Undergraduate Texts in Mathematics

Gérard Iooss
Daniel D. Joseph

Elementary Stability and Bifurcation Theory

Springer-Verlag
New York Heidelberg Berlin

Gérard Iooss

Faculté des Sciences
Institut des Mathématiques
et Sciences Physiques
Université de Nice
Parc Valrose, Nice 06034
FRANCE

Daniel D. Joseph

Department of Aerospace Engineering
and Mechanics
University of Minnesota
Minneapolis, MN 55455
USA

AMS Subject Classification (1980): 34-01, 34, A34, 34D30, 34D99, 34C99

With 47 illustrations.

Library of Congress Cataloging in Publication Data

Iooss, Gérard.
Elementary stability and bifurcation theory.

(Undergraduate texts in mathematics)
Bibliography: p.
Includes index.
1. Differential equations—Numerical solutions.
2. Evolution equations—Numerical solutions.
3. Stability. 4. Bifurcation theory. I. Joseph,
Daniel D., joint author. II. Title.
QA372.I68 515.3′5 80-20782

Printed in the United States of America.

9 8 7 6 5 4 3 2 1

ISBN 0-387-90526-X Springer-Verlag New York
ISBN 3-540-90526-X Springer-Verlag Berlin Heidelberg

Everything should be made as simple as possible,
but not simpler.

ALBERT EINSTEIN

Contents

List of Frequently Used Symbols

All symbols are fully defined at the place where they are first introduced. As a convenience to the reader we have collected some of the more frequently used symbols in several places. The largest collection is the one given below. Shorter lists, for later use can be found in the introductions to Chapters X and XI.

$\stackrel{\text{def}}{=}$	equality by definition
$\in$	"$a \in A$" means "a belongs to the set A" or "a is an element of A"
$\mathbb{N}$	the set of nonnegative integers (0 included)
$\mathbb{N}^*$	the set of strictly positive integers (0 excluded)
$\mathbb{Z}$	the set of positive and negative integers including 0
$\mathbb{R}$	the set of real numbers (the real line)
$\mathbb{R}^n$	the set of ordered n-tuples of real numbers. $\mathbf{a} \in \mathbb{R}^n$ may be represented as $\mathbf{a} = (a_1, \ldots, a_n)$. Moreover, $\mathbb{R}^n$ is a Euclidian space with the scalar product $\langle \mathbf{a}, \mathbf{b} \rangle = \sum_{i=1}^{n} a_i b_i$ where $\mathbf{a} = (a_1, \ldots, a_n), \mathbf{b} = (b_1, \ldots, b_n)$. $\mathbb{R}^1 = \mathbb{R}$; $\mathbb{R}^2$ is the plane
$\mathbb{C}$	the set of complex numbers
$\mathbb{C}^n$	the set of ordered n-tuples of complex numbers. The scalar product in $\mathbb{C}^n$ is denoted as in $\mathbb{R}^n$, but $\langle \mathbf{a}, \mathbf{b} \rangle = \sum_{i=1}^{n} a_i \overline{b_i} = \overline{\langle \mathbf{b}, \mathbf{a} \rangle}$.

$\mathscr{C}^n(\mathscr{V})$ — the set of n-times continuously differentiable functions on a domain $\mathscr{V}$. We may furthermore specify the domain E where these functions take their values by writing $\mathscr{C}^n(\mathscr{V}; E)$.

$\|\mathbf{u}\|$ — the norm of $\mathbf{u}$. For instance, if $\mathbf{u} \in \mathbb{C}^n$ we have $\|\mathbf{u}\| = \langle \mathbf{u}, \mathbf{u} \rangle^{1/2}$; if $\mathbf{u} \in \mathscr{C}(\mathscr{V})$, $\|\mathbf{u}\| = \text{l.u.b}_{x \in \mathscr{V}} \|\mathbf{u}(x)\|$, where $\|\mathbf{u}(x)\|$ is the norm of $\mathbf{u}(x)$ in the domain of values for $\mathbf{u}$; $\|\mathbf{u}\| = 0$ implies that $\mathbf{u} = 0$.

$\mathbf{A}(\cdot)$ — a linear operator:

$$\mathbf{A}(\alpha\mathbf{u} + \beta\mathbf{v}) = \alpha\mathbf{A}(\mathbf{u}) + \beta\mathbf{A}(\mathbf{v}).$$

$\mathbf{B}(\cdot, \cdot)$ — a bilinear operator:

$$\mathbf{B}(\alpha_1\mathbf{u}_1 + \alpha_2\mathbf{u}_2, \beta_1\mathbf{v}_1 + \beta_1\mathbf{v}_2) = \alpha_1\beta_1\mathbf{B}(\mathbf{u}_1, \mathbf{v}_1) + \alpha_1\beta_2\mathbf{B}(\mathbf{u}_1, \mathbf{v}_2) + \alpha_2\beta_1\mathbf{B}(\mathbf{u}_2, \mathbf{v}_1) + \alpha_2\beta_2\mathbf{B}(\mathbf{u}_2, \mathbf{v}_2)$$

$\mathbf{C}(\cdot, \cdot, \cdot)$ — a trilinear operator

$\mathbf{N}(\cdot)$ — a general nonlinear operator with no constant term and no linear term in a neighborhood of 0:

$$\mathbf{N}(\mathbf{u}) \overset{\text{def}}{=} B(\mathbf{u}, \mathbf{u}) + C(\mathbf{u}, \mathbf{u}, \mathbf{u}) + O(\|\mathbf{u}\|^4)$$

Sometimes we assign a slightly different meaning to **A**, **B**, **C**:

$$(\mathbf{A} \cdot \mathbf{u})_i = A_{ij}u_j = A_{i1}u_1 + A_{i2}u_2 + \cdots + A_{in}u_n$$

$$(\mathbf{B} \cdot \mathbf{u} \cdot \mathbf{v})_i = B_{ijk}u_j v_k$$

$$(\mathbf{C} \cdot \mathbf{u} \cdot \mathbf{v} \cdot \mathbf{w})_i = C_{ijkl}u_j v_k w_l$$

where we use the summation convention for repeated indices and where

(A_{ij}) is the matrix of a linear operator

(B_{ijk}) is the matrix of a bilinear operator

(C_{ijkl}) is the matrix of a trilinear operator

$\mathbf{F}(t, \mu, \mathbf{U})$ — a nonlinear operator—see the opening paragraph of Chapter I

$\mathbf{f}(t, \mu, \mathbf{u})$ — reduction of **F** to "local form," see §I.3

$\mathbf{F}_u$, $\mathbf{F}_{uu}$, etc. — derivatives of **F**; see §I.6–7

$\mathbf{F}_u(t, \mu, \mathbf{U}_0|\cdot)$ — the linear operator associated with the derivative of **F** at $\mathbf{U} = \mathbf{U}_0$

$\mathbf{F}_u(t, \mu, \mathbf{U}_0|\mathbf{v})$ — first derivative of $\mathbf{F}(t, \mu, \mathbf{U})$, evaluated at $\mathbf{U} = \mathbf{U}_0$, acting on $\mathbf{v}$

$\sigma = \xi + i\eta$ — an eigenvalue of a linear operator arising in the study of stability of $\mathbf{u} = 0$

When $\mathbf{u} = 0$ corresponds to a time-periodic $\mathbf{U}(t) = \mathbf{U}(t + T)$, then σ is a *Floquet exponent*

$\lambda = e^{\sigma T}$ — a *Floquet multiplier*; see preceding entry and §VII.6.2

$\gamma = \xi + i\eta$ — an eigenvalue of a linear operator arising in the study of bifurcating solution. We use the same notation, ξ and η, for the real and imaginary part of σ and γ and depend on the context to define the difference.

ω; $T = 2\pi/\omega$ — frequency ω and period T

ε — amplitude of a bifurcating solution defined in various ways: under (II.2), (V.2), (VI.72), $(\text{VII.6})_2$, (VIII.22), Figure X.1.

$\langle \mathbf{a}, \mathbf{b} \rangle$ — notation for a scalar product $\overline{\langle \mathbf{a}, \mathbf{b} \rangle} = \langle \mathbf{b}, \mathbf{a} \rangle$ with the usual conventions. For vectors in $\mathbb{C}^n$, $\langle \mathbf{a}, \mathbf{b} \rangle = \mathbf{a} \cdot \bar{\mathbf{b}}$. For vector fields in $\mathscr{V}$, $\langle \mathbf{a}, \mathbf{b} \rangle = \int_{\mathscr{V}} \mathbf{a}(\mathbf{x})\bar{\mathbf{b}}(\mathbf{x})\, d\mathscr{V}$. See (IV.7), under (VI.4), §VI.6, under $(\text{VI.134})_2$, and under (VI.144)

$[\mathbf{a}, \mathbf{b}]$ — another scalar product for 2π-periodic functions, defined above (VIII.15)

$[\mathbf{a}, \mathbf{b}]_{nT}$ — see (IX.16)

Some operators whose domains are 2π-periodic functions of s:

$$J(\cdot, \varepsilon), \qquad J(\cdot, 0) = J_0 \text{ (VII.38)}$$

$$\mathbb{J}_0 \text{ (VIII.15)}; \qquad \mathbb{J}_0^* \text{ (VIII.19)}; \qquad \mathbb{J}(\varepsilon) \text{ above (VIII.37)}.$$

Similar operators for nT-periodic functions are defined under notation for Chapter IX, at the beginning of Chapter IX.

Order Symbols. we say that $f(\varepsilon) = O(\varepsilon^n)$ if

$$\frac{f(\varepsilon)}{\varepsilon^n} \text{ is bounded when } \varepsilon \to 0$$

we say that $f(\varepsilon) = o(\varepsilon^n)$

$$\lim_{\varepsilon \to 0} \frac{f(\varepsilon)}{\varepsilon^n} = 0.$$

Introduction

In its most general form bifurcation theory is a theory of equilibrium solutions of nonlinear equations. By equilibrium solutions we mean, for example, steady solutions, time-periodic solutions, and quasi-periodic solutions. The purpose of this book is to teach the theory of bifurcation of equilibrium solutions of evolution problems governed by nonlinear differential equations. We have written this book for the broadest audience of potentially interested learners: engineers, biologists, chemists, physicists, mathematicians, economists, and others whose work involves understanding equilibrium solutions of nonlinear differential equations.

To accomplish our aims, we have thought it necessary to make the analysis

1. general enough to apply to the huge variety of applications which arise in science and technology, and
2. simple enough so that it can be understood by persons whose mathematical training does not extend beyond the classical methods of analysis which were popular in the 19th Century.

Of course, it is not possible to achieve generality and simplicity in a perfect union but, in fact, the general theory is simpler than the detailed theory required for particular applications. The general theory abstracts from the detailed problems only the essential features and provides the student with the skeleton on which detailed structures of the applications must rest.

It is generally believed that the mathematical theory of bifurcation requires some functional analysis and some of the methods of topology and dynamics. This belief is certainly correct, but in a special sense which it is useful to specify as motivation for the point of view which we have adopted in this work.

The main application of functional analysis of problems of bifurcation is the justification of the reduction of problems posed in spaces of high or infinite dimension to one and two dimensions. These low-dimensional problems are associated with eigenfunction projections, and in some special cases, like those arising in degenerate problems involving symmetry-breaking steady bifurcations, analysis of problems of low dimension greater than two is required. But the one- and two-dimensional projections are the most important. They fall under the category of problems mathematicians call bifurcation at a simple eigenvalue.

The existence and nature of bifurcation and the stability of the bifurcating solutions are completely determined by analysis of the nonlinear ordinary differential and algebraic equations which arise from the methods of reduction by projections. The simplest way, then, to approach the teaching of the subject is to start with the analysis of low-dimensional problems and only later to demonstrate how the lower-dimensional problems may be projected out of high-dimensional problems. In the first part of the analysis we require only classical methods of analysis of differential equations and functions. In the second part of the analysis, which is treated in Chapters VI and VIII, we can proceed in a formal way without introducing the advanced mathematical tools which are required for the ultimate justification of the formal analysis. It goes almost without saying that we believe that all statements which we make are mathematically justified in published works which are cited and left for further study by courageous students.

It is perhaps useful to emphasize that we confine our attention to problems which can be reduced to one or two dimensions. In this setting we can discuss the following types of bifurcation: bifurcation of steady solutions in one dimension (Chapter II) and for general problems which can be projected into one dimension (Chapter VI); isolated solutions which perturb bifurcation in one dimension (Chapter III) and for general problems which can be projected into one dimension (Chapter VI); bifurcation of steady solutions from steady solutions in two dimensions (Chapters IV and V) and for general problems which can be projected into two dimensions (Chapter VI); bifurcation of time-periodic solutions from steady ones in two dimensions (Chapter VII) and for general problems which can be projected into two dimensions (Chapter VIII); the bifurcation of subharmonic solutions from T-periodic ones in the case of T-periodic forcing (Chapter IX), the bifurcating torus of "asymptotically quasi-periodic" solutions which bifurcate from T-periodic ones in the case of T-periodic forcing (Chapter X), the bifurcation of subharmonic solutions and tori from self-excited periodic solutions (the autonomous case, Chapter XI). It is not possible to do much better in an elementary book because even apparently benign systems of three nonlinear ordinary differential equations give rise to very complicated dynamics with turbulentlike attracting sets which defy description in simple terms. In one dimension all solutions lie on the real line, in two dimensions all solutions of the initial value problem lie in the plane and their trajectories cannot intersect transversally because the solutions are unique. This severe restriction on solutions of two-dimensional problems has already much less force in three dimensions where nonintersecting trajectories can ultimately generate attracting sets of considerable complexity (for example, see E. N. Lorenz, Deterministic nonperiodic flow, *J. Atmos. Sci.* **20**, 130 (1963)).

We regard this book as a text for the teaching of the principles of bifurcation. Our aim was to give a complete theory for all problems which in a sense, through projections, could be said to be set in two dimensions. To do this we had to derive a large number of new research results. In fact new results appear throughout this book but most especially in the problem of bifurcation of periodic solutions which is studied in Chapters X and XI. Students who wish to continue their studies after mastering the elementary theory may wish to consult some of the references listed at the end of Chapter I.

There are many very good and important papers among the thousands published since 1963. We have suppressed our impulse to make systematic reference to these papers because we wish to emphasize only the elementary parts of the subject. It may be helpful, however, to note that some papers use the "method of Liapunov–Schmidt" to decompose the space of solutions and equations into a finite-dimensional and an infinite-dimensional part. The infinite part can be solved and the resulting finite dimensional problem has all the information about bifurcation. Other papers use the "center manifold" to reduce the problems to finite dimensions. This method uses the fact that in problems like those in this book, solutions are attracted to the center manifold, which is finite dimensional. Both methods are good for proving existence theorems. Though they can also be used to construct solutions, they in fact involve redundant computations. These methods are systematically avoided in this book. Instead, we apply the *implicit function theorem* to justify the direct, sequential computation of *power series solutions* in an amplitude ε, using the *Fredholm alternative*, as the most economic way to determine qualitative properties of the bifurating solutions and to compute them.

Acknowledgments

This book was begun in 1978 during a visit of G. Iooss to the University of Minnesota made possible by a grant from the Army Research Office in Durham. Continued support of the research of D. D. Joseph by the Fluid Mechanics program of the N.S.F. is most gratefully acknowledged.

CHAPTER I

Equilibrium Solutions of Evolution Problems

We are going to study equilibrium solutions of evolution equations of the form

$$\frac{d\mathbf{U}}{dt} = \mathbf{F}(t, \mu, \mathbf{U}), \tag{I.1}$$

where $t \geq 0$ is the time and μ is a parameter which lies on the real line $-\infty < \mu < \infty$. The unknown in (I.1) is $\mathbf{U}(t)$. $\mathbf{F}(t, \mu, \mathbf{U})$ is a given nonlinear function or operator.* When $\mathbf{F}$ is independent of t we omit t and write $\mathbf{F}(\mu, \mathbf{U})$. (I.1) governs the evolution of $\mathbf{U}(t)$ from its *initial value* $\mathbf{U}(0) = \mathbf{U}_0$. An equilibrium solution is a solution to which $\mathbf{U}(t)$ evolves after the transient effects associated with the initial values, have died away. It is necessary to state more precisely what is meant by $\mathbf{U}(t)$, $\mathbf{F}(t, \mu, \mathbf{U})$, and an equilibrium solution. This statement requires some preliminary explanations and definitions.

I.1 One-Dimensional, Two-Dimensional, *n*-Dimensional, and Infinite-Dimensional Interpretations of (I.1)

In one-dimensional problems $U(t)$ is a scalar, $-\infty < U < \infty$, and $F(t, \mu, U)$ is a scalar-valued function of (t, μ, U). For example, in a coarse approximation

* We assume here that $\mathbf{F}$ depends on the present value of $\mathbf{U}(t)$ and not on its history. For more general possibilities see Notes for I.

of an ecological logistics problem, U could be the density of mosquitoes in Minnesota and μ the available food supply. The rate of increase of the density of mosquitoes is given by the nonlinear function $F(\mu, U)$. The mosquito population increases when $F > 0$, decreases when $F < 0$ and is in equilibrium when $F(\mu, U) = 0$. In the equilibrium distribution the food supply μ and population density U are related by $F = 0$. There can be many equilibrium distributions; for example,

$$F(\mu, U) = (a_1(\mu) - U)(a_2(\mu) - U)\dots(a_n(\mu) - U) \tag{I.2}$$

could have n equilibrium distributions each one corresponding to a zero of a factor: $a_l(\mu) = U$. The determination of equilibrium distributions does not tell us what density of mosquitoes to expect given the availability of human flesh because some equilibrium distributions are unstable and will not persist under perturbation. So we have to find the equilibrium distributions and to determine their stability.

In two-dimensional problems $\mathbf{U}(t)$ is a two-dimensional vector with components $(U_1(t), U_2(t))$, and $\mathbf{F}(t, \mu, \mathbf{U})$ is vector-function whose components $[F_1(t, \mu, U_1, U_2), F_2(t, \mu, U_1, U_2)]$ are nonlinear functions of the components of $\mathbf{U}$. The same notations are adopted for n-dimensional problems with $n > 2$; in this case the vectors have n components.

We follow the usual mathematical conventions and define

$$(\mathbb{R}^1, \mathbb{R}^2, \mathbb{R}^n) = (\text{the real line, the plane, } n\text{-dimensional space}).$$

Scalars take on values in $\mathbb{R}^1$ and n-component vectors take on values in $\mathbb{R}^n$. It is customary to simplify the notation for $\mathbb{R}^1$ by dropping the superscript: $\mathbb{R}^1 = \mathbb{R}$.

It is also conventional in mathematics to speak of infinite-dimensional problems but, in general, something more than and something different from $n \to \infty$ is meant. By an infinite-dimensional problem we mean that $\mathbf{U} = \mathbf{U}(x_1, \dots, x_n, t)$ is a field on a n-dimensional (usually ≤ 3 dimensions) region $\mathscr{V}$ of $(x_1, \dots, x_n)$-space and that $\mathbf{F}(t, \mu, \mathbf{U})$ is an operator involving operations on the spatial variables $x_1, x_2, \dots, x_n$ which carry vector fields in $\mathscr{V}$ into vector fields in $\mathscr{V}$. Partial differential equations and integral equations fit this description. For partial differential equations it is necessary to supplement (I.1) with boundary conditions. For example, in problems of reaction and diffusion involving n different species fields $C_i(\mathbf{x}, t)$ in a temperature field $T(\mathbf{x}, t)$ defined on the region $\mathscr{V}$ of three-dimensional physical space the evolution of the $(n + 1)$-dimensional vector field $\mathbf{U}(\mathbf{x}, t) = (C_1(\mathbf{x}, t), C_2(\mathbf{x}, t), \dots, C_n(\mathbf{x}, t), T(\mathbf{x}, t)) = (U_1(\mathbf{x}, t), U_2(\mathbf{x}, t), \dots, U_{n+1}(\mathbf{x}, t))$ is governed by

$$\frac{\partial \mathbf{U}}{\partial t} = \mathbf{F}(t, \mu, \mathbf{U}); \quad \text{i.e.,} \quad \frac{\partial U_\alpha}{\partial t} = F_\alpha(t, \mu, \mathbf{U}), \qquad \alpha = 1, 2, \dots, n + 1 \tag{I.3}$$

where

$$F_\alpha(t, \mu, \mathbf{U}) = \nabla \cdot (D_{\alpha\beta}\nabla)U_\beta + g_\alpha(\mu, \mathbf{U}) + h_\alpha(\mathbf{x}, t, \mu) \tag{I.4}^*$$

$(D_{\alpha\beta}) = (n + 1) \times (n + 1)$ matrix of diffusion coefficients;

$g_\alpha(\mu, \mathbf{U})$ is a nonlinear function of μ and $\mathbf{U}$ and $g_\alpha(\mu, 0) = 0$; (I.5)

and

$h_\alpha(\mathbf{x}, t, \mu)$ is a prescribed function of μ and t.

On the boundary $\partial\mathscr{V}$ of $\mathscr{V}$ with outward normal $\mathbf{n}$ some linear combination of the normal derivatives and values of the components of $\mathbf{U}$ are prescribed:

$$(\mathbf{n} \cdot \nabla)M_{\alpha\beta}(\mathbf{x}, t, \mu)U_\beta + N_{\alpha\beta}(\mathbf{x}, t, \mu)U_\beta = P_\alpha(\mathbf{x}, t, \mu), \tag{I.6}$$

where $M_{\alpha\beta}$ and $N_{\alpha\beta}$ are square matrices and $P_\alpha(\mathbf{x}, t, \mu)$ is prescribed. This problem is infinite-dimensional because it is defined for each of the infinitely many places $\mathbf{x}$ of $\mathscr{V}$.

Another example is the Navier–Stokes equations for a homogeneous incompressible fluid. Here (I.1) can be taken as the equation for the vorticity $\boldsymbol{\omega} = \text{curl } \mathbf{V}$, where $\mathbf{V}(\mathbf{x}, t)$ is the velocity, ν is the kinematic viscosity and

$$\begin{aligned} \frac{\partial \boldsymbol{\omega}}{\partial t} &= \nu\nabla^2\boldsymbol{\omega} + (\boldsymbol{\omega} \cdot \nabla)\mathbf{V} - (\mathbf{V} \cdot \nabla)\boldsymbol{\omega} + \mathbf{p}(\mathbf{x}, t, \mu), \\ \boldsymbol{\omega} &= \text{curl } \mathbf{V}, \\ \text{div } \mathbf{V} &= 0, \end{aligned} \tag{I.7}$$

where $\mathbf{p}(\mathbf{x}, t, \mu)$ is a prescribed forcing term. The solutions $(\mathbf{V}, \boldsymbol{\omega})$ of (I.7) together with boundary conditions prescribing $\mathbf{V}$, say

$$\mathbf{V}(\mathbf{x}, t) = \boldsymbol{\psi}(\mathbf{x}, t, \mu) \quad \text{for} \quad \mathbf{x} \in \partial\mathscr{V}, \tag{I.8}$$

determine $\mathbf{V}(\mathbf{x}, t)$ in $\mathscr{V}$.

It is perhaps useful here to state that in many situations the higher-dimensional problems can be reduced to one- or two-dimensional ones (see, e.g., Chapters VI and VIII).

I.2 Forced Solutions; Steady Forcing and *T*-Periodic Forcing; Autonomous and Nonautonomous Problems

Now we adopt the convention, which clearly applies to the examples given in §I.1, that $\mathbf{U} \equiv 0$ is not a solution of the evolution problem associated with (I.1). The function $\mathbf{U} = 0$ cannot solve this problem because $\mathbf{U} \neq 0$ is

* We use the convention of repeated indices. A repeated index is to be summed over its range: e.g., $D_{\alpha\beta}u_\beta = D_{\alpha1}u_1 + D_{\alpha2}u_2 + D_{\alpha3}u_3$.

forced by nonzero forcing data. In the examples mentioned in §I.1, the forcing data is $a_1(\mu)a_2(\mu)\ldots a_n(\mu) \neq 0$ in (I.2), $h_\alpha(\mathbf{x}, t, \mu)$ and $P_\alpha(\mathbf{x}, t, \mu) \neq 0$ in (I.3–6), and $\mathbf{p}(x, t, \mu)$ and $\boldsymbol{\psi}(\mathbf{x}, t, \mu) \neq 0$ in (I.7, 8). If we ignore the boundary conditions in problems of partial differential equations then the forcing data is given by $\mathbf{F}(t, \mu, 0) \neq 0$.

We are going to restrict our attention to problems in which the forcing data

$$\mathbf{F}(t, \mu, 0) \stackrel{\text{def}}{=} \mathbf{F}(\mu, 0) \neq 0 \quad \text{and} \quad \mathbf{F}(\mu, \mathbf{U}) \text{ is independent of } t \tag{I.9}$$

or

$$\mathbf{F}(t, \mu, 0) = \mathbf{F}(t + T, \mu, 0) \neq 0 \quad \text{and } \mathbf{F}(t, \mu, \mathbf{U}) \text{ is } T\text{-periodic.} \tag{I.10}$$

When $\mathbf{F}(\mu, U)$ is independent of t, the problem

$$\frac{d\mathbf{U}}{dt} = \mathbf{F}(\mu, \mathbf{U}) \tag{I.11}$$

is said to be *autonomous*. When $\mathbf{F}(t, \mu, \mathbf{U})$ is periodic in t with period T, the problem

$$\frac{d\mathbf{U}}{dt} = \mathbf{F}(t, \mu, \mathbf{U}) = \mathbf{F}(t + T, \mu, \mathbf{U}) \tag{I.12}$$

is said to be *nonautonomous*, T-periodic. We shall usually omit the words "T-periodic" in describing nonautonomous problems since only the T-periodic ones are considered in this book.

I.3 Reduction to Local Form

We make the assumption that for μ in a certain interval of $\mathbb{R}^1$ there are equilibrium solutions of (I.11) and (I.12) which imitate evolution properties of the forcing data. So there is a steady solution $\tilde{\mathbf{U}}(\mu)$ (I.11) and a T-periodic solution $\tilde{\mathbf{U}}(t, \mu) = \tilde{\mathbf{U}}(t + T, \mu)$ of (I.12).

Consider an arbitrary disturbance, $\mathbf{u}$ of $\tilde{\mathbf{U}}$. The equations which govern this disturbance are

$$\frac{d\mathbf{u}}{dt} = \mathbf{F}(\mu, \tilde{\mathbf{U}} + \mathbf{u}) - \mathbf{F}(\mu, \tilde{\mathbf{U}}) \stackrel{\text{def}}{=} \mathbf{f}(\mu, \mathbf{u}) \tag{I.13}$$

in the autonomous case, and

$$\frac{d\mathbf{u}}{dt} = \mathbf{F}(t, \mu, \tilde{\mathbf{U}} + \mathbf{u}) - \mathbf{F}(t, \mu, \tilde{\mathbf{U}}) \stackrel{\text{def}}{=} \mathbf{f}(t, \mu, \mathbf{u}) \tag{I.14}$$

where $\mathbf{f}(t, \mu, \mathbf{u}) = \mathbf{f}(t + T, \mu, \mathbf{u})$ in the nonautonomous case; $\mathbf{u}$ identically zero is a solution of (I.13) and (I.14). Problems (I.13) and (I.14), in which $\mathbf{u} = 0$ is a solution, are said to be *reduced to local form*.

There is no great loss of generality involved in the reduction to local form. It is a valid reduction for those values of μ for which $\tilde{\mathbf{U}}(\mu)$ and $\tilde{\mathbf{U}}(t, \mu)$ exist.

Equations (I.13) and (I.14) are identical except for the presence of t in $\mathbf{f}(t, \mu, \mathbf{u})$ in (I.14). But the behavior of solutions of these two problems is very different. This is no surprise. The difference arises from a big difference in the nature of forcing data which drives the dynamical equations from the outside.

I.4 Equilibrium Solutions

We have already defined two types of equilibrium solutions: (1) steady solutions of autonomous problems and (2) T-periodic solutions of nonautonomous problems.

One of the main features of bifurcation is the appearance of solutions which break the symmetry pattern of the forcing data. For example, we may get (3) a τ-periodic solution $\mathbf{U}(t) = \mathbf{U}(t + \tau)$ or $\mathbf{u}(t) = \mathbf{u}(t + \tau)$ of the steady problem (I.11) or (I.13), respectively. We may get (4) subharmonic solutions $\mathbf{U}(t) = \mathbf{U}(t + nT)$ or $\mathbf{u}(t) = \mathbf{u}(t + nT)$, where $n = 1, 2, 3, \ldots$ of nonautonomous, T-periodic problems (I.12) or (I.14), respectively. We may also get (5) subharmonic bifurcating solutions of τ-periodic solutions of autonomous problems. Suppose there is a τ-periodic solution of (I.13)

$$\frac{d\mathbf{u}}{dt} = \mathbf{f}(\mu, \mathbf{u}(\mu, t)), \qquad \mathbf{u}(\mu, t) = \mathbf{u}(\mu, t + \tau). \tag{I.15}$$

Then $\mathbf{f}$ is autonomous even though $\mathbf{u}$ depends on t. A disturbance $\mathbf{v}$ of $\mathbf{u}(\mu, t)$ satisfies

$$\frac{d(\mathbf{u} + \mathbf{v})}{dt} = \mathbf{f}(\mu, \mathbf{u}(\mu, t) + \mathbf{v}(t)). \tag{I.16}$$

If there are periodic solutions of (I.16), $\mathbf{u}(\mu, t) + \mathbf{v}(t) = \mathbf{u}(\mu, t + \tilde{\tau}) + \mathbf{v}(t + \tilde{\tau})$ where $\tilde{\tau} \to n\tau$, $n = 1, 2, 3, \ldots,$ as $\mathbf{v} \to 0$ then $\mathbf{u} + \mathbf{v}$ is said to be subharmonic. Finally we can get bifurcation of periodic solutions of autonomous and nonautonomous problems into "asymptotically quasi-periodic" solutions. These solutions are sometimes said to live on a bifurcating torus and they are discussed in Chapter X.

We do not give a general definition of equilibrium solutions. Instead, by "equilibrium solution" we mean one of the six types listed above.

I.5 Equilibrium Solutions and Bifurcating Solutions

Bifurcating solutions are equilibrium solutions which form intersecting branches in a suitable space of functions. For example, when U lies in $\mathbb{R}^1$ the bifurcating steady solutions form intersecting branches of the curve $F(\mu, U) = 0$ in the μ, U plane. When $\mathbf{U}$ lies in $\mathbb{R}^2$ the bifurcating solutions form connected intersecting surfaces or curves in the three-dimensional (μ, U_1, U_2) space. We say that one equilibrium solution bifurcates from another at $\mu = \mu_0$ if there are two distinct equilibrium solutions $\mathbf{U}^{(1)}(\mu, t)$ and $\mathbf{U}^{(2)}(\mu, t)$ of the evolution problem, continuous in μ, and such that $\mathbf{U}^{(1)}(\mu_0, t) = \mathbf{U}^{(2)}(\mu_0, t)$.

Not all equilibrium solutions arise from bifurcation. Isolated solutions and disjoint branches of solutions are common in nonlinear problems (see Figure II.7).

I.6 Bifurcating Solutions and the Linear Theory of Stability

To get the linearized theory we subject an equilibrium solution to a small initial perturbation. If the perturbation grows the equilibrium solution is unstable, and if it eventually decays the equilibrium solution is stable to small disturbances. It may be unstable to larger disturbances, but if it is stable to small disturbances then there is no other equilibrium solution of the evolution problem close to the given one. Since solutions which bifurcate from the given one branch off the given one in a continuous fashion it is often (but not always) true that a necessary condition for bifurcation is the instability of the equilibrium solution to indefinitely small disturbances. (This necessary condition is true for bifurcation at a simple eigenvalue.) The stability theory for indefinitely small disturbances is linear because quadratic terms in the disturbance equations are negligible compared to linear ones.

Suppose, for example, that $\tilde{\mathbf{U}}(t, \mu)$ is a solution of (I.12) and $\delta\mathbf{v}$ is a disturbance of $\tilde{\mathbf{U}}$ where δ is a constant. Then

$$\delta \frac{d\mathbf{v}}{dt} = \mathbf{F}(t, \mu, \tilde{\mathbf{U}}(t, \mu) + \delta\mathbf{v}(t)) - \mathbf{F}(t, \mu, \tilde{\mathbf{U}})$$

so

$$\frac{d\mathbf{v}}{dt} \cong \left[\frac{d}{d\delta}\mathbf{F}(t, \mu, \tilde{\mathbf{U}} + \delta\mathbf{v})\right]_{\delta=0} \stackrel{\text{def}}{=} \mathbf{F}_U(t, \mu, \tilde{\mathbf{U}} \mid \mathbf{v})$$

where $\mathbf{F}_U(t, \mu, \tilde{\mathbf{U}} \mid \cdot)$ is a linear operator, linear in the variable after the vertical bar, called the derivative or linearization of $\mathbf{F}$ evaluated at $\tilde{\mathbf{U}}(t, \mu)$. In the same way $\mathbf{f}_u(t, \mu, 0 \mid \cdot)$ is the derivative of $\mathbf{f}$ evaluated at the solution $\mathbf{u} = 0$ and

$$\frac{d\mathbf{v}}{dt} = \mathbf{f}_u(t, \mu, 0 \mid \mathbf{v}) \tag{I.17$_1$}$$

defines the linearized equation reduced to local form. To simplify notation we write

$$\mathbf{f}_u(t, \mu \mid \cdot) \stackrel{\text{def}}{=} \mathbf{f}_u(t, \mu, 0 \mid \cdot). \tag{I.17$_2$}$$

The solution $\mathbf{u} = 0$ of (I.14) is said to be *asymptotically stable* if $\mathbf{v} \to 0$ as $t \to \infty$ (see §II.7).

I.7 Notation for the Functional Expansion of $\mathbf{F}(t, \mu, \mathbf{U})$

It is frequently useful to expand the nonlinear operator $\mathbf{F}(t, \mu, \mathbf{U})$ as a Taylor series around the vector $\mathbf{U}_0$. Thus

$$\begin{aligned}\mathbf{F}(t, \mu, \mathbf{U}_0 + \mathbf{v}) = {} & \mathbf{F}(t, \mu, \mathbf{U}_0) + \mathbf{F}_U(t, \mu, \mathbf{U}_0 | \mathbf{v}) \\ & + \frac{1}{2}\mathbf{F}_{UU}(t, \mu, \mathbf{U}_0 | \mathbf{v} | \mathbf{v}) + \frac{1}{3!}\mathbf{F}_{UUU}(t, \mu, \mathbf{U}_0 | \mathbf{v} | \mathbf{v} | \mathbf{v}) \\ & + O(\|\mathbf{v}\|^4),\end{aligned} \tag{I.18}$$

where, for example,

$$\begin{aligned}\mathbf{F}_{UU}(t, \mu, \mathbf{U}_0 | \mathbf{a} | \mathbf{b}) & = \mathbf{F}_{UU}(t, \mu, \mathbf{U}_0 | \mathbf{b} | \mathbf{a}) \\ & \stackrel{\text{def}}{=} \left.\frac{\partial^2 \mathbf{F}(t, \mu, \mathbf{U}_0 + \delta_1 \mathbf{a} + \delta_2 \mathbf{b})}{\partial\delta_1\, \partial\delta_2}\right|_{\delta_1 = \delta_2 = 0}\end{aligned} \tag{I.19}$$

is a bilinear operator carrying vectors into vectors. $\mathbf{F}_{UUU}(t, \mu, \mathbf{U}_0 | \mathbf{v} | \mathbf{v} | \mathbf{v})$ is generated from a trilinear operator in the same way. The multilinear operators are obviously symmetric with respect to the argument vectors to the right of the vertical bars.

When $\mathbf{U}(t) \in \mathbb{R}^n$ we may express the functional derivatives in terms of matrices

$$\begin{aligned}\{\mathbf{F}_U(t, \mu, \mathbf{U}_0 | \mathbf{v})\}_i & = \{\mathbf{A}(t, \mu, \mathbf{U}_0) \cdot \mathbf{v}\}_i = A_{ij}(t, \mu, \mathbf{U}_0) v_j, \\ \frac{1}{2}\{\mathbf{F}_{UU}(t, \mu, \mathbf{U}_0 | \mathbf{v} | \mathbf{v})\} & = \{\mathbf{B}(t, \mu, \mathbf{U}_0) \cdot \mathbf{v} \cdot \mathbf{v}\} = B_{ijk}(t, \mu, \mathbf{U}_0) v_j v_k, \\ \frac{1}{3!}\{F_{UUU}(t, \mu, \mathbf{U}_0 | \mathbf{v} | \mathbf{v} | \mathbf{v})\}_I & = \{\mathbf{C}(t, \mu, \mathbf{U}_0)\mathbf{v} \cdot \mathbf{v} \cdot \mathbf{v}\}_i \\ & = C_{ijkl}(t, \mu, \mathbf{U}_0) v_j v_k v_l\end{aligned} \tag{I.20}$$

where indices range from 1 through n, summation of repeated indices is implied, and B_{ijk} and C_{ijkl} are symmetric with respect to interchange of the subscripts following i.

The same considerations apply when the problem is reduced to local form. In this case (see $(\text{I.17})_2$) we have

$$\frac{d\mathbf{u}}{dt} = \mathbf{f}(t, \mu, \mathbf{u}) = \mathbf{f}_u(t, \mu | \mathbf{u}) + \frac{1}{2}\mathbf{f}_{uu}(t, \mu | \mathbf{u} | \mathbf{u}) + \frac{1}{3!}\mathbf{f}_{uuu}(t, \mu | \mathbf{u} | \mathbf{u} | \mathbf{u}) + \cdots \tag{I.21}$$

and in $\mathbb{R}^n$

$$\mathbf{f}(t, \mu, \mathbf{u}) = \mathbf{A}(t, \mu) \cdot \mathbf{u} + \mathbf{B}(t, \mu) \cdot \mathbf{u} \cdot \mathbf{u} + \mathbf{C} \cdot \mathbf{u} \cdot \mathbf{u} \cdot \mathbf{u} + \cdots. \tag{I.22}$$

Notes

The theory of bifurcation applies generally to nonlinear problems, not only when bifurcating solutions are equilibrium solutions of evolution problems like (I.1), but also in the case of integral equations, nonlinear algebraic and functional equations, integro-differential and functional-differential equations, especially those of retarded type in which memory effects are important; for example

$$\frac{\partial \mathbf{u}}{\partial t} = \int_{-\infty}^{t} G(t - \tau)\mathbf{F}(\tau, \mu, \mathbf{u}(\tau))\, d\tau.$$

The theory given in this book is a guide to how to study these other problems; in many cases only slight and obvious changes are required.

The time-derivative in (I.1) is important in the definition of equilibrium solutions and the discussion of their stability. For example, in the next chapter we shall show that the theory of bifurcation of plane curves $F(\mu, \varepsilon) = 0$ is the same as the study of singular points of these curves. The study of singular points may be connected with stability but the connection is incidental and not intrinsic. The problem of stability depends on whether the system is dissipative or conservative. Conservative systems are more difficult in the sense that small perturbations of them never decay. In this book we treat only dissipative systems.

There are many works and some monographs devoted to problems of bifurcation. The word bifurcation or *Abzweigung* seems to have been introduced by C. Jacobi, Über die Figur des Gleichgewichts, *Pogg. Ann.*, **32**, 229 (1834), in his study of the bifurcation of the MacLaurin spheroidal figures of equilibrium of self-gravitating rotating bodies. The French word *bifurcation* was introduced by H. Poincaré, Sur l'équilibre d'une masse fluide animée d'un mouvement de rotation, *Acta Math.*, **7**, 259–380 (1885). There are many books and monographs devoted to problems of bifurcation and stability. Most of these are not elementary or, if elementary, they are too biased toward particular applications whose study, however meritorious, involves many details of application which are not intrinsic or central to the problems of bifurcation and stability. A partial list of review articles, collections of papers, books, and monographs which may help students after they have mastered the elementary theory is given below.

Arnold, V. I. *Complements on the Theory of Ordinary Differential Equations*. Moscow: Nauka, 1978 (in Russian).

Amann, H., Bazley, N., Kirchgässner, K. *Applications of Nonlinear Analysis in the Physical Science*. Boston–London–Melbourne: Pitman, 1981.

Gurel O., and Rossler, O., eds. Bifurcation theory and its applications in scientific disciplines. *Annals of the New York Academy of Sciences*, **316**, 1979.

Haken, H., ed. *Synergetics*. Berlin–Heidelberg–New York: Springer-Verlag, 1977.

Iooss, G. *Bifurcation of Maps and Applications*. Lecture Notes, Mathematical Studies. Amsterdam: North-Holland, 1979.

Joseph, D. D., *Stability of Fluid Motions, I and II*. Springer Tracts in Natural Philosophy. Vol. 27 and 28. Berlin–Heidelberg–New York: Springer-Verlag, 1976.

Keller, J. and Antman, S., eds. *Bifurcation Theory and Nonlinear Eigenvalue Problems*. New York: W. A. Benjamin, 1969.

Krasnosel'ski, M. A., *Topological Methods in the Theory of Nonlinear Integral Equations*. New York: Macmillan, 1964.

Marsden, J. and McCracken, M. *The Hopf Bifurcation and Its Applications.* Lecture notes in Applied Mathematical Sciences, Vol. 18. Berlin–Heidelberg–New York: Springer-Verlag, 1976.

Pimbley, G. H. *Eigenfunction Branches of Nonlinear Operators and Their Bifurcations.* Lecture Notes in Mathematics No. 104. Berlin–Heidelberg–New York: Springer-Verlag, 1969.

Rabinowitz, P., ed. *Applications of Bifurcation Theory.* New York: Academic Press, 1977.

Sattinger, D. H. *Topics in Stability and Bifurcation Theory.* Lecture Notes in Mathematics No. 309. Berlin–Heidelberg–New York: Springer-Verlag, 1972.

Sattinger, D. H. *Group Theoretic Methods in Bifurcation Theory.* Lecture Notes in Mathematics No. 762. Berlin–Heidelberg–New York, Springer-Verlag 1980.

Stakgold, I. Branching of solutions of nonlinear equations. *SIAM Review B*, 289 (1971).

Vainberg, M. M., and Trenogin, V. A., The methods of Lyapunov and Schmidt in the theory of nonlinear equations and their further development. *Russ. Math. Surveys* **17** (2): 1 (1962).

CHAPTER II

Bifurcation and Stability of Steady Solutions of Evolution Equations in One Dimension

We consider an evolution equation in $\mathbb{R}^1$ of the form

$$\frac{du}{dt} = F(\mu, u) \tag{II.1}$$

where $F(\cdot, \cdot)$ has two continuous derivatives with respect to μ and u. It is conventional in the study of stability and bifurcation to arrange things so that

$$F(\mu, 0) = 0 \quad \text{for all real numbers } \mu. \tag{II.2}$$

But we shall not require (II.2). Instead we require that equilibrium solutions of (II.1) satisfy $u = \varepsilon$, independent of t and

$$F(\mu, \varepsilon) = 0. \tag{II.3}$$

The study of bifurcation of equilibrium solutions of the autonomous problem (II.1) is equivalent to the study of singular points of the curves (II.3) in the (μ, ε) plane.

II.1 The Implicit Function Theorem

The implicit function theorem is a basic mathematical result used in bifurcation theory. The simplest version of this theorem may be stated as follows:

Let $F(\mu_0, \varepsilon_0) = 0$ and let F be continuously differentiable in some open region containing the point (μ_0, ε_0) of the (μ, ε) plane. Then, if $F_\varepsilon(\mu_0, \varepsilon_0) \neq 0$, there exist $\alpha > 0$ and $\beta > 0$ such that:

(i) *The equation $F(\mu, \varepsilon) = 0$ has a unique solution $\varepsilon = \varepsilon(\mu)$ when $\mu_0 - \alpha < \mu < \mu_0 + \alpha$ such that $\varepsilon_0 - \beta < \varepsilon < \varepsilon_0 + \beta$.*

(ii) *The function* $\varepsilon(\cdot)$ *is continuously differentiable when* $\mu_0 - \alpha < \mu < \mu_0 + \alpha$.

(iii) $\varepsilon_\mu(\mu) = -F_\mu(\mu, \varepsilon(\mu))/F_\varepsilon(\mu, \varepsilon(\mu))$.

Remark 1. We can solve for $\mu(\varepsilon)$ if $F_\mu(\mu_0, \varepsilon_0) \neq 0$.

Remark 2. If F is analytic so is $\mu(\varepsilon)$ or $\varepsilon(\mu)$.

Remark 3. Suppose we wish to solve the equation

$$F[\mu, \varepsilon^{(1)}, \ldots, \varepsilon^{(n)}] = 0$$

for μ. If $F(\mu_0, \varepsilon_0^{(1)}, \ldots, \varepsilon_0^{(n)}) = 0$ and $F_\mu(\mu_0, \varepsilon_0^{(1)}, \ldots, \varepsilon_0^{(n)}) \neq 0$, the implicit function theorem holds with $\varepsilon_0^{(k)} - \beta_k < \varepsilon^{(k)} < \varepsilon_0^{(k)} + \beta_k$, $k = 1, \ldots, n$ and we obtain a unique function $\mu = \mu(\varepsilon^{(1)}, \ldots, \varepsilon^{(n)})$ in the interval $\mu_0 - \alpha < \mu < \mu_0 + \alpha$.

Remark 4. The proof of the implicit function theorem is given in nearly every book on advanced calculus and is omitted here.

II.2 Classification of Points on Solution Curves

In our study of equilibrium solutions (II.3) it is desirable to introduce the following classification of points.

(i) A *regular point* of $F(\mu, \varepsilon) = 0$ is one for which the implicit function theorem works:

$$F_\mu \neq 0 \quad \text{or} \quad F_\varepsilon \neq 0. \tag{II.4}$$

If (II.4) holds, then we can find a unique curve $\mu = \mu(\varepsilon)$ or $\varepsilon = \varepsilon(\mu)$ through the point.

(ii) A *regular turning point* is a point at which $\mu_\varepsilon(\varepsilon)$ changes sign and $F_\mu(\mu, \varepsilon) \neq 0$.

(iii) A *singular point* of the curve $F(\mu, \varepsilon) = 0$ is a point at which

$$F_\mu = F_\varepsilon = 0. \tag{II.5}$$

(iv) A *double point* of the curve $F(\mu, \varepsilon) = 0$ is a singular point through which pass two and only two branches of $F(\mu, \varepsilon) = 0$ possessing distinct tangents. We shall assume that all second derivatives of F do not simultaneously vanish at a double point.

(v) A *singular turning (double) point* of the curve $F(\mu, \varepsilon) = 0$ is a double point at which μ_ε changes sign on one branch.

(vi) A *cusp point* of the curve $F(\mu, \varepsilon) = 0$ is a point of second order contact between two branches of the curve. The two branches of the curve have the same tangent at a cusp point.

(vii) A *conjugate point* is an isolated singular point solution of $F(\mu, \varepsilon) = 0$.

(viii) A *higher-order singular point* of the curve $F(\mu, \varepsilon) = 0$ is a singular point at which all three second derivatives of $F(\mu, \varepsilon)$ are null.

Remarks. The elementary theory of singular points of plane curves is discussed in many books on classical analysis; for example, see R. Courant, *Differential and Integral Calculus*, Vol. II, Chap. III (New York: Interscience, 1956). To complete the study of bifurcation in $\mathbb{R}^1$ we shall also need to study the stability of the bifurcating solutions (see Sections II.8–II.14 extending results presented by D. D. Joseph, Factorization theorems and repeated branching of solution at a simple eigenvalue, *Annals of the New York Academy of Sciences*, **316**, 150–167 (1979)).

II.3 The Characteristic Quadratic. Double Points, Cusp Points, and Conjugate Points

It is necessary to be precise about double points. Suppose (μ_0, ε_0) is a singular point. Then equilibrium curves passing through the singular points satisfy

$$2F(\mu, \varepsilon) = F_{\mu\mu}\delta\mu^2 + 2F_{\varepsilon\mu}\delta\varepsilon\delta\mu + F_{\varepsilon\varepsilon}\delta\varepsilon^2 + o[(|\delta\mu| + |\delta\varepsilon|)^2] = 0 \quad \text{(II.6)}$$

where $\delta\mu = \mu - \mu_0$, $\delta\varepsilon = \varepsilon - \varepsilon_0$ and $F_{\mu\mu} = F_{\mu\mu}(\mu_0, \varepsilon_0)$, etc. In the limit, as $(\mu, \varepsilon) \to (\mu_0, \varepsilon_0)$ the equation (II.6) for the curves $F(\mu, \varepsilon) = 0$ reduces to the quadratic equation

$$F_{\mu\mu}\,d\mu^2 + 2F_{\varepsilon\mu}\,d\varepsilon\,d\mu + F_{\varepsilon\varepsilon}d\varepsilon^2 = 0. \quad \text{(II.7)}$$

for the tangents to the curve. We find that

$$\begin{bmatrix} \mu_\varepsilon^{(1)}(\varepsilon_0) \\ \mu_\varepsilon^{(2)}(\varepsilon_0) \end{bmatrix} = -\frac{F_{\varepsilon\mu}}{F_{\mu\mu}}\begin{bmatrix} 1 \\ 1 \end{bmatrix} + \sqrt{\frac{D}{F_{\mu\mu}^2}}\begin{bmatrix} 1 \\ -1 \end{bmatrix} \quad \text{(II.8)}$$

or

$$\begin{bmatrix} \varepsilon_\mu^{(1)}(\mu_0) \\ \varepsilon_\mu^{(2)}(\mu_0) \end{bmatrix} = -\frac{F_{\varepsilon\mu}}{F_{\varepsilon\varepsilon}}\begin{bmatrix} 1 \\ 1 \end{bmatrix} - \sqrt{\frac{D}{F_{\varepsilon\varepsilon}^2}}\begin{bmatrix} 1 \\ -1 \end{bmatrix} \quad \text{(II.9)}$$

where

$$D = F_{\varepsilon\mu}^2 - F_{\mu\mu}F_{\varepsilon\varepsilon}. \quad \text{(II.10)}$$

If $D < 0$ there are no real tangents through (μ_0, ε_0) and the point (μ_0, ε_0) is an isolated (conjugate) point solution of $F(\mu, \varepsilon) = 0$.

We shall consider the case when (μ_0, ε_0) is *not* a higher-order singular point. Then (μ_0, ε_0) is a double point if and only if $D > 0$. If two curves pass through the singular point and $D = 0$ then the slope at the singular point of higher-order contact is given by (II.8) or (II.9). If $D > 0$ and $F_{\mu\mu} \neq 0$, then there are two tangents with slopes $\mu_\varepsilon^{(1)}(\varepsilon_0)$ and $\mu_\varepsilon^{(2)}(\varepsilon_0)$ given by (II.8). If $D > 0$ and $F_{\mu\mu} = 0$, then $F_{\varepsilon\mu} \neq 0$ and

$$d\varepsilon[2\,d\mu F_{\varepsilon\mu} + d\varepsilon F_{\varepsilon\varepsilon}] = 0 \tag{II.11}$$

and there are two tangents $\varepsilon_\mu(\mu_0) = 0$ and $\mu_\varepsilon(\varepsilon_0) = -F_{\varepsilon\varepsilon}/2F_{\varepsilon\mu}$. If $\varepsilon_\mu(\mu_0) = 0$ then $F_{\mu\mu}(\mu_0, \varepsilon_0) = 0$. So all possibilities are covered in the following two cases:

(A) $D > 0$, $F_{\mu\mu} \neq 0$ with tangents $\mu_\varepsilon^{(1)}(\varepsilon_0)$ and $\mu_\varepsilon^{(2)}(\varepsilon_0)$.

(B) $D > 0$, $F_{\mu\mu} = 0$ with tangents $\varepsilon_\mu(\mu_0) = 0$ and $\mu_\varepsilon(\varepsilon_0) = -F_{\varepsilon\varepsilon}/2F_{\varepsilon\mu}$.

II.4 Double-Point Bifurcation and the Implicit Function Theorem

Solutions (μ, ε) of $F(\mu, \varepsilon) = 0$ are said to undergo double-point bifurcation at (μ_0, ε_0) if two curves with distinct tangents pass through (μ_0, ε_0). We suppose $D > 0$ and use the implicit function theorem to find the curves. Consider case (A) specified in the last paragraph of §II.3 and define a to be determined function $v(\varepsilon)$ satisfying the equation $\mu - \mu_0 = v(\varepsilon)(\varepsilon - \varepsilon_0)$ and such that

$$v_0 \stackrel{\text{def}}{=} v(\varepsilon_0) = \mu_\varepsilon(\varepsilon_0)$$

where $\mu_\varepsilon(\varepsilon_0)$ has one of the two values $\mu_\varepsilon^{(1)}, \mu_\varepsilon^{(2)}$ given by (II.8) as the solution of the characteristic quadratic equation. Now define

$$\begin{aligned} G(v, \varepsilon) &\stackrel{\text{def}}{=} \frac{2F(\mu, \varepsilon)}{(\varepsilon - \varepsilon_0)^2} \\ &= F_{\mu\mu}v^2 + 2F_{\varepsilon\mu}v + F_{\varepsilon\varepsilon} \\ &\quad + \tfrac{1}{3}\{F_{\varepsilon\varepsilon\varepsilon} + 3F_{\varepsilon\varepsilon\mu}v + 3F_{\varepsilon\mu\mu}v^2 + F_{\mu\mu\mu}v^3\}(\varepsilon - \varepsilon_0) + o(|\varepsilon - \varepsilon_0|). \end{aligned} \tag{II.12}$$

We have defined G so that

$$G(v_0, \varepsilon_0) = F_{\mu\mu}v_0^2 + 2F_{\varepsilon\mu}v_0 + F_{\varepsilon\varepsilon} = 0$$

for both choices of v_0. Moreover, differentiation of (II.12) using (II.8) shows that

$$G_v(v_0, \varepsilon_0) = 2(\mu_\varepsilon(\varepsilon_0)F_{\mu\mu} + F_{\varepsilon\mu}) = \pm 2\sqrt{D}\,\operatorname{sgn} F_{\mu\mu} \neq 0. \tag{II.13}$$

So the existence of two functions $v^{(1)}(\varepsilon)$ and $v^{(2)}(\varepsilon)$ with $v^{(1)}(\varepsilon_0) = \mu_\varepsilon^{(1)}(\varepsilon_0)$ and $v^{(2)}(\varepsilon_0) = \mu_\varepsilon^{(2)}(\varepsilon_0)$ is guaranteed by the implicit function theorem.

We leave the strict proof of bifurcation for case (B) using the implicit function theorem as an exercise for the reader.

II.5 Cusp-Point Bifurcation and Characteristic Quadratics

We now suppose that $F(\cdot, \cdot)$ has four continuous partial derivatives and show what happens at a cusp point of second-order contact. When $\mu = \mu(\varepsilon)$ all derivatives $\mathscr{F}(\varepsilon) \equiv F(\mu(\varepsilon), \varepsilon) \equiv 0$ vanish. Then we have

$$\frac{d^2\mathscr{F}}{d\varepsilon^2} = F_{\varepsilon\varepsilon} + 2\mu_\varepsilon F_{\varepsilon\mu} + \mu_\varepsilon^2 F_{\mu\mu} + \mu_{\varepsilon\varepsilon} F_\mu = 0, \tag{II.14}$$

$$\begin{aligned}\frac{d^3\mathscr{F}}{d\varepsilon^3} &= F_{\varepsilon\varepsilon\varepsilon} + 3\mu_\varepsilon F_{\varepsilon\varepsilon\mu} + 3\mu_\varepsilon^2 F_{\varepsilon\mu\mu} + \mu_\varepsilon^3 F_{\mu\mu\mu} \\ &\quad + 3\mu_{\varepsilon\varepsilon} F_{\varepsilon\mu} + 3\mu_{\varepsilon\varepsilon}\mu_\varepsilon F_{\mu\mu} + \mu_{\varepsilon\varepsilon\varepsilon} F_\mu = 0,\end{aligned} \tag{II.15}$$

$$\begin{aligned}\frac{d^4\mathscr{F}}{d\varepsilon^4} &= F_{\varepsilon\varepsilon\varepsilon\varepsilon} + 4\mu_\varepsilon F_{\varepsilon\varepsilon\varepsilon\mu} + 6\mu_\varepsilon^2 F_{\varepsilon\varepsilon\mu\mu} + 4\mu_\varepsilon^3 F_{\varepsilon\mu\mu\mu} \\ &\quad + \mu_\varepsilon^4 F_{\mu\mu\mu\mu} + 4\mu_{\varepsilon\varepsilon\varepsilon} F_{\varepsilon\mu} + 4\mu_{\varepsilon\varepsilon\varepsilon}\mu_\varepsilon F_{\mu\mu} + 3\mu_{\varepsilon\varepsilon}^2 F_{\mu\mu} \\ &\quad + 6\mu_{\varepsilon\varepsilon} F_{\varepsilon\varepsilon\mu} + 12\mu_\varepsilon\mu_{\varepsilon\varepsilon} F_{\varepsilon\mu\mu} + 6\mu_{\varepsilon\varepsilon}\mu_\varepsilon^2 F_{\mu\mu\mu} \\ &\quad + \mu_{\varepsilon\varepsilon\varepsilon\varepsilon} F_\mu = 0.\end{aligned} \tag{II.16}$$

When $\varepsilon = \varepsilon(\mu)$, $\mathscr{f}(\mu) \equiv F(\mu, \varepsilon(\mu)) \equiv 0$ and

$$\begin{aligned}\frac{d\mathscr{f}^2}{d\mu^2} &= F_{\mu\mu} + 2\varepsilon_\mu F_{\varepsilon\mu} + \varepsilon_\mu^2 F_{\varepsilon\varepsilon} + \varepsilon_{\mu\mu} F_\varepsilon = 0, \\ \frac{d^3\mathscr{f}}{d\mu^3} &= F_{\mu\mu\mu} + 3\varepsilon_\mu F_{\varepsilon\mu\mu} + 3\varepsilon_\mu^2 F_{\varepsilon\varepsilon\mu} + \varepsilon_\mu^3 F_{\varepsilon\varepsilon\varepsilon} \\ &\quad + 3\varepsilon_{\mu\mu} F_{\varepsilon\mu} + 3\varepsilon_{\mu\mu}\varepsilon_\mu F_{\varepsilon\varepsilon} + \varepsilon_{\mu\mu\mu} F_\varepsilon = 0,\end{aligned} \tag{II.17}$$

$$\begin{aligned}\frac{d^4\mathscr{f}}{d\mu^4} &= F_{\mu\mu\mu\mu} + 4\varepsilon_\mu F_{\varepsilon\mu\mu\mu} + 6\varepsilon_\mu^2 F_{\varepsilon\varepsilon\mu\mu} + 4\varepsilon_\mu^3 F_{\varepsilon\varepsilon\varepsilon\mu} \\ &\quad + \varepsilon_\mu^4 F_{\varepsilon\varepsilon\varepsilon\varepsilon} + 4\varepsilon_{\mu\mu\mu} F_{\varepsilon\mu} + 4\varepsilon_{\mu\mu\mu}\varepsilon_\mu F_{\varepsilon\varepsilon} + 3\varepsilon_{\mu\mu}^2 F_{\varepsilon\varepsilon} \\ &\quad + 6\varepsilon_{\mu\mu} F_{\varepsilon\mu\mu} + 12\varepsilon_\mu\varepsilon_{\mu\mu} F_{\varepsilon\varepsilon\mu} + 6\varepsilon_{\mu\mu}\varepsilon_\mu^2 F_{\varepsilon\varepsilon\varepsilon} \\ &\quad + \varepsilon_{\mu\mu\mu\mu} F_\varepsilon = 0.\end{aligned} \tag{II.18}$$

At a cusp point $F = F_\varepsilon = F_\mu = D = 0$. In case (A), $F_{\mu\mu} \neq 0$ $\mu_\varepsilon(\varepsilon_0) = -F_{\varepsilon\mu}/F_{\mu\mu}$, (II.14) is satisfied identically, (II.15) becomes

$$F_{\varepsilon\varepsilon\varepsilon} + 3\mu_\varepsilon(\varepsilon_0)F_{\varepsilon\varepsilon\mu} + 3\mu_\varepsilon^2(\varepsilon_0)F_{\varepsilon\mu\mu} + \mu_\varepsilon^3 F_{\mu\mu\mu} = 0$$

and the coefficient of $\mu_{\varepsilon\varepsilon\varepsilon}$ in (II.16) vanishes, leaving a quadratic equation for the curvature $\mu_{\varepsilon\varepsilon}$ at ε_0:

$$\mu_{\varepsilon\varepsilon}^2 + 2\left(\frac{\mu_{\varepsilon\varepsilon}\xi}{F_{\mu\mu}}\right) + \left(\frac{\zeta}{F_{\mu\mu}}\right) = 0, \tag{II.19}$$

where

$$\xi = F_{\varepsilon\varepsilon\mu} + 2\mu_\varepsilon F_{\varepsilon\mu\mu} + \mu_\varepsilon^2 F_{\mu\mu\mu}$$

and

$$3\zeta = F_{\varepsilon\varepsilon\varepsilon\varepsilon} + 4\mu_\varepsilon F_{\varepsilon\varepsilon\varepsilon\mu} + 6\mu_\varepsilon^2 F_{\varepsilon\varepsilon\mu\mu} + 4\mu_\varepsilon^3 F_{\varepsilon\mu\mu\mu} + \mu_\varepsilon^4 F_{\mu\mu\mu\mu}.$$

Equation (II.19) has two roots

$$\begin{bmatrix} \mu_{\varepsilon\varepsilon}^{(1)}(\varepsilon_0) \\ \mu_{\varepsilon\varepsilon}^{(2)}(\varepsilon_0) \end{bmatrix} = -\frac{\xi}{F_{\mu\mu}}\begin{bmatrix} 1 \\ 1 \end{bmatrix} + \sqrt{\frac{\mathscr{D}_1}{F_{\mu\mu}^2}}\begin{bmatrix} 1 \\ -1 \end{bmatrix}, \tag{II.20}$$

where

$$\mathscr{D}_1 = \xi^2 - F_{\mu\mu}\zeta. \tag{II.21}$$

In case (B), $F_{\mu\mu} = 0$, and since $D = 0$, $F_{\varepsilon\mu} = 0$, $F_{\varepsilon\varepsilon} \neq 0$ and $\varepsilon_\mu(\mu_0) = 0$. (II.17) then shows that $F_{\mu\mu\mu} = 0$ and (II.18) reduces to a quadratic equation for the curvature $\varepsilon_{\mu\mu}(\mu_0)$:

$$\varepsilon_{\mu\mu}^2 + 2\left(\frac{\varepsilon_{\mu\mu}F_{\varepsilon\mu\mu}}{F_{\varepsilon\varepsilon}}\right) + \left(\frac{F_{\mu\mu\mu\mu}}{3F_{\varepsilon\varepsilon}}\right) = 0. \tag{II.22}$$

Equation (II.22) has two roots,

$$\begin{bmatrix} \varepsilon_{\mu\mu}^{(1)}(\mu_0) \\ \varepsilon_{\mu\mu}^{(2)}(\mu_0) \end{bmatrix} = -\frac{F_{\varepsilon\mu\mu}}{F_{\varepsilon\varepsilon}}\begin{bmatrix} 1 \\ 1 \end{bmatrix} + \sqrt{\frac{\mathscr{D}_2}{F_{\varepsilon\varepsilon}^2}}\begin{bmatrix} 1 \\ -1 \end{bmatrix} \tag{II.23}$$

where

$$\mathscr{D}_2 = F_{\varepsilon\mu\mu}^2 - \left(\frac{F_{\varepsilon\varepsilon}F_{\mu\mu\mu\mu}}{3}\right). \tag{II.24}$$

At a point of second-order contact the two curves have common tangents and different real-valued curvatures. It follows that $\mathscr{D}_1 > 0$ or $\mathscr{D}_2 > 0$ at a point of second-order contact.

The implicit function theorem may be used to show that the curvatures defined by (II.20) and (II.23) belong to real curves passing through the cusp point.

II.6 Triple-Point Bifurcation

We turn next to the case in which all second-order derivatives of $F(\cdot, \cdot)$ are null at a singular point. Confining our attention to the case in which $F_{\mu\mu\mu} \neq 0$ we may write (II.15) as

$$(\mu_\varepsilon - \mu_\varepsilon^{(1)})(\mu_\varepsilon - \mu_\varepsilon^{(2)})(\mu_\varepsilon - \mu_\varepsilon^{(3)}) = \mu_\varepsilon^3 + 3\mu_\varepsilon^2 \frac{F_{\varepsilon\mu\mu}}{F_{\mu\mu\mu}} + 3\mu_\varepsilon \frac{F_{\varepsilon\varepsilon\mu}}{F_{\mu\mu\mu}} + \frac{F_{\varepsilon\varepsilon\varepsilon}}{F_{\mu\mu\mu}} = 0. \quad \text{(II.25)}$$

where $\mu_\varepsilon^{(1)}$, $\mu_\varepsilon^{(2)}$ and $\mu_\varepsilon^{(3)}$ are values of $\mu_\varepsilon(\varepsilon)$ at $\varepsilon = \varepsilon_0$. It follows from (II.25) that

$$\frac{F_{\varepsilon\varepsilon\mu}}{F_{\mu\mu\mu}} = \frac{1}{3}(\mu_\varepsilon^{(1)}\mu_\varepsilon^{(2)} + \mu_\varepsilon^{(1)}\mu_\varepsilon^{(3)} + \mu_\varepsilon^{(2)}\mu_\varepsilon^{(3)}), \quad \text{(II.26)}$$

$$\frac{F_{\varepsilon\mu\mu}}{F_{\mu\mu\mu}} = -\frac{1}{3}(\mu_\varepsilon^{(1)} + \mu_\varepsilon^{(2)} + \mu_\varepsilon^{(3)}) \quad \text{(II.27)}$$

and

$$\frac{F_{\varepsilon\varepsilon\varepsilon}}{F_{\mu\mu\mu}} = -\mu_\varepsilon^{(1)}\mu_\varepsilon^{(2)}\mu_\varepsilon^{(3)}.$$

If the three roots of (II.25) are real and distinct, three bifurcating solutions pass through the singular point (μ_0, ε_0). If two roots are complex, then there is no bifurcation. The formulas (II.26, 27) are useful in relating the stability of bifurcation to the shape of the bifurcating curves at a triple point.

II.7 Conditional Stability Theorem

Some of the solutions which bifurcate are stable and some are unstable. To study the stability of the solution $u = \varepsilon$ we very often study the linearized equation

$$Z_t = F_\varepsilon(\mu, \varepsilon)Z, \quad \text{(II.28)}$$

the general solution of which is

$$Z = e^{\sigma t}Z_0, \quad \text{(II.29)}$$

where

$$\sigma = F_\varepsilon(\mu, \varepsilon). \quad \text{(II.30)}$$

Since all solutions of (II.28) are in the form (II.29) we find that disturbances Z or ε grow when $\sigma > 0$ and decay when $\sigma < 0$. The linearized theory then implies that $(\mu(\varepsilon), \varepsilon)$ satisfying $F(\mu, \varepsilon) = 0$ is stable when $\sigma < 0$ and is unstable when $\sigma > 0$.

Now we shall demonstrate that the conclusion of linearized theory holds for the nonlinear equations provided that the disturbance is not too large. Let v be a disturbance of ε, $u = \varepsilon + v$, where

$$\frac{dv}{dt} = F(\mu(\varepsilon), \varepsilon + v) - F(\mu(\varepsilon), \varepsilon)$$
$$= F_\varepsilon(\mu(\varepsilon), \varepsilon)v + R(\varepsilon, v), \tag{II.31}$$

where

$$|R(\varepsilon, v)| \le K|v|^2 \tag{II.32}$$

when $|v|$ is small enough. We want to show that near the origin v is like $Z(t) = e^{\sigma t}Z_0$, $\sigma = F_\varepsilon(\mu(\varepsilon), \varepsilon)$ and tends to zero exponentially or increases exponentially according as $\sigma < 0$ or $\sigma > 0$. We may write (II.31) as

$$\frac{d}{dt}(ve^{-\sigma t}) = R(\varepsilon, v)e^{-\sigma t}. \tag{II.33}$$

Hence

$$v(t)e^{-\sigma t} = v(0) + \int_0^t R(\varepsilon, v(s))e^{-\sigma s}\, ds \tag{II.34}$$

and, using (II.32), we find that

$$|v(t) - v(0)e^{\sigma t}| \le K\int_0^t e^{\sigma(t-s)}|v(s)|^2\, ds. \tag{II.35}$$

We want to show that (II.35) implies that $v(t) \to 0$ as $t \to \infty$ if $\sigma(\varepsilon) < 0$ and $|v(0)|$ is small enough. Suppose $\sigma < 0$. Then there is $\eta > 0$ such that $\sigma(\varepsilon) + \eta < 0$. Now, for the time being, assume that

$$|v(t)| \le \frac{\eta}{K} \quad \text{for all } t \ge 0. \tag{II.36}$$

Combining (II.35) and (II.36) we find that

$$|v(t) - v(0)e^{\sigma t}| \le \eta e^{\sigma t}\int_0^t e^{-\sigma s}|v(s)|\, ds. \tag{II.37}$$

From (II.37) we deduce that

$$|v(t)| \le |v(0)|e^{\sigma t} + \eta e^{\sigma t}\int_0^t e^{-\sigma s}|v(s)|\, ds. \tag{II.38}$$

Setting

$$y(t) \stackrel{\text{def}}{=} \int_0^t e^{-\sigma s}|v(s)|\, ds \tag{II.39}$$

we have (II.38) in the form

$$0 \le \dot{y}(t) \le |v(0)| + \eta y(t),\ y(0) = 0. \tag{II.40}$$

Multiplication of (II.40) by the integrating factor $e^{-\eta t}$ followed by integration leads to

$$y(t)e^{-\eta t} \le \{1 - e^{-\eta t}\}|v(0)|/\eta. \tag{II.41}$$

Returning now to (II.37) with (II.41) we find that

$$|v(t) - v(0)e^{\sigma t}| \le |v(0)|e^{(\sigma+\eta)t}. \tag{II.42}$$

Equation (II.42) shows that $v(t) \to 0$ exponentially when $\sigma(\varepsilon) < 0$ and that $|v(t)| \le \eta/K$ is satisfied for all $t \ge 0$ when $|v(0)|$ is small enough.

We have shown that $(\mu(\varepsilon), \varepsilon)$ is exponentially stable when $\sigma(\varepsilon) = F_\varepsilon(\mu(\varepsilon), \varepsilon) < 0$ and $|v(0)|$ is small enough. The condition on $|v(0)|$ is the reason that $(\mu(\varepsilon), \varepsilon)$ is said to be conditionally stable. If $|v(0)|$ were unrestricted we would have unconditional or global stability. Global stability is a rare property since it implies that at a fixed value of μ there is just one steady solution $u = \varepsilon$ which attracts all solutions of (II.1) at that fixed μ. Frequently more than one solution exists at the same μ and each stable equilibrium solution attracts its own restricted set of u's (see Fig. II.7).

We next demonstrate that the null-solution equation (II.31) is unstable when $\sigma > 0$. Equation (II.34) is still valid. Hence if we assume that

$$|v(t)| \le \varepsilon$$

for all $t \ge 0$, and let $t \to \infty$ in (II.34), we obtain

$$v(0) = -\int_0^\infty R(\varepsilon, v(s))e^{-\sigma s}\, ds.$$

Then we may rewrite (II.34) as

$$v(t) = -\int_t^\infty R(\varepsilon, v(s))e^{\sigma(t-s)}\, ds.$$

Now, using the estimate (II.32), we have

$$|v(t)| \le K\varepsilon^2 \int_t^\infty e^{\sigma(t-s)}\, ds = \frac{K\varepsilon^2}{\sigma},$$

and ε has to satisfy

$$\varepsilon \le \frac{K\varepsilon^2}{\sigma}. \tag{II.43}$$

Choose $v(0) \neq 0$, and $\varepsilon < \sigma/K$, contradicting (II.43). It is then impossible to maintain $|v(t)| \leq \varepsilon$ for all t. Hence the solution $v(t)$ leaves the fixed interval at a time $t_0 < +\infty$. This completes the proof of the equivalence between the linear result and the nonlinear one for the stability of the null solution of (II.31).

Remarks. The proof of the conditional stability theorem in $\mathbb{R}^1$ follows the proof of the classical theorem of Lyapunov for systems in $\mathbb{R}^n$ (for example, see Coddington and Levinson, *Theory of Ordinary Differential Equations* (New York: McGraw-Hill, 1965), Chapter 3).

II.8 The Factorization Theorem in One Dimension

Theorem 1 (Factorization Theorem). *For every equilibrium solution $F(\mu, \varepsilon) = 0$ for which $\mu = \mu(\varepsilon)$ we have*

$$\sigma(\varepsilon) = F_\varepsilon(\mu(\varepsilon), \varepsilon) = -\mu_\varepsilon(\varepsilon) F_\mu(\mu(\varepsilon), \varepsilon) \stackrel{\text{def}}{=} -\mu_\varepsilon \hat{\sigma}(\varepsilon). \qquad \text{(II.44)}$$

The proof of Theorem 1 follows from the equation

$$\frac{dF(\mu(\varepsilon), \varepsilon)}{d\varepsilon} = F_\varepsilon(\mu(\varepsilon), \varepsilon) + \mu_\varepsilon(\varepsilon) F_\mu(\mu(\varepsilon), \varepsilon) = 0.$$

This type of factorization may be proved for the stability of bifurcating solutions in spaces more complicated than $\mathbb{R}^1$. But the theorem is most easily understood in $\mathbb{R}^1$. One of the main implications of the factorization theorem is that $\sigma(\varepsilon)$ *must change sign as ε is varied across a regular turning point*. This implies that the solution $u = \varepsilon$, $\mu = \mu(\varepsilon)$ is stable on one side of a regular turning point and is unstable on the other side (Figure II.1).

Corollary 1. (A) *Any point (μ_0, ε_0) of the curve $\mu = \mu(\varepsilon)$ for which $\hat{\sigma}(\varepsilon_0) = 0$, is a singular point.* **(B)** *Any point (μ_0, ε_0) of the curve $\varepsilon(\mu)$ for which $\sigma(\mu_0) = 0$ is a singular point.*

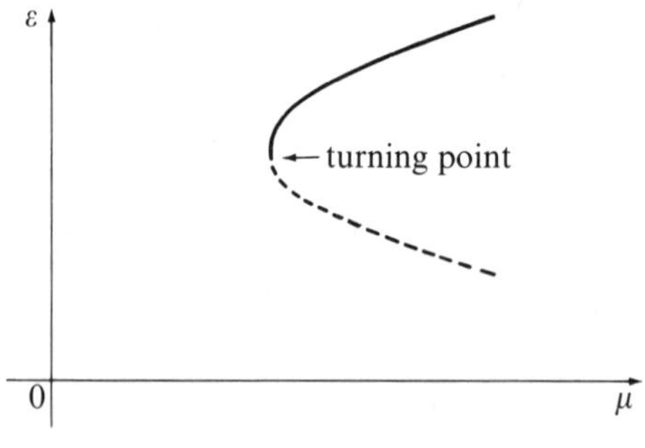

Figure II.1 Exchange of stability at a regular turning point

The proof of (A) follows from (II.44) and the proof of (B) from

$$\sigma(\mu) = F_\varepsilon(\mu, \varepsilon(\mu)), \qquad \frac{dF}{d\mu} = F_\mu + \varepsilon_\mu F_\varepsilon = 0. \tag{II.45}$$

II.9 Equivalence of Strict Loss of Stability and Double-Point Bifurcation

We connect the study of stability to the study of bifurcation under the "strict crossing" hypothesis introduced by Hopf* and used in almost all studies of bifurcation and stability. This hypothesis restricts the study of bifurcation to *double points*; cusp points and higher-order singular points are excluded.

Corollary 2. *Suppose that* (μ_0, ε_0) *is a singular point and* (A) $\sigma_\varepsilon(\varepsilon_0) \neq 0$ *or* (B) $\sigma_\mu(\mu_0) \neq 0$. *Then* (μ_0, ε_0) *is a double point.*

In case (A) we find from (II.44) that at the singular point $(\mu(\varepsilon_0), \varepsilon_0)$

$$\sigma_\varepsilon(\varepsilon_0) = F_{\varepsilon\varepsilon} + \mu_\varepsilon F_{\varepsilon\mu} = -\mu_\varepsilon^2 F_{\mu\mu} - \mu_\varepsilon F_{\varepsilon\mu} \neq 0. \tag{II.46}$$

Equation (II.46) shows that the characteristic quadratic equation (II.7) holds at $(\mu(\varepsilon_0), \varepsilon_0)$. Since there is a curve through this point, $D \geq 0$ and we need to show that $D \neq 0$. We shall assume that $D = F_{\varepsilon\mu}^2 - F_{\mu\mu}F_{\varepsilon\varepsilon} = 0$ and show that this assumption contradicts (II.46). We first note that (II.46) implies that not all three of the second derivatives of F are null at $(\mu(\varepsilon_0), \varepsilon_0)$. If $F_{\mu\mu}F_{\varepsilon\varepsilon} \neq 0$ and $D = 0$ then (II.8) becomes $\mu_\varepsilon(\varepsilon_0) = -F_{\varepsilon\mu}/F_{\mu\mu}$ and (II.46) may be written as $F_{\varepsilon\varepsilon} - (F_{\varepsilon\mu}^2/F_{\mu\mu}) = -D/F_{\mu\mu} \neq 0$. So $D \neq 0$ after all. If $F_{\mu\mu}F_{\varepsilon\varepsilon} = 0$ and $D = 0$ then $F_{\varepsilon\mu} = 0$ and (II.46) may be written as $\sigma_\varepsilon = F_{\varepsilon\varepsilon} = -\mu_\varepsilon^2 F_{\mu\mu} \neq 0$. So $D \neq 0$ after all.

In case (B) we solve $F(\mu, \varepsilon) = 0$ for $\varepsilon(\mu)$. At the singular point (μ_0, ε_0), we have strict loss of stability because $\sigma_\mu = F_{\varepsilon\mu} + F_{\varepsilon\varepsilon}\varepsilon_\mu = F_{\varepsilon\mu} = \sqrt{D}\,\operatorname{sgn} F_{\varepsilon\mu}$.

II.10 Exchange of Stability at a Double Point

It is possible to make precise statements about the stability of solutions near double points. All of the possibilities for the stability of double-point bifurcation can be described by the cases (A) and (B) which were fully specified under (II.11). In case (A) two curves $\mu^{(1)}(\varepsilon)$ and $\mu^{(2)}(\varepsilon)$ pass through the

* E. Hopf, Abzweigung einer periodischen Lösung von einer stationären Lösung eines Differentialsystems, *Berichten der Mathematisch-Physischen Klasse der Sächsischen Akademie der Wissenschaften zu Leipzig XCIV*, 1–22 (1942). An English translation of this paper by L. N. Howard can be found in the book by Marsden and McCracken (see Notes to Chapter I).

double point (μ_0, ε_0). In case (B) two curves, $\varepsilon^{(1)}(\mu)$ (with $\varepsilon_\mu^{(1)}(\mu_0) = 0$) and $\mu^{(2)}(\varepsilon)$, pass through the double point. The eigenvalue $\sigma^{(1)}$ belongs to the curve with superscript (1) and $\sigma^{(2)}$ to the curve with superscript (2).

Theorem 2. *Suppose (μ_0, ε_0) is a double point. Then, in case* (A),

$$\sigma^{(1)}(\varepsilon) = -\mu_\varepsilon^{(1)}(\varepsilon)\{\hat{s}\sqrt{D}(\varepsilon - \varepsilon_0) + o(\varepsilon - \varepsilon_0)\} \tag{II.47}$$

and

$$\sigma^{(2)}(\varepsilon) = \mu_\varepsilon^{(2)}(\varepsilon)\{\hat{s}\sqrt{D}(\varepsilon - \varepsilon_0) + o(\varepsilon - \varepsilon_0)\} \tag{II.48}$$

where $\hat{s} = F_{\mu\mu}/|F_{\mu\mu}|$ and D and $F_{\mu\mu}$ are evaluated at $\varepsilon = \varepsilon_0$. And in case (B),

$$\sigma^{(1)}(\mu) = s\sqrt{D}(\mu - \mu_0) + o(\mu - \mu_0) \tag{II.49}$$

and

$$\sigma^{(2)}(\varepsilon) = -s\mu_\varepsilon^{(2)}(\varepsilon)\{\sqrt{D}(\varepsilon - \varepsilon_0) + o(\varepsilon - \varepsilon_0)\} \tag{II.50}$$

where $s = F_{\varepsilon\mu}/|F_{\varepsilon\mu}|$

Proof. If $\mu = \mu(\varepsilon)$ we have (II.44) in the form,

$$\begin{aligned}\sigma(\varepsilon) &= -\mu_\varepsilon(\varepsilon)F_\mu(\mu(\varepsilon), \varepsilon)\\ &= -\mu_\varepsilon(\varepsilon)\{[F_{\mu\mu}(\mu_0, \varepsilon_0)\mu_\varepsilon(\varepsilon_0) + F_{\varepsilon\mu}(\mu_0, \varepsilon_0)](\varepsilon - \varepsilon_0)\\ &\quad + o(\varepsilon - \varepsilon_0)\}.\end{aligned} \tag{II.51}$$

The formulas (II.47) and (II.48) arise from (II.51) when $\mu_\varepsilon(\varepsilon_0)$ is replaced with the values given by (II.8). If $\varepsilon = \varepsilon(\mu)$ with $\varepsilon_\mu(\mu_0) = 0$ then $F_{\mu\mu}(\mu_0, \varepsilon_0) = 0$, $F_{\varepsilon\mu}^2(\mu_0, \varepsilon_0) = D$, and

$$\begin{aligned}\sigma(\mu) &= F_\varepsilon(\mu, \varepsilon(\mu)) = F_{\varepsilon\mu}(\mu_0, \varepsilon_0)(\mu - \mu_0) + o(\mu - \mu_0)\\ &= s\sqrt{D}(\mu - \mu_0) + o(\mu - \mu_0).\end{aligned}$$

Theorem 2 gives an exhaustive classification relating the stability of solutions near a double point to the slope of the bifurcation curves near that point. The result may be summarized as follows. Suppose $|\varepsilon - \varepsilon_0| > 0$ is small. Then (II.47) and (II.48) show that $\sigma^{(1)}(\varepsilon)$ and $\sigma^{(2)}(\varepsilon)$ have the same (different) sign if $\mu_\varepsilon^{(1)}(\varepsilon)$ and $\mu_\varepsilon^{(2)}(\varepsilon)$ have different (the same) sign. A similar conclusion can be drawn from (II.49) and (II.50). The possible distributions of stability of solutions is sketched in Figure II.2 (dotted lines mean unstable).

Theorem 3. *Assume that all singular points of solutions of $F(\mu, \varepsilon) = 0$ are double points. The stability of such solutions must change at each regular turning point and at each singular point (which is not a turning point), and only at such points.*

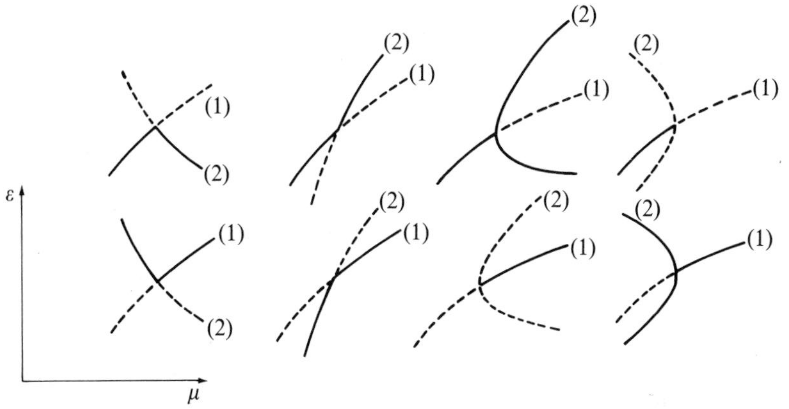

Figure II.2 Stability of solutions in the neighborhood of double-point bifurcation. The double point in each of the eight sketches is (μ_0, ε_0). The top four sketches have $\hat{s} = -1$ and the bottom four have $\hat{s} = 1$. Stability is determined by the sign of the eigenvalue given by (II.47) and (II.48)

II.11 Exchange of Stability at a Double Point for Problems Reduced to Local Form

The analysis of double-point bifurcation is even easier when one first makes the reduction (I.14) to local form. It may be helpful to make a few remarks about the bifurcation diagrams which follow from analysis of (I.14). Nearly all the literature starts from a setup in which $u = 0$ is a solution of the evolution problem. If $F(\mu, 0) = 0$ for all μ then $F_\mu(0, 0) = F_{\mu\mu}(0, 0) = 0$ and the strict loss of stability of the solution $u = 0$ as μ is increased past zero is

$$\sigma_\mu^{(1)}(0) = F_{\mu\varepsilon}(0, 0) \neq 0, \quad \text{say} > 0. \tag{II.52}$$

Then $D = F_{\varepsilon\mu}^2 > 0$ and

$$\sigma^{(2)}(\varepsilon) = -\mu_\varepsilon^{(2)}(\varepsilon)\sigma_\mu^{(1)}(0)\{\varepsilon + o(\varepsilon)\}. \tag{II.53}$$

The bifurcation diagrams which follow from these results and the conventional statements which we make about them are given by the diagrams and caption to Figure II.3.

A marvelous demonstration which can help to fix the ideas embodied in theorem 3 has been found by T. B. Benjamin. Benjamin's demonstration is an example of the buckling of a simple structure under the action of gravity. His apparatus is a board with two holes through which a wire is passed. The wire forms an arch above the board whose arc length is l. The wire which is actually used in Benjamin's demonstration is like a bicycle brake cable: it is wound like a tight coil spring and covered with a plastic sheath. The demonstration apparatus is sketched in Figure II.4.

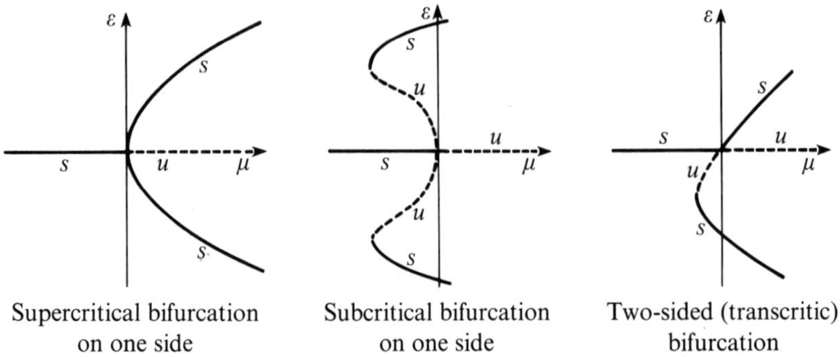

Figure II.3 Stability of solutions bifurcating from $\varepsilon = 0$. Supercritical solutions have $|\varepsilon| > 0$ for values of μ (>0 in the diagram) for which $\varepsilon = 0$ is unstable. Subcritical solutions have $|\varepsilon| > 0$ for values of μ for which $\varepsilon = 0$ is stable

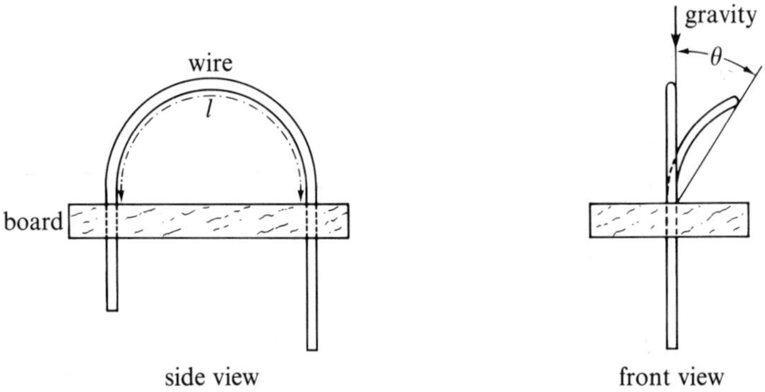

Figure II.4 Benjamin's apparatus for demonstrating the buckling of a wire arch under gravity loading. The bifurcation diagram which fits this system is shown in Figure II.5. When l is small the only stable solution of (II.54) is the upright one ($\theta = 0$). When $l > l_c$ is large the upright position is unstable and the arch falls to the left or to the right as shown in the front view. The bent position of the wire is also stable when $l < l_c$. When $l_0 < l < l_c$ there are three stable steady solutions, the upright one ($\theta = 0$) and the left or right bent one ($|\theta| \neq 0$)

We imagine that the equation of motion for the wire arch is

$$\frac{d\theta}{dt} = F(l, \theta). \tag{II.54}$$

The steady solutions of (II.54) are imagined to be in the form $F(l(\theta), \theta) = 0$ shown in Figure II.5. Here $\theta = 0$ is one solution (the upright one) and $l(\theta)$ is another solution (the bent arch). In fact there is a one-to-one correspondence between Benjamin's demonstration and the bifurcation diagram (II.5); nothing is seen in the demonstration that does not appear in the diagram and there is nothing in the diagram that is not in the demonstration. The

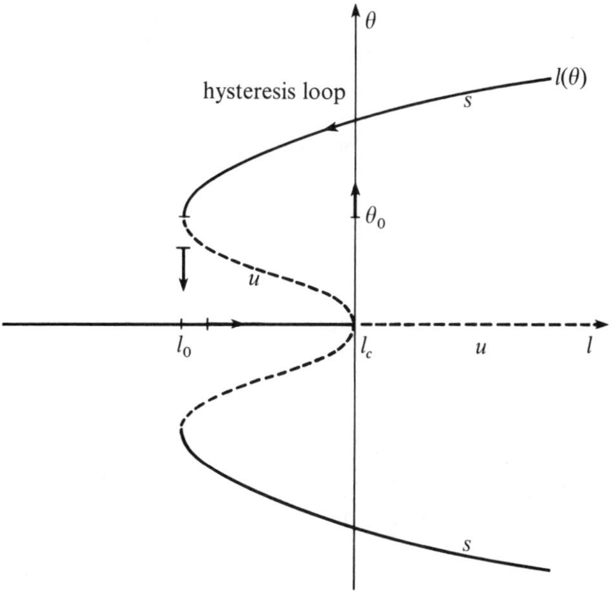

Figure II.5 Bifurcation diagram for the buckling of the wire arch. When l is small the only equilibrium of (II.54) is the upright one ($\theta = 0$). The solution $\theta = 0$ loses stability when $\mu = l - l_c$ is increased past zero. A new solution $\mu(\theta) = l(\theta) - l_c$ corresponding to the bent arch then undergoes double-point bifurcation at a singular turning point $(l, \theta) = (l_c, 0)$. The system is symmetric in θ. When $l > l_c$ only the left and right bent equilibrium configurations are stable. The points $(l, \theta) = (l_0, \pm\theta_0)$ are regular turning points. When $l_0 \leq l \leq l_c$ there are three stable solutions $\theta = 0$ and the symmetric left and right bent positions. In this region the system exhibits hysteresis. If the length l of the arch of the wire above the board is decreased while the wire is bent the bent configuration will continue to be observed until $l = l_0$. When $l = l_0$ the bifurcating bent position is a regular turning point. When $l < l_0$ only $\theta = 0$ is stable. So when l is reduced below l_0 the arch snaps through to the upright solution. Now if we increase l the arch stays in the vertical position until $l = l_c$. When $l > l_c$ the upright solution loses stability and the arch falls back into the left or right stable bent position

interpretation of events in the demonstration is given in the caption for Figure II.5.

Double-point bifurcation is the most common form of bifurcation which can occur at a singular point. Other types of bifurcation, cusp points, triple points, etc., are less common because they require some relationship between higher-order derivatives of $F(\mu, \varepsilon)$. Such situations are sometimes called nongeneric bifurcation. There is a technical mathematical sense for the word *generic* (having to do with dense open coverings), but most of the time the word is used as a fancy alternative for the plain english word *typical*. Analysis of typical problems does not help you if your problem is not typical. For example, it is surely wise to base calculations of the gravitational attraction between massy points on Newton's law of the inverse square rather than on some imagined generic law, say inverse square plus epsilon, leading to an even stranger epsilon-not-zero world than the epsilon–zero world we now know. In the same sense if your problem is such that $D = 0$ when all second derivatives are not null you will eventually get cusp-point bifurcation no matter how typical double-point bifurcation may be.

II.12 Exchange of Stability at a Cusp Point

Restricting our attention to a point of second-order contact, we expand the factor $F_\mu(\mu(\varepsilon), \varepsilon)$ into a series of powers of $(\varepsilon - \varepsilon_0)$ and find that in case A

$$\begin{bmatrix} \sigma^{(1)}(\varepsilon) \\ \sigma^{(2)}(\varepsilon) \end{bmatrix} = -\tfrac{1}{2}\hat{s}\sqrt{\mathscr{D}_1}\begin{bmatrix} \mu_\varepsilon^{(1)}(\varepsilon) \\ -\mu_\varepsilon^{(2)}(\varepsilon) \end{bmatrix}[(\varepsilon - \varepsilon_0)^2 + O(\varepsilon - \varepsilon_0)^3] \quad \text{(II.55)}$$

where $\hat{s} = \operatorname{sgn} F_{\mu\mu}$. In case (B), we expand $\sigma(\mu) = F_\varepsilon(\mu, \varepsilon(\mu))$ into a series of powers in $(\mu - \mu_0)$ and find that

$$\begin{bmatrix} \sigma^{(1)}(\mu) \\ \sigma^{(2)}(\mu) \end{bmatrix} = -\tfrac{1}{2}s\sqrt{\mathscr{D}_2}\begin{bmatrix} 1 \\ -1 \end{bmatrix}[(\mu - \mu_0)^2 + O(\mu - \mu_0)^3] \quad \text{(II.56)}$$

where $s = \operatorname{sgn} F_{\varepsilon\varepsilon}$. It follows from (II.55) and (II.56) that the stability of any branch passing through a cusp point of second order changes sign if and only if $\mu_\varepsilon(\varepsilon)$ does. The possible distributions of stability at a cusp point are exhibited in Figure II.6.

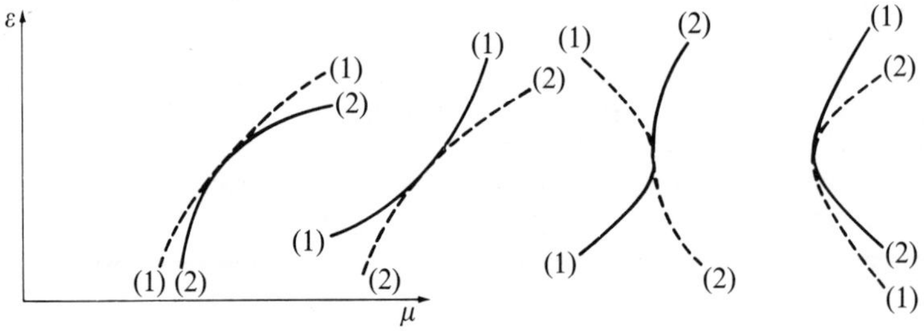

Figure II.6 Stability of solutions bifurcating at a cusp point of second order

II.13 Exchange of Stability at a Triple Point

The stability of the branches (II.25) may be determined from the sign of

$$\begin{aligned} \sigma(\varepsilon) &= -\mu_\varepsilon(\varepsilon)F_\mu(\mu(\varepsilon), \varepsilon) \\ &= -\tfrac{1}{2}\mu_\varepsilon(\varepsilon)F_{\mu\mu\mu}\{\tfrac{1}{3}(\mu_\varepsilon^{(1)}\mu_\varepsilon^{(2)} + \mu_\varepsilon^{(1)}\mu_\varepsilon^{(3)} + \mu_\varepsilon^{(2)}\mu_\varepsilon^{(3)}) \\ &\quad - \tfrac{2}{3}\mu_\varepsilon(\varepsilon_0)(\mu_\varepsilon^{(1)} + \mu_\varepsilon^{(2)} + \mu_\varepsilon^{(3)}) + \mu_\varepsilon^2(\varepsilon_0)\}(\varepsilon - \varepsilon_0)^2 \\ &\quad + O(\varepsilon - \varepsilon_0)^3 \end{aligned}$$

where we have used (II.26–27) to express the expansion of $F_\mu(\mu(\varepsilon), \varepsilon)$ in powers of $\varepsilon - \varepsilon_0$. This expression may be evaluated on each of the three branches as follows

$$\begin{bmatrix} \sigma^{(1)}(\varepsilon) \\ \sigma^{(2)}(\varepsilon) \\ \sigma^{(3)}(\varepsilon) \end{bmatrix} = -\tfrac{1}{6}F_{\mu\mu\mu}\begin{bmatrix} \mu_\varepsilon^{(1)}(\varepsilon)(\mu_\varepsilon^{(1)} - \mu_\varepsilon^{(2)})(\mu_\varepsilon^{(1)} - \mu_\varepsilon^{(3)}) \\ \mu_\varepsilon^{(2)}(\varepsilon)(\mu_\varepsilon^{(1)} - \mu_\varepsilon^{(2)})(\mu_\varepsilon^{(3)} - \mu_\varepsilon^{(2)}) \\ \mu_\varepsilon^{(3)}(\varepsilon)(\mu_\varepsilon^{(1)} - \mu_\varepsilon^{(3)})(\mu_\varepsilon^{(2)} - \mu_\varepsilon^{(3)}) \end{bmatrix}(\varepsilon - \varepsilon_0)^2 + O(\varepsilon - \varepsilon_0)^3 \tag{II.57}$$

where it may be assumed without loss of generality that $\mu_\varepsilon^{(1)} > \mu_\varepsilon^{(2)} > \mu_\varepsilon^{(3)}$. The distribution of stability of the three distinct branches is easily determined from (II.57). For example, the sign of

$$\frac{6\sigma^{(j)}(\varepsilon)}{\mu_\varepsilon^{(j)}(\varepsilon)F_{\mu\mu\mu}}$$

is $(-1)^j$. We leave further deductions about bifurcation and stability at a singular point where the second derivatives are all null as an exercise for the interested reader. It will suffice here to remark that the stability of a branch passing through such a point can change if and only if $\mu_\varepsilon(\varepsilon)$ changes sign there.

II.14 Global Properties of Stability of Isolated Solutions

All the results which we have asserted so far can be shown to apply to problems of partial differential equations, like the Navier–Stokes equations, under a condition, to be explained in Chapter VI, called bifurcation at simple eigenvalues. Theorem 2 applies in these more general problems because all the branches are connected; they are really branches in a higher-dimensional space whose projections are represented as plane curves.

It is necessary here to emphasize that it is not necessary for equilibrium solutions of evolution equations to be connected by bifurcations. There are isolated solutions, which are as common as rain, which are not connected to other solutions through bifurcation. Such isolated solutions of $F(\mu, \varepsilon) = 0$ occur even in one-dimensional problems (see Figure II.7 for one typical example). In the one-dimensional case it is possible to prove that the stability of solutions which pierce the line $\mu =$ constant is of alternating sign, as shown in Figure II.7. This result, however, is strictly one-dimensional and does not apply to one-dimensional projections of higher-dimensional problems, in which curves of solutions which appear to intersect when projected onto the plane of the bifurcation diagram actually do not intersect in the higher-dimensional space. The strictly one-dimensional result to which

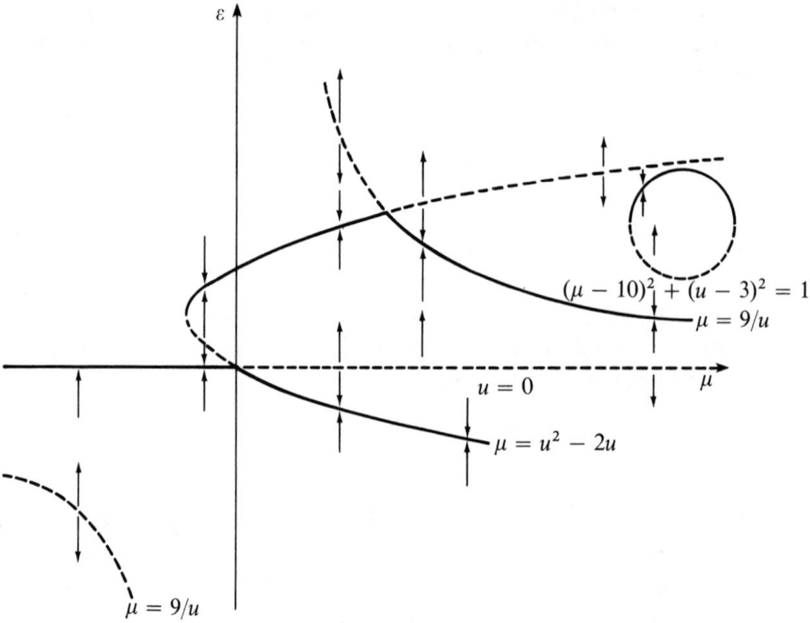

Figure II.7 Bifurcation, stability, and domains of attraction of equilibrium solutions of

$$\frac{du}{dt} = u(9 - \mu u)(\mu + 2u - u^2)([\mu - 10]^2 + [u - 3]^2 - 1). \tag{II.58}$$

The equilibrium solution $\mu = 9/u$ in the third quadrant and the circle are isolated solutions which cannot be obtained by bifurcation analysis

we have just alluded gives a complete description of the domains of initial values attracted by a steady solution.

To have the strong $\mathbb{R}^1$ result that the stability of solutions is of alternating sign we must assume that F satisfies some reasonable regularity conditions. For example, if for a fixed μ, the solutions ε of $F(\mu, \varepsilon) = 0$ are isolated, then

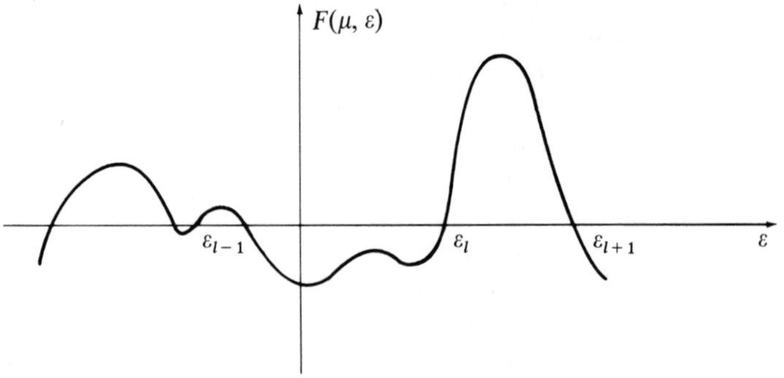

Figure II.8 Variation of F satisfying good conditions on the line μ = constant. We see immediately that the sign of the slopes $F_\varepsilon(\mu, \varepsilon_l)$ alternate

they are countable and we may write them ε_l, where $\varepsilon_{l-1} < \varepsilon_l < \varepsilon_{l+1}$ and l is in $\mathbb{Z}$ (positive or negative integers). Now we assume that the line $\mu =$ constant does not meet any singular point of F and that $F_\varepsilon(\mu, \varepsilon_l) \neq 0$ for all l. This situation is sketched in Figure II.8.

The significance of this result is dramatized by the sketch of the domain of attraction of equilibrium solutions of (II.58) in Figure II.7.

CHAPTER III

Imperfection Theory and Isolated Solutions Which Perturb Bifurcation

Isolated solutions are probably very common in dynamical problems. One way to treat them is as a perturbation of problems which do bifurcate. This method of studying isolated solutions which are close to bifurcating solutions is known as imperfection theory. Some of the basic ideas involved in imperfection theory can be understood by comparing the bending of an initially straight column with an initially imperfect, say bent, column (see Figure III.1). The first column will remain straight under increasing end loadings P until a critical load P_c is reached. The column then undergoes supercritical, one-sided, double-point bifurcation (Euler buckling). In this perfect (plane) problem there is no way to decide if the column will buckle to the left or to the right. The situation is different for the initially bent column. The sidewise deflection starts as soon as the bent column is loaded and it deflects in the direction $x < 0$ of the initial bending. If the initial bending is small the deflection will resemble that of the perfect column. There will be a small, nonzero deflection with increasing load until a neighborhood of P_c is reached; then the deflection will increase rapidly with increasing load. When P is large it will be possible to push the deflected bent column into a stable "abnormal" position ($x > 0$) opposite to the direction of initial bending.

To understand the isolated solutions which perturb bifurcation it is desirable to examine the possibilities with some generality. It is possible to do this simply, again by studying steady solutions of one-dimensional problems.

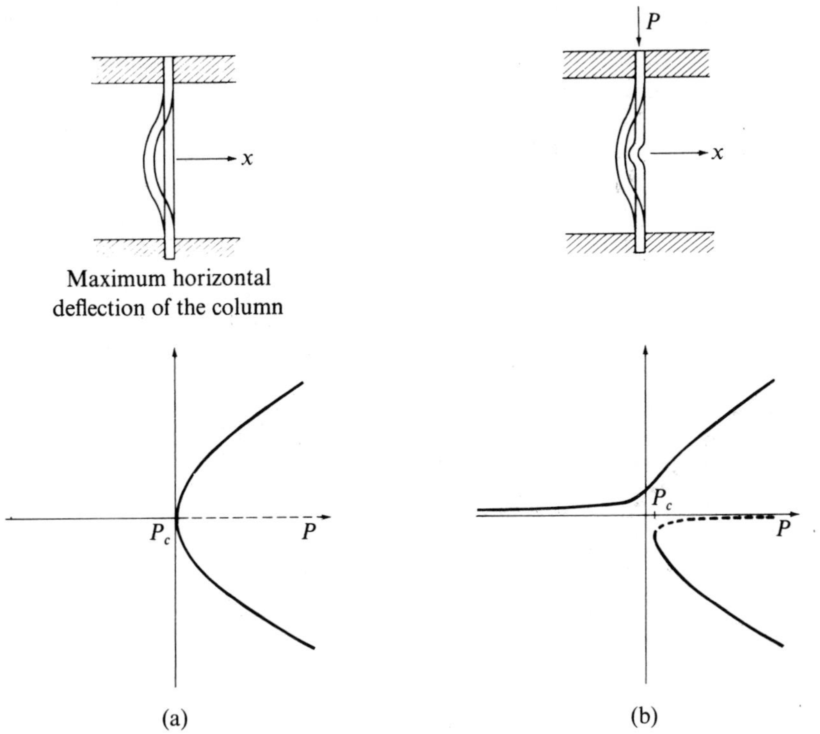

Figure III.1 (a) Buckling of a straight column. Double-point supercritical bifurcation. (b) Bending of a bent column. Isolated solutions which perturb double-point bifurcation

III.1 The Structure of Problems Which Break Double-Point Bifurcation

Consider an evolution equation in one dimension

$$\frac{dx}{dt} = \tilde{F}(\mu, x, \delta) \tag{III.1}$$

where δ and μ are parameters, $\tilde{F}$ has at least three continuous derivatives with respect to each of its three variables in a neighborhood of the point $(\mu, x, \delta) = (0, 0, 0)$. To simplify notation we drop the tilde overbar on $\tilde{F}$ and on the partial derivatives of $\tilde{F}$ when these quantities are evaluated at the point $(0, 0, 0)$. For example,

$$F \stackrel{\text{def}}{=} \tilde{F}(0, 0, 0),$$

$$F_\mu \stackrel{\text{def}}{=} \tilde{F}_\mu(0, 0, 0), \quad \text{etc.}$$

It is assumed that $(\mu, x) = (0, 0)$ is a double point of $\tilde{F}(\mu, x, 0) = 0$. At such a point we have

$$\begin{aligned} F &= 0 \\ F_x &= 0 \\ F_\mu &= 0 \\ D = F_{x\mu}^2 - F_{\mu\mu}F_{xx} &> 0. \end{aligned} \tag{III.2}$$

We are interested in the steady solutions of

$$\tilde{F}(\mu, \varepsilon, \delta) = 0 \tag{III.3}$$

which break the solutions which bifurcate at the double point into isolated solutions for $\delta \neq 0$. To break bifurcation it is enough that

$$F_\delta \neq 0. \tag{III.4}$$

III.2 The Implicit Function Theorem and the Saddle Surface Breaking Bifurcation

Let us derive the form of the isolated solutions which break the bifurcation. The implicit function theorem, $(\text{III.2})_1$, and (III.4) imply that there is a function $\delta = \Delta(\mu, \varepsilon)$ such that $\Delta(0, 0) = 0$ and

$$\tilde{F}(\mu, \varepsilon, \Delta(\mu, \varepsilon)) = 0. \tag{III.5}$$

It follows from (III.5) that

$$F_\mu + F_\delta \Delta_\mu = 0 \tag{III.6}$$

and

$$F_\varepsilon + F_\delta \Delta_\varepsilon = 0. \tag{III.7}$$

Since $F_\mu = F_\varepsilon = 0$ and $F_\delta \neq 0$ at the double point we may conclude that

$$\Delta_\mu = \Delta_\varepsilon = 0. \tag{III.8}$$

Equations (III.8) show that the surface $\delta = \Delta(\mu, \varepsilon)$ is tangent to the plane $\delta = 0$ in the three-dimensional space with coordinates $(\mu, \varepsilon, \delta)$ at the point $(0, 0, 0)$. We shall show that this point is a saddle. For this it suffices to demonstrate that in addition to (III.8) we have

$$\Delta_{\mu\varepsilon}^2 - \Delta_{\mu\mu}\Delta_{\varepsilon\varepsilon} > 0. \tag{III.9}$$

Equation (III.9) follows from the three second partial derivatives of (III.5),

$$\begin{aligned} F_{\mu\mu} + F_\delta \Delta_{\mu\mu} &= 0, \\ F_{\varepsilon\varepsilon} + F_\delta \Delta_{\varepsilon\varepsilon} &= 0, \\ F_{\mu\varepsilon} + F_\delta \Delta_{\mu\varepsilon} &= 0, \end{aligned} \tag{III.10}$$

and the inequality $D > 0$, which holds at a double point $(\text{III.2})_4$.

Since $\Delta(\mu, \varepsilon)$ is as smooth as $\tilde{F}(\mu, \varepsilon, \delta)$ we may represent $\Delta(\mu, \varepsilon)$ as a series

$$\begin{aligned} \delta = \Delta(\mu, \varepsilon) = a\varepsilon^2 + 2b\varepsilon\mu + c\mu^2 + d\varepsilon^3 \\ + e\varepsilon^2\mu + f\varepsilon\mu^2 + g\mu^3 + o((|\mu| + |\varepsilon|)^3)^* \end{aligned} \tag{III.11}$$

where

$$\begin{aligned} a &= -\frac{F_{\varepsilon\varepsilon}}{2F_\delta} \\ b &= -\frac{F_{\varepsilon\mu}}{2F_\delta} \\ c &= -\frac{F_{\mu\mu}}{2F_\delta} \\ d &= -\frac{[F_{\varepsilon\varepsilon\varepsilon} - 3F_{\varepsilon\varepsilon}F_{\varepsilon\delta}/F_\delta]}{3!F_\delta} \\ e &= -\frac{[F_{\mu\varepsilon\varepsilon} - (2F_{\varepsilon\delta}F_{\varepsilon\mu} + F_{\delta\mu}F_{\varepsilon\varepsilon})/F_\delta]}{2F_\delta} \\ f &= -\frac{[F_{\mu\mu\varepsilon} - (2F_{\mu\varepsilon}F_{\mu\delta} + F_{\mu\mu}F_{\varepsilon\delta})/F_\delta]}{2F_\delta} \\ g &= -\frac{[F_{\mu\mu\mu} - 3F_{\mu\mu}F_{\mu\delta}/F_\delta]}{3!F_\delta}. \end{aligned} \tag{III.12}$$

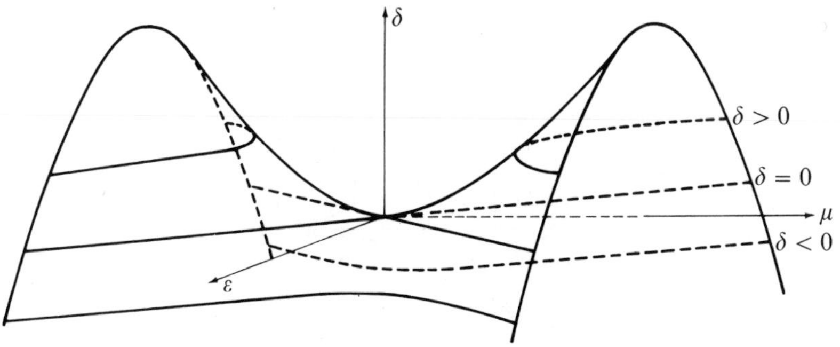

Figure III.2 Saddle surface $\delta = \Delta(\mu, \varepsilon)$

* $((|\mu| + |\varepsilon|)^3)$ goes to zero faster than $(|\mu| + |\varepsilon|)^3$ as $\mu \to 0$ and $\varepsilon \to 0$.

Our problem now is to solve (III.11) with coefficients (III.12) for $\mu(\varepsilon, \delta)$ (or $\varepsilon(\mu, \delta)$) for a fixed value of δ. The intersection of the surface $\delta = \Delta(\mu, \varepsilon)$ and the planes $\delta =$ constant determines these curves (see Figure III.2).

III.3 Examples of Isolated Solutions Which Break Bifurcation

It is of interest to give some simple typical examples of the isolated solutions which are generated by breaking double-point bifurcation with the parameter δ. For small values of δ we get a local representation of the isolated solution by the lowest-order truncation of (III.11)

$$\delta = a\varepsilon^2 + 2b\varepsilon\mu + c\mu^2. \tag{III.13}$$

This local section of the surface $\delta = \Delta(\mu, \varepsilon)$ is a hyperbola (see Figure III.3).

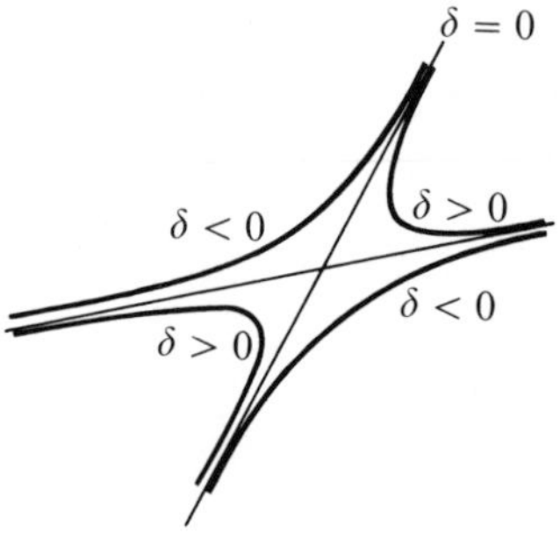

Figure III.3 Projection of the curves (III.13) onto the plane $\delta = 0$

Example III.1. Two-sided bifurcation (Figure III.4):

$$\tilde{F}(\mu, \varepsilon, \delta) = \varepsilon(\varepsilon - \mu) + \delta = 0. \tag{III.14}$$

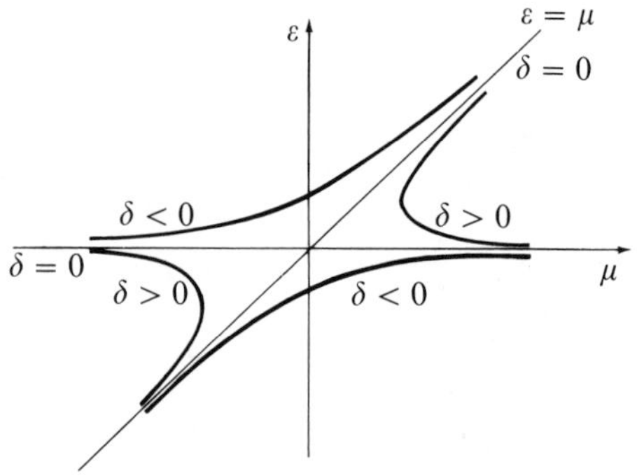

Figure III.4 Projection of the curves (III.14) onto the plane $\delta = 0$

EXAMPLE III.2. Two-sided bifurcation with a turning point (Figure III.5):

$$\tilde{F}(\mu, \varepsilon, \delta) = \varepsilon(\mu - \varepsilon - \varepsilon^2) + \delta = 0. \tag{III.15}$$

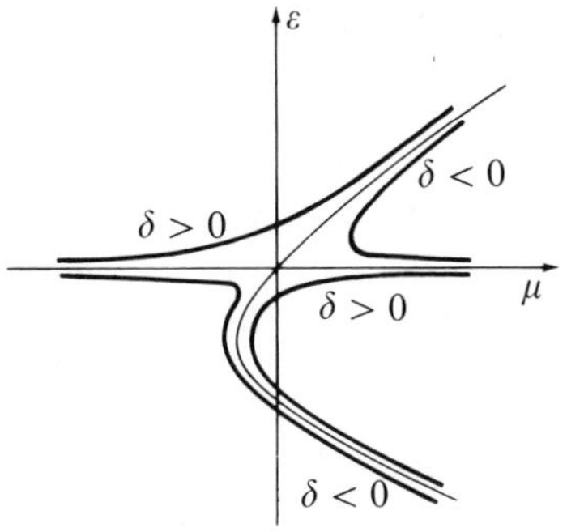

Figure III.5 Projection of the curves (III.15) onto the plane $\delta = 0$

EXAMPLE III.3. One-sided supercritical bifurcation (Figure III.6):

$$\tilde{F}(\mu, \varepsilon, \delta) = \varepsilon(\varepsilon^2 - \mu) + \delta = 0. \tag{III.16}$$

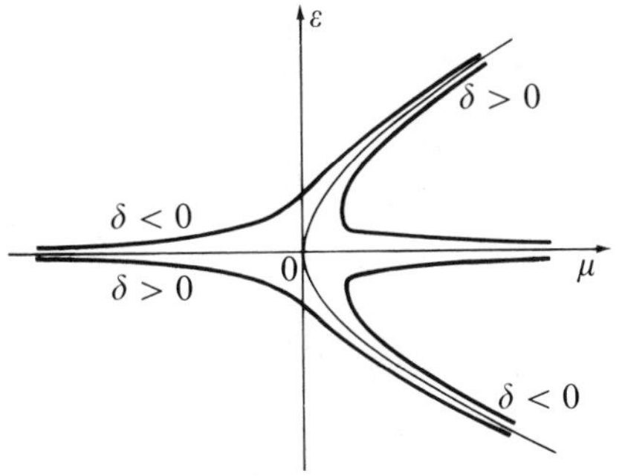

Figure III.6 Projection of the curves (III.16) onto the plane $\delta = 0$

III.4 Iterative Procedures for Finding Solutions

In the next step of our analysis we give a systematic procedure to develop the function $\mu(\varepsilon, \delta)$. Two cases are considered:

(i) $\tilde{F}(\mu, \varepsilon, \delta)$ is in general form and $F_{\mu\mu} \neq 0$.
(ii) $\tilde{F}(\mu, \varepsilon, \delta)$ is the local form (see §I.3) that is, $\tilde{F}(\mu, 0, 0) = 0$ for all μ in an interval around zero.

Case (i). $F_{\mu\mu} \neq 0$. We introduce a new parameter

$$\tilde{\delta} = \frac{\delta}{\varepsilon^2} \tag{III.17}$$

having the same sign as δ. The coefficients for the series representation

$$\mu(\varepsilon, \delta) = \mu_1(\tilde{\delta})\varepsilon + \mu_2(\tilde{\delta})\varepsilon^2 + o(|\varepsilon|^2) \tag{III.18}$$

of the isolated solutions which break double-point bifurcation may be computed by identification through terms of order $O(\varepsilon^2)$. If $\tilde{F}(\cdot, \cdot, \cdot)$ is analytic in the neighborhood of (0, 0, 0) then all the terms of the Taylor series for (III.18) may be computed by identification. If $\tilde{F}$ is sufficiently smooth but not analytic, we may compute a unique asymptotic representation of the form (III.18) in finite terms (a Taylor polynomial). To get μ_1 and μ_2, insert (III.17) and (III.18) into (III.11) and identify the coefficients of identical powers of ε:

$$\tilde{\delta} = a + 2b\mu_1 + c\mu_1^2, \tag{III.19}$$

$$0 = 2b\mu_2 + 2c\mu_1\mu_2 + d + e\mu_1 + f\mu_1^2 + g\mu_1^3. \tag{III.20}$$

Equation (III.19) has two roots

$$\begin{aligned} \mu_1^+(\tilde{\delta}) &= -\frac{b}{c} + \frac{1}{c}\sqrt{\frac{D}{4F_\delta^2} + \tilde{\delta}c} \\ \mu_1^-(\tilde{\delta}) &= -\frac{b}{c} - \frac{1}{c}\sqrt{\frac{D}{4F_\delta^2} + \tilde{\delta}c}. \end{aligned} \tag{III.21}$$

For each of the two roots μ_1^+ and μ_1^- there is a unique μ_2^+ and μ_2^- given by (III.20) provided that

$$\frac{D}{4F_\delta^2} + \tilde{\delta}c \neq 0, \tag{III.22}$$

and there are unique isolated solutions breaking bifurcation given by

$$\begin{aligned} \mu^+(\varepsilon, \delta) &= \mu_1^+(\tilde{\delta})\varepsilon + \mu_2^+\varepsilon^2 + o(|\varepsilon|^2) \\ \mu^-(\varepsilon, \delta) &= \mu_1^-(\tilde{\delta})\varepsilon + \mu_2^-\varepsilon^2 + o(|\varepsilon|^2), \end{aligned} \tag{III.23}$$

where

$$\begin{bmatrix} \mu_1^+(\tilde{\delta}) \\ \mu_1^-(\tilde{\delta}) \end{bmatrix} \varepsilon = -\frac{F_{\varepsilon\mu}}{F_{\mu\mu}}\varepsilon \begin{bmatrix} 1 \\ 1 \end{bmatrix} + h(\varepsilon, \delta) \begin{bmatrix} 1 \\ -1 \end{bmatrix} \stackrel{\text{def}}{=} \begin{bmatrix} \hat{\mu}_1^+(\varepsilon, \delta) \\ \hat{\mu}_1^-(\varepsilon, \delta) \end{bmatrix} \tag{III.24}$$

and

$$h(\varepsilon, \delta) = -\frac{1}{F_{\mu\mu}}\sqrt{D\varepsilon^2 - 2F_\delta F_{\mu\mu}\delta} \cdot (\operatorname{sgn}(\varepsilon F_\delta)),$$

$$\begin{aligned} \operatorname{sgn}(\varepsilon F_\delta) &= +1 \quad \text{if } \varepsilon F_\delta > 0, \\ &= -1 \quad \text{if } \varepsilon F_\delta < 0. \end{aligned}$$

Hence $(\hat{\mu}_1^+(\varepsilon, \delta), \hat{\mu}_1^-(\varepsilon, \delta))$ is a first approximation to $\mu^+(\varepsilon, \delta)$ and $\mu^-(\varepsilon, \delta)$. To find the second order approximation we solve (III.20) for μ_2 and find that

$$\begin{bmatrix} \mu^+(\varepsilon, \delta) \\ \mu^-(\varepsilon, \delta) \end{bmatrix} = \begin{bmatrix} \hat{\mu}_1^+(\varepsilon, \delta) \\ \hat{\mu}_1^-(\varepsilon, \delta) \end{bmatrix} - \frac{1}{2ch(\varepsilon, \delta)} \left\{ d\varepsilon^3 \begin{bmatrix} 1 \\ -1 \end{bmatrix} + e\varepsilon^2 \begin{bmatrix} \hat{\mu}_1^+(\varepsilon, \delta) \\ -\hat{\mu}_1^-(\varepsilon, \delta) \end{bmatrix} \right.$$

$$\left. + f\varepsilon \begin{bmatrix} (\hat{\mu}_1^+(\varepsilon, \delta))^2 \\ -(\hat{\mu}_1^-(\varepsilon, \delta))^2 \end{bmatrix} + g \begin{bmatrix} (\hat{\mu}_1^+(\varepsilon, \delta))^3 \\ -(\hat{\mu}_1^-(\varepsilon, \delta))^3 \end{bmatrix} \right\} + o(|\varepsilon|^2) \quad \text{(III.25)}$$

for a fixed $\tilde{\delta} = \delta/\varepsilon^2$, when $\varepsilon \to 0$.

Case (ii). $\tilde{F}(\mu, 0, 0) = 0$ for μ in an interval around zero. In this case $\varepsilon = 0$ is a solution of the bifurcation problem and $c = g = 0$. It is easy to verify that in this case

$$\Delta(\mu, 0) = 0,$$

so that ε is a factor of the series on the right of (III.11). To find the curve $\mu = \mu(\varepsilon, \delta)$ which breaks bifurcation, we again introduce a parameter $\hat{\delta}$

$$\delta = \varepsilon\hat{\delta} = \Delta(\mu, \varepsilon) = \varepsilon\hat{\Delta}(\mu, \varepsilon), \quad \text{(III.26)}$$

$$\hat{\delta} = \hat{\Delta}(\mu, \varepsilon) = a\varepsilon + 2b\mu + d\varepsilon^2 + e\varepsilon\mu + f\mu^2 + o[(|\varepsilon| + |\mu|)^2]. \quad \text{(III.27)}$$

We can solve (III.27) with respect to μ by the series method used in case (i) or by the method of successive approximations described below:

$$\mu = \frac{1}{2b}\{\hat{\delta} - a\varepsilon - d\varepsilon^2 - e\varepsilon\mu - f\mu^2\} + o[(|\varepsilon| + |\mu|)^2]. \quad \text{(III.28)}$$

The first approximation is given by

$$\mu \sim \mu^{(1)} = \frac{1}{2b}\{\hat{\delta} - a\varepsilon\} = \frac{1}{2b}\left\{\frac{\delta}{\varepsilon} - a\varepsilon\right\}. \quad \text{(III.29)}$$

The denominator $b = -F_{\varepsilon\mu}/F_\delta = -\sqrt{D}/F_\delta \neq 0$. Hence (III.29) gives two isolated solutions which break double-point bifurcation. For example, if $a = 0$ as in (III.16) (Example III.3), we get two bifurcating solutions when $\delta = 0$: $\varepsilon = 0$, and $\mu = 0$. The isolated solutions which perturb these bifurcating solutions when $\delta \neq 0$ are given by the hyperbola $\mu = \delta/2b\varepsilon$ (Figure III.7). The second approximation is given by

$$\mu \sim \mu^{(2)} = \frac{1}{2b}\{\hat{\delta} - a\varepsilon - d\varepsilon^2 - e\varepsilon\mu^{(1)} - f\mu^{(1)^2}\}$$

$$= \frac{1}{2b}\left\{\frac{\delta}{\varepsilon} - a\varepsilon - d\varepsilon^2 - \frac{e}{2b}(\delta - a\varepsilon^2) - \frac{f}{4b^2}\left(\frac{\delta}{\varepsilon} - a\varepsilon\right)^2\right\}. \quad \text{(III.30)}$$

For example, if $a = 0$ we get two bifurcating solutions when $\delta = 0$. These bifurcating solutions are given locally by

$$\varepsilon = 0 \quad \text{and} \quad \mu = -\frac{d}{2b}\varepsilon^2 \quad \text{(III.31)}$$

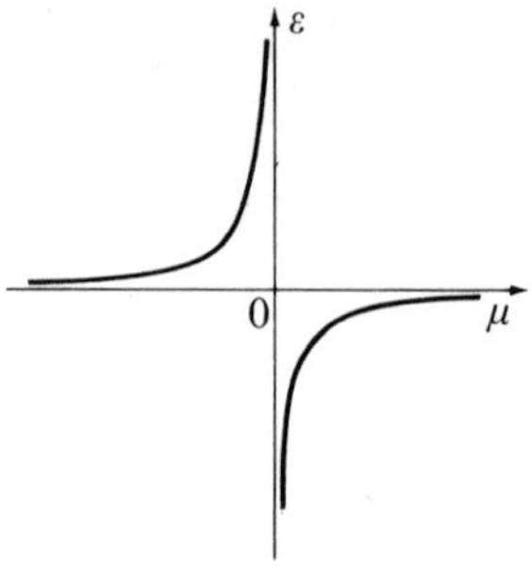

Figure III.7 Hyperbola which breaks double-point bifurcation in the first approximation when $a = 0$

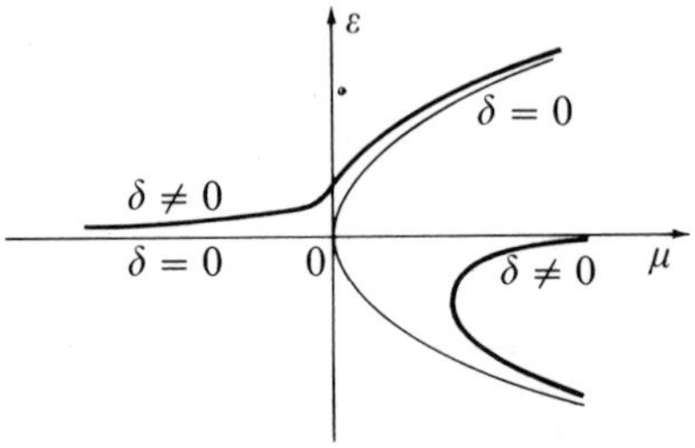

Figure III.8 Second approximation for Figure III.7

corresponding to one-sided bifurcation if $d/2b \neq 0$ ($d/2b < 0$ in Figure III.8). The isolated solutions which break bifurcation when $\delta \neq 0$ are given by (III.30). In the supercritical case the picture shown in Figure III.7 is corrected by the second approximation in the manner shown in Figure III.8.

It is necessary to add that cases (i) and (ii) exclude certain possibilities; for example, $F_{\mu\mu} = 0$, $F_{\mu\mu\mu} \neq 0$, which in any event could be obtained under case (i) in some new coordinates (μ', ε') obtained under orthogonal transformation of the (μ, ε) plane. The required orthogonal transformation suppresses the mixed product ($\mu\varepsilon$ in the hyperbola (III.13)) and brings us back to the case $F_{\mu'\mu'} \neq 0$.

III.5 Stability of Solutions Which Break Bifurcation

The stability of isolated solutions on the curve $\mu(\varepsilon, \delta)$ may be obtained from the factorization theorem. Perturbing the solutions $\mu = \mu(\varepsilon, \delta)$, $x = \varepsilon$ of (III.1) with small disturbances proportional to $e^{\gamma t}$ we find that

$$\gamma(\varepsilon) = F_\varepsilon(\mu(\delta, \varepsilon), \varepsilon, \delta) = -\mu_\varepsilon(\delta, \varepsilon)F_\mu(\mu(\delta, \varepsilon), \varepsilon, \delta). \qquad \text{(III.32)}$$

We can prove that:

(i) Stable branches with $\mu_\varepsilon(\varepsilon, 0)$ of one sign perturb with δ to stable branches with $\mu_\varepsilon(\varepsilon, \delta)$ of the same sign;

(ii) the stability of any branch $\mu = \mu(\varepsilon, \delta)$ changes sign at each and every regular turning point.

Property (i) follows from continuity and property (ii) from (III.32). Typical examples of the stability principles (i) and (ii) are shown in Figure III.9.

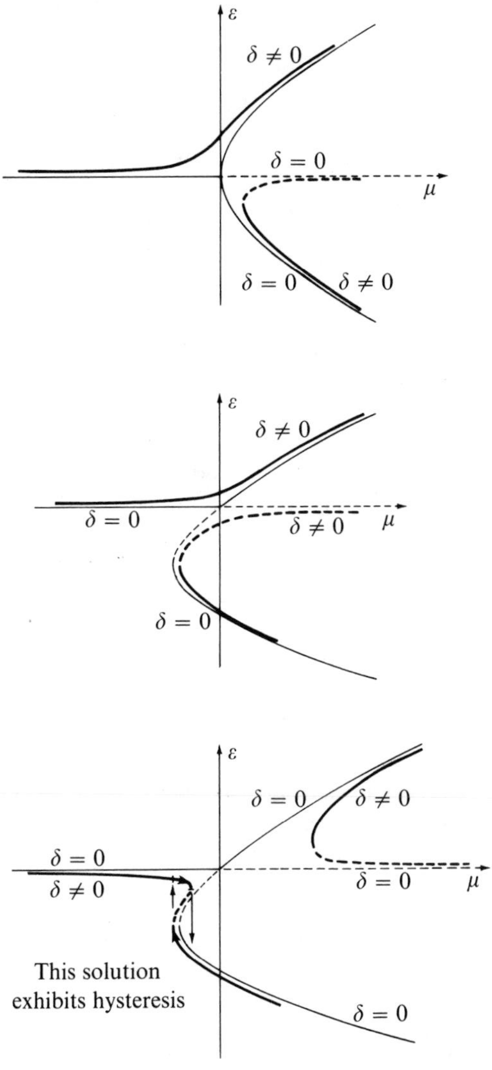

Figure III.9 Stability of isolated solutions perturbing bifurcation

III.6 Isolas

We now relax the assumption introduced in §III.1 and assume that (III.2) and (III.4) hold with $D < 0$. This assumption means that the singular point of $\tilde{F}(\mu, \varepsilon, 0)$ is isolated (a conjugate point) and there is no bifurcation when $\delta = 0$. We can proceed as in § III.2 and compute $\delta = \Delta(\mu, \varepsilon) = O(|\mu| + |\varepsilon|)^2$ where

$$\Delta_{\mu\varepsilon}^2 - \Delta_{\mu\mu}\Delta_{\varepsilon\varepsilon} < 0.$$

The principal part of $\delta = \Delta(\mu, \varepsilon)$; i.e., $\delta = \frac{1}{2}\Delta_{\varepsilon\varepsilon}\varepsilon^2 + \Delta_{\varepsilon\mu}\varepsilon\mu + \frac{1}{2}\Delta_{\mu\mu}\mu^2$ determines sections of an elliptic paraboloid instead of the hyperbolic paraboloid studied in § III.2. When δ has the correct sign the curves in the planes $\delta =$ constant are *closed* (isolas) and they shrink to zero with δ. There are no solutions when δ has the other sign.

EXAMPLE III.4

$$\tilde{F}(\mu, \varepsilon, \delta) \overset{\text{def}}{=} \mu^2 + \mu\varepsilon + \varepsilon^2 - \delta + O\{\delta^2 + |\delta|(|\varepsilon| + |\mu|) + (|\varepsilon| + |\mu|)^3\} = 0$$

determines closed curves close to ellipses when δ is positive and small (see Figure III.10).

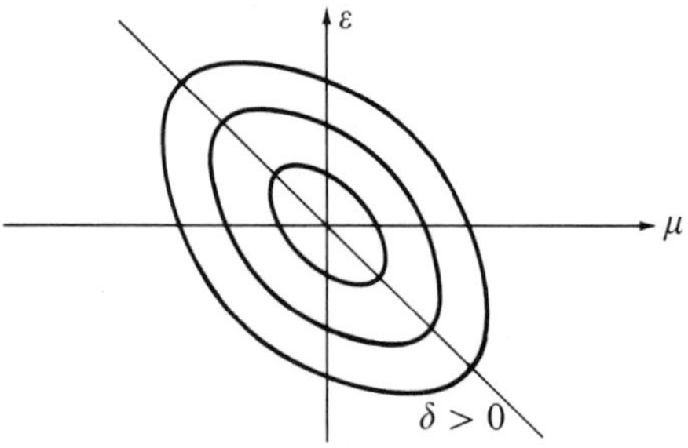

Figure III.10 Level lines of isolas

EXERCISE

III.1 (Imperfection theory for a "bifurcation at infinity" ($\mu \to \infty$).) Let us consider the two following examples

$$\frac{dx}{dt} = x\left(\frac{1}{\mu} - x - x^2\right) + \delta \tag{III.33}$$

$$\frac{dx}{dt} = x\left(\frac{1}{\mu} - x^2\right) - \delta, \tag{III.34}$$

where $\mu > 0$ or $\mu < 0$.

Show that the steady solutions and their stabilities are as in Figures III.11 and III.12. (Compare with Figures III.5 and III.6.

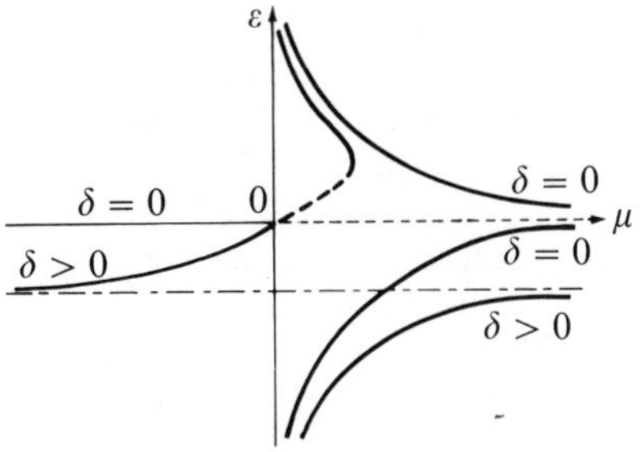

Figure III.11 Case (III.34)

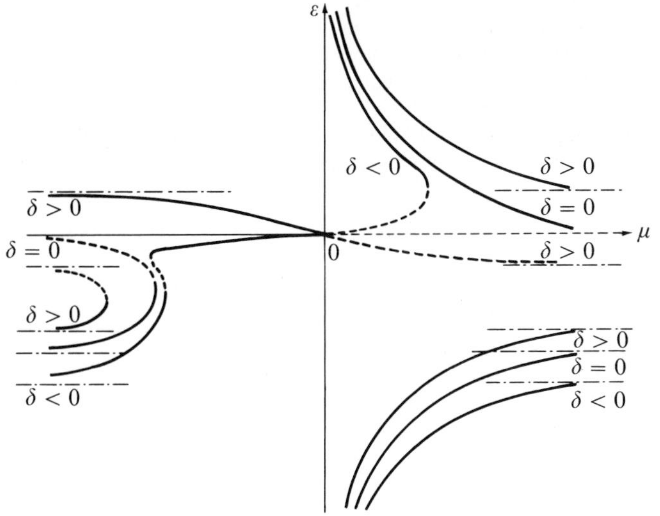

Figure III.12 Case (III.33)

Problems of bifurcation from infinity have been studied by Rosenblat and Davis, *SIAM J. Appl. Math*, **1**, 1–20 (1979).)

Notes

Imperfection theory can be traced back at least to Koiter (1945) in problems in elastic stability and Zochner (1933) in problems involving liquid crystals. The imperfection theory of Matkowsky and Reiss (1977) is close to the one given here but their aims were such as to lead them to treat the problem, which is analytic when F is, as a singular perturbation. We have preferred to stress the analytic nature of the problem, which is implied by the implicit function theorem, to define analytic iterative procedures for obtaining the curves. Imperfection theory can be regarded as a special case arising from the singularity theory of R. Thom (1968) when there is a single control parameter (δ). In this simplest case of singularity theory a canonical cubic

$$\delta = 2b\varepsilon\mu + d\varepsilon^3 \tag{III.35}$$

governing the breakup of one-sided bifurcation is important. This curve is the lowest-order approximation of the curve (III.30) relating δ to ε on the plane $\mu =$ constant. Examination of the terms on the right of (III.30) shows that $\delta \sim O(\varepsilon^3)$ and the terms neglected in going from (III.30) to (III.35) are $O(\varepsilon^4)$. The graph of (III.35) is like that shown in (III.8). In our theory we find that (III.7) is the first approximation to the curves breaking bifurcation and that it is never necessary to consider a cubic equation. The recent work of Golubitsky and Schaeffer (1979) relaxes some of the assumptions of Thom's theory and treats the problem of the breaking of bifurcation by equivalence classes of control parameters from a general, but more or less advanced, standpoint.

Golubitsky, M. and Schaeffer, D. A Theory for imperfect bifurcation via singularity theory. *Comm. Pure Appl. Math*, **32**, 1–78 (1979).

Koiter, W. T., On the stability of elastic equilibrium (in Dutch), Amsterdam: H. J. Paris, 1945; translated into English as NASA TTF-10833 (1967).

Matkowsky, B. J. and Reiss, E. Singular perturbation of bifurcations, *SIAM J. Appl. Math.*, **33**, 230–255 (1977).

Thom, R., Topological methods in biology, *Topology*, **8**, 313–335 (1968).

Zochner, H. The effect of a magnetic field on the Nematic state. *Trans. Faraday Soc.*, **29**, 945–957 (1933).

CHAPTER IV

Stability of Steady Solutions of Evolution Equations in Two Dimensions and n Dimensions

We noted in the introduction that the solutions of three nonlinear ordinary differential equations can be turbulent-like and outside the scope of elementary analysis. In fact, the most complete results known in bifurcation theory are for problems which can be reduced to one or two dimensions. So we shall start our analysis with two-dimensional autonomous problems, reduced to local form (I.21):

$$\frac{d\mathbf{u}}{dt} = \mathbf{f}(\mu, \mathbf{u}) \tag{IV.1}$$

where

$$f_i(\mu, \mathbf{u}) = A_{ij}(\mu)u_j + B_{ijk}(\mu)u_j u_k + C_{ijkl}(\mu)u_j u_k u_l + O(\|u\|^4). \tag{IV.2}$$

The same equations (IV.1) and (IV.2) hold in $\mathbb{R}^n$. In general, the subscripts range over $(1, 2, \ldots, n)$; in $\mathbb{R}^2$, $n = 2$.

To test the stability of the steady solution $\tilde{\mathbf{U}}(\mu)$ corresponding to the zero solution $\mathbf{u} = 0$ of (IV.1), we examine the evolution of a disturbance $\mathbf{v}$ of $\mathbf{u} = 0$ which, in the linearized approximation, satisfies

$$\frac{d\mathbf{v}}{dt} = \mathbf{f}_u(\mu, 0 \,|\, \mathbf{v}) = \mathbf{A}(\mu) \cdot \mathbf{v} \tag{IV.3}$$

or, in index notation,

$$\frac{dv_i}{dt} = A_{ij}(\mu)v_j. \tag{IV.4}$$

The stability to small disturbances of the solution $\mathbf{u} = 0$ is controlled by the eigenvalues of $\mathbf{A}(\mu)$ (see § IV.3). We are especially interested in the case in which $\mathbf{A}(\mu)$ is a 2×2 matrix (see § IV.2). But it is best to start more generally.

IV.1 Eigenvalues and Eigenvectors of a $n \times n$ Matrix

Let $\mathbf{A}(\mu)$ be an $n \times n$ matrix with real-valued components $A_{ij}(\mu)$. Let $\mathbf{x}$, $\mathbf{y}$ be n-component vectors with possibly complex components.

The system of linear homogeneous equations

$$\mathbf{A} \cdot \mathbf{x} = \sigma \mathbf{x}, \qquad A_{ij}x_j = \sigma x_j \tag{IV.5}$$

gives a nonzero solution $\mathbf{x}$ if and only if $\sigma = \sigma_l$ is a root of the polynomial

$$P(\sigma) = \det[\mathbf{A} - \sigma \mathbf{I}] = (-1)^n(\sigma - \sigma_1)(\sigma - \sigma_2)\cdots(\sigma - \sigma_n) = 0,$$

where $\mathbf{I}$ is the unit matrix with components δ_{ij}, σ_l ($l = 1, \ldots, n$) is an eigenvalue of $\mathbf{A}$, and $\mathbf{x}$ solving $\mathbf{A} \cdot \mathbf{x} = \sigma_l \mathbf{x}$ is an eigenvector.

IV.2 Algebraic and Geometric Multiplicity—The Riesz Index

We define μ_l as the number of repeated values of σ_l in $P(\sigma) = 0$; then μ_l is called the *algebraic multiplicity* of σ_l. It is the order of the zero σ_l of $P(\sigma) = (\sigma - \sigma_l)^{\mu_l}\tilde{P}(\sigma) = 0$, $\tilde{P}(\sigma_l) \neq 0$; σ_l is a *simple eigenvalue* of $\mathbf{A}$ if $\mu_l = 1$.

We define n_l as the number of linearly independent eigenvectors belonging to σ_l; then n_l is called the *geometric multiplicity* of σ_l.

There are always n complex values of σ for which the polynomial (of degree n) $P(\sigma) = 0$. Of course some (or all) of these values may be repeated. There is one and only one eigenvector belonging to each simple eigenvalue. If an eigenvalue is repeated then there is at least one eigenvector and, at most, μ_l linearly independent eigenvectors; that is

$$\mu_l \geq n_l.$$

The Riesz index γ_l of the eigenvalue σ_l may be defined as the lowest integer γ_l such that the two systems

$$(\mathbf{A} - \sigma_l \mathbf{I})^{\gamma_l}\mathbf{x} = 0, \qquad (\mathbf{A} - \sigma_l \mathbf{I})^{\gamma_l + 1}\mathbf{x} = 0, \tag{IV.6}$$

have the same solutions $\mathbf{x}$. If $\mu_l = n_l$ then $\gamma_l = 1$ and σ_l is said to be *semi-simple*. If the Riesz index is greater than one there are fewer eigenvectors than repeated roots and it is necessary to introduce the notion of generalized eigenvectors (see § IV.4).*

IV.3 The Adjoint Eigenvalue Problem

We now define the usual scalar product

$$\langle \mathbf{x}, \mathbf{y} \rangle \stackrel{\text{def}}{=} \mathbf{x} \cdot \bar{\mathbf{y}} = \langle \overline{\mathbf{y}, \mathbf{x}} \rangle, \tag{IV.7}$$

where the overbar designates complex conjugation. $\mathbf{A}^*$ is the adjoint of $\mathbf{A}$ if $\langle \mathbf{x}, \mathbf{A} \cdot \mathbf{y} \rangle = \langle \mathbf{A}^* \cdot \mathbf{x}, \mathbf{y} \rangle$ for all $\mathbf{x}, \mathbf{y} \in \mathbb{C}^n$. Since

$$\langle \mathbf{y}, \mathbf{A} \cdot \mathbf{x} \rangle = y_i A_{ij} \bar{x}_j = (\mathbf{A}^{\mathrm{T}})_{ji} y_i \bar{x}_j = \langle \mathbf{A}^{\mathrm{T}} \cdot \mathbf{y}, \mathbf{x} \rangle,$$

we conclude that $\mathbf{A}^* = \mathbf{A}^{\mathrm{T}}$, where $\mathbf{A}^{\mathrm{T}}$ is the transpose of the matrix $\mathbf{A}$. If the elements of $\mathbf{A}$ were complex we would find that $\mathbf{A}^* = \overline{\mathbf{A}^{\mathrm{T}}}$.

We now note that

$$\langle \mathbf{y}, (\mathbf{A} - \sigma \mathbf{I}) \cdot \mathbf{x} \rangle = \langle (\mathbf{A}^{\mathrm{T}} - \bar{\sigma} \mathbf{I}) \cdot \mathbf{y}, \mathbf{x} \rangle = 0$$

for all $\mathbf{y} \in \mathbb{C}^n$ when x solves (IV.5) and for all $x \in \mathbb{C}^n$ when $\mathbf{y}$ solves $\mathbf{A}^{\mathrm{T}}\mathbf{y} = \bar{\sigma}\mathbf{y}$ or

$$\mathbf{A}^{\mathrm{T}} \cdot \bar{\mathbf{y}} = \sigma \bar{\mathbf{y}}. \tag{IV.8}$$

Since

$$\det (\mathbf{A}^{\mathrm{T}} - \sigma \mathbf{I}) = \det (\mathbf{A} - \sigma \mathbf{I}) = P(\sigma),$$

the adjoint eigenvalue problem determines the same set of eigenvalues.

Suppose $\mathbf{x}_J$ and $\mathbf{y}_J$ belong to the eigenvalue σ_J. Then, comparing (IV.5) and (IV.8) we get

$$\begin{aligned}(\sigma_I - \sigma_J)\langle \mathbf{x}_I, \mathbf{y}_J \rangle &= (\sigma_I - \sigma_J)\mathbf{x}_I \cdot \bar{\mathbf{y}}_J \\ &= (\mathbf{A} \cdot \mathbf{x}_I) \cdot \bar{\mathbf{y}}_J - \mathbf{x}_I \cdot (\mathbf{A}^{\mathrm{T}} \cdot \bar{\mathbf{y}}_J) \\ &= \langle \mathbf{A} \cdot \mathbf{x}_I, \mathbf{y}_J \rangle - \langle \mathbf{x}_I, \mathbf{A}^{\mathrm{T}} \cdot \mathbf{y}_J \rangle = 0,\end{aligned}$$

so that any eigenvector of A belonging to the eigenvalue σ_I is orthogonal to any eigenvector of $\mathbf{A}^{\mathrm{T}}$ belonging to an eigenvalue $\bar{\sigma}_J$ such that

$$\sigma_I \neq \sigma_J.$$

* Generalized eigenvectors are frequently associated with nondiagonalizable matrices in Jordan form. We shall give the theory of these in $\mathbb{R}^2$ (this chapter) and in $\mathbb{R}^n$ (Appendix IV.1). Generalized eigenvectors are important in the theory of linear ordinary differential equations. They correspond to "secular" solutions, polynomials in t times exponentials (for example, see E. Coddington, E. and N. Levinson, *Theory of Ordinary Differential Equations* (New York: McGraw-Hill, 1955, Chapter 3).

We have, therefore,

$$\langle \mathbf{x}_I, \mathbf{y}_J \rangle = 0. \tag{IV.9}$$

If $\sigma_I = \sigma_J$ is semi-simple with algebraic multiplicity $\mu_I = m$, there are $\mu_I = m$ linearly independent eigenvectors $\mathbf{x}_{Ij}$ and adjoint eigenvectors $\mathbf{y}_{Ij}$ which may be selected so that

$$\langle \mathbf{x}_{Ii}, \mathbf{y}_{Ij} \rangle = \delta_{ij} \quad \text{for } i, j = 1, 2, \ldots, m. \tag{IV.10}$$

So biorthonormal bases may be selected on the subspace spanned by the $\mu_l = m$ eigenvectors belonging to semi-simple eigenvalues. It is not possible to select biorthonormal bases of eigenvectors if σ_l has Riesz index larger than one (see Appendix IV.1).

If $\mathbf{A} = \mathbf{A}^{\mathrm{T}}$ the eigenvalue problem is self-adjoint. The eigenvalues of self-adjoint operators are real and semi-simple (see Appendix IV.1).

IV.4 Eigenvalues and Eigenvectors of a 2 × 2 Matrix

$$\mathbf{A} = \begin{bmatrix} a & b \\ c & d \end{bmatrix}, \qquad \mathbf{A}^{\mathrm{T}} = \begin{bmatrix} a & c \\ b & d \end{bmatrix}.$$

IV.4.1 Eigenvalues

$$P(\sigma) = \det \begin{bmatrix} a - \sigma & b \\ c & d - \sigma \end{bmatrix} = \sigma^2 - \sigma(a + d) + ad - bc = 0,$$

$$\begin{aligned} \sigma_1 &= \frac{a + d}{2} + \sqrt{\Delta} \\ \sigma_2 &= \frac{a + d}{2} - \sqrt{\Delta} \end{aligned} \tag{IV.11}$$

where the discriminant Δ is defined by

$$\Delta \stackrel{\text{def}}{=} \frac{(a - d)^2}{4} + bc = \frac{(a + d)^2}{4} - ad + bc = \tfrac{1}{4}(\operatorname{tr} \mathbf{A})^2 - \det \mathbf{A},$$

and

$$\operatorname{tr} \mathbf{A} \stackrel{\text{def}}{=} a + d.$$

IV.4.2 Eigenvectors

$$\mathbf{x}_1 = \begin{bmatrix} x_{11} \\ x_{12} \end{bmatrix} \quad \text{where} \quad \begin{aligned} (a - \sigma_1)x_{11} + bx_{12} &= 0, \\ cx_{11} + (d - \sigma_1)x_{12} &= 0; \end{aligned}$$

$$\mathbf{x}_2 = \begin{bmatrix} x_{21} \\ x_{22} \end{bmatrix} \quad \text{where} \quad \begin{aligned} (a - \sigma_2)x_{21} + bx_{22} &= 0, \\ cx_{21} + (d - \sigma_2)x_{22} &= 0; \end{aligned}$$

$$\bar{\mathbf{y}}_1 = \begin{bmatrix} \bar{y}_{11} \\ \bar{y}_{12} \end{bmatrix} \quad \text{where} \quad \begin{aligned} (a - \sigma_1)\bar{y}_{11} + c\bar{y}_{12} &= 0, \\ b\bar{y}_{11} + (d - \sigma_1)\bar{y}_{12} &= 0; \end{aligned}$$

$$\bar{\mathbf{y}}_2 = \begin{bmatrix} \bar{y}_{21} \\ \bar{y}_{22} \end{bmatrix} \quad \text{where} \quad \begin{aligned} (a - \sigma_2)\bar{y}_{21} + c\bar{y}_{22} &= 0, \\ b\bar{y}_{21} + (d - \sigma_2)\bar{y}_{22} &= 0. \end{aligned}$$

IV.4.3 Algebraically Simple Eigenvalues

Case 1. $\Delta > 0$. $\sigma_1 \neq \sigma_2$ are real. There are two real eigenvectors and two adjoint eigenvectors.

Case 2. $\Delta < 0$. $\sigma_2 = \bar{\sigma}_1$. There are two eigenvectors and they are conjugate. The same is true of the adjoint problem.

IV.4.4 Algebraically Double Eigenvalues

Case 3. $\Delta = 0$. Then $\sigma_1 = \sigma_2 = (a + d)/2$ is an eigenvalue of algebraic multiplicity two. The eigenvector problems are given by

$$\begin{aligned} qx_1 + bx_2 &= 0, \\ cx_1 - qx_2 &= 0, \end{aligned} \tag{IV.12}$$

$$\begin{aligned} q\bar{y}_1 + c\bar{y}_2 &= 0, \\ b\bar{y}_1 - q\bar{y}_2 &= 0, \end{aligned} \tag{IV.13}$$

where $q = (a - d)/2$ and $\Delta = q^2 + bc = 0$.

IV.4.4.1 *Riesz Index 1*

$$q = b = c = 0. \tag{IV.14}$$

Then

$$\mathbf{A} = \begin{bmatrix} a & 0 \\ 0 & a \end{bmatrix} = a\mathbf{I}$$

and every vector $\mathbf{x}$ is an eigenvector belonging to $\sigma_1 = \sigma_2 = a$. We can select two orthonormal ones. So a is a double semi-simple eigenvalue.

IV.4.4.2 *Riesz Index 2*

$$q^2 + bc = 0, \qquad |q| + |b| + |c| \neq 0.$$

$\sigma = \sigma_1 = \sigma_2 = (a + d)/2$ is algebraically double and geometrically simple.

There is one and only one eigenvector satisfying $(\mathbf{A} - \sigma\mathbf{I}) \cdot \mathbf{x} = 0$ and an arbitrary normalizing condition. The components x_1, x_2 of $\mathbf{x}$ satisfy (IV.12) and if $q \neq 0$, or $c \neq 0$, or $b \neq 0$ then $\mathbf{x}$ is given by

$$\mathbf{x} = \begin{bmatrix} -b/q \\ 1 \end{bmatrix} x_2 = \begin{bmatrix} q/c \\ 1 \end{bmatrix} x_2 = \begin{bmatrix} 1 \\ -q/b \end{bmatrix} x_1 = \begin{bmatrix} 1 \\ c/q \end{bmatrix} x_1.$$

Similarly, there is one and only one adjoint eigenvector satisfying $(\mathbf{A}^{\mathrm{T}} - \sigma\mathbf{I}) \cdot \bar{\mathbf{y}} = 0$ and an arbitrary normalizing condition. The components $\bar{y}_1, \bar{y}_2$ of $\bar{\mathbf{y}}$ satisfy (IV.13), and if $b \neq 0$, or $q \neq 0$, or $c \neq 0$ then $\bar{\mathbf{y}}$ is given by

$$\bar{\mathbf{y}} = \begin{bmatrix} q/b \\ 1 \end{bmatrix} \bar{y}_2 = \begin{bmatrix} -c/q \\ 1 \end{bmatrix} \bar{y}_2 = \begin{bmatrix} 1 \\ b/q \end{bmatrix} \bar{y}_1 = \begin{bmatrix} 1 \\ -q/c \end{bmatrix} \bar{y}_1.$$

The scalar product of $\mathbf{x}$ with its adjoint *cannot* be normalized because

$$\langle \mathbf{x}, \mathbf{y} \rangle = \mathbf{x} \cdot \bar{\mathbf{y}} = 0. \tag{IV.15}$$

We next consider generalized eigenvectors $\boldsymbol{\zeta}$ satisfying

$$(\mathbf{A} - \sigma\mathbf{I}) \cdot \boldsymbol{\zeta} = \mathbf{x}. \tag{IV.16}$$

Since $\mathbf{x}$ is an eigenvector we have

$$(\mathbf{A} - \sigma\mathbf{I})^2 \cdot \boldsymbol{\zeta} = 0. \tag{IV.17}$$

Every vector in $\mathbb{R}^2$ satisfies (IV.17) because

$$(\mathbf{A} - \sigma\mathbf{I})^2 = \begin{bmatrix} q & b \\ c & -q \end{bmatrix}^2 = \begin{bmatrix} 0 & 0 \\ 0 & 0 \end{bmatrix}.$$

But the $\boldsymbol{\zeta}$ satisfying (IV.16) must have

$$\begin{aligned} q\zeta_1 + b\zeta_2 &= x_1 \\ &= -\frac{b}{q} x_2 \quad \text{if } q \neq 0 \\ c\zeta_1 - q\zeta_2 &= x_2. \end{aligned}$$

Similarly there is a generalized adjoint eigenvector $\bar{\boldsymbol{\zeta}}^*$ satisfying

$$(\mathbf{A}^{\mathrm{T}} - \sigma\mathbf{I}) \cdot \bar{\boldsymbol{\zeta}}^* = \bar{\mathbf{y}}, \qquad (\mathbf{A}^{\mathrm{T}} - \sigma\mathbf{I})^2 \cdot \bar{\boldsymbol{\zeta}}^* = 0$$

and

$$q\bar{\zeta}_1^* + c\bar{\zeta}_2^* = \bar{y}_1,$$
$$b\bar{\zeta}_1^* - q\bar{\zeta}_2^* = \bar{y}_2.$$

Given $\bar{\mathbf{y}}$, we may normalize by specifying some value, say one, for the scalar product of the generalized eigenvector and the adjoint eigenvector,

$$\langle \boldsymbol{\zeta}, \mathbf{y} \rangle = 1. \tag{IV.18}$$

Since

$$\begin{aligned} \langle \boldsymbol{\zeta}, \mathbf{y} \rangle &= \langle \boldsymbol{\zeta}, (\mathbf{A}^{\mathrm{T}} - \bar{\sigma}\mathbf{I}) \cdot \boldsymbol{\zeta}^* \rangle \\ &= \langle (\mathbf{A} - \sigma\mathbf{I}) \cdot \boldsymbol{\zeta}, \boldsymbol{\zeta}^* \rangle \\ &= \langle \mathbf{x}, \boldsymbol{\zeta}^* \rangle \\ &= \zeta_1 \bar{y}_1 + \zeta_2 \bar{y}_2 \\ &= -\frac{x_2 \bar{y}_2}{q} = \frac{x_1 \bar{y}_2}{b} = \frac{x_2 \bar{y}_1}{c} = 1, \end{aligned} \tag{IV.19}$$

we may set $x_1 = b/\bar{y}_2$ for arbitrary values of $\bar{y}_2 \neq 0$ if $b \neq 0$.

In the case which we adopt as canonical, $c = q = 0$, $b \neq 0$, we have (IV.19), $x_2 = \bar{y}_1 = 0$, $\zeta_2 = x_1/b$ and $\bar{\zeta}_1^* = 1/x_1$, $\bar{y}_2 = b/x_1$, $\bar{\zeta}_2^* = -\zeta_1 b/x_1^2$ (we may choose $\zeta_1 = 0$).

The results discussed for $\mathbb{R}^2$ under § IV.4.4.2 are special cases in the general theory for $\mathbb{R}^n$, $n \geq 2$, of eigenvalues which are not semi-simple. The general theory is discussed in the appendix to this chapter.

IV.5 The Spectral Problem and Stability of the Solution $\mathbf{u} = 0$ in $\mathbb{R}^n$

Set $\mathbf{v} = e^{\sigma t}\mathbf{x}$ in (IV.3). We find that

$$\mathbf{A}(\mu) \cdot \mathbf{x} = \sigma \mathbf{x}, \tag{IV.20}$$

where

$$\sigma(\mu) = \xi(\mu) + i\eta(\mu) \tag{IV.21}$$

is an eigenvalue of $\mathbf{A}(\mu)$ if $\mathbf{x} \neq 0$. We say that $\mathbf{u} = 0$ is stable by the criterion of the spectral problem if $\xi(\mu) < 0$ for all eigenvalues $\sigma(\mu)$, and is unstable if there is a value σ solving (IV.20) with $\mathbf{x} \neq 0$ for which $\xi(\mu) > 0$. The adjoint eigenvalue problem is given by (IV.8).

IV.6 Nodes, Saddles, and Foci

In $\mathbb{R}^2$, $\mathbf{x}$ is a two-component vector and $\mathbf{A}(\mu)$ is a 2×2 matrix. Let us suppose that σ_1 and σ_2 are simple eigenvalues of $\mathbf{A}(\mu)$. There are two eigenvectors. Suppose $\mathbf{x}_1$ is an eigenvector of σ_1, $\mathbf{x}_2$ is an eigenvector of σ_2 and $\mathbf{v}_J = e^{\sigma_J t}\mathbf{x}_J$, $J = 1, 2$. Then

$$\frac{d\mathbf{v}_J}{dt} = \sigma_J \mathbf{v}_J. \tag{IV.22}$$

From (IV.22) we may construct a picture of the geometric properties of equilibrium points in the plane. There are three cases to consider.

(1) $\sigma_J(\mu) = \xi_J(\mu)$ are real and $\xi_1(\mu)$ and $\xi_2(\mu)$ are of the same sign. By the construction $\mathbf{v}_1$ and $\mathbf{v}_2$ are real independent vectors. If $\xi_1(\mu)$ and $\xi_2(\mu)$ are negative $\mathbf{v}_1$ and $\mathbf{v}_2$ tend to 0 and 0 is called a stable node. In the other case $\mathbf{v}_1$ and $\mathbf{v}_2$ leave 0 and 0 is called an unstable node (see Figure IV.1).

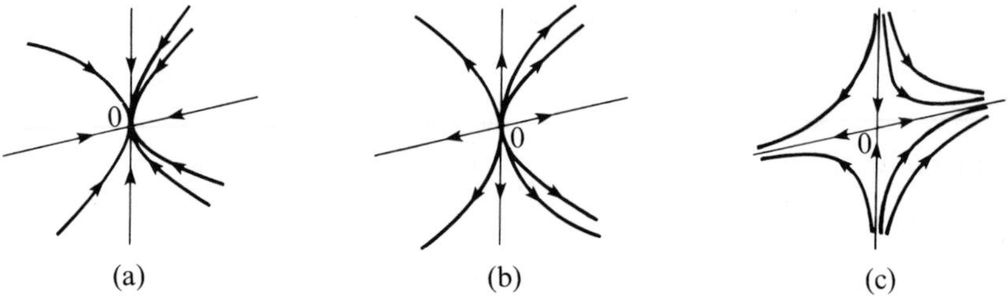

Figure IV.1 Trajectories near an equilibrium point in $\mathbb{R}^2$ when $\sigma(\mu)$ is real-valued (a) Stable node; (b) Unstable node; (c) Saddle

(2) $\sigma_J(\mu) = \xi_J(\mu)$ and $\xi_1(\mu)$ and $\xi_2(\mu)$ have different signs. Then 0 is a saddle point, one of the two trajectories goes to 0 and the other escapes from 0 (see Figure IV.1). A saddle is always unstable.

(3) The two eigenvalues are complex:

$$\sigma_1 = \xi(\mu) + i\eta(\mu) = \bar{\sigma}_2,$$

and the eigenvectors satisfy $\mathbf{v}_1 = \bar{\mathbf{v}}_2$.

The general solution of (IV.3) takes the form

$$\mathbf{v}(t) = \operatorname{Re}(\alpha e^{\sigma_1 t}\mathbf{v}_1) = \tfrac{1}{2}[\alpha e^{\sigma_1 t}\mathbf{v}_1 + \bar{\alpha} e^{\sigma_2 t}\mathbf{v}_2],$$

which may be also written as

$$\mathbf{v}(t) = e^{\xi(\mu)t}[\operatorname{Re}(\alpha\mathbf{v}_1)\cos(\eta(\mu)t) - \operatorname{Im}(\alpha\mathbf{v}_1)\sin(\eta(\mu)t)]. \tag{IV.23}$$

The trajectories (IV.23) are shown in Figure IV.2. The point $\mathbf{v} = 0$ is a stable focus if $\xi(\mu) < 0$ and an unstable focus if $\xi(\mu) > 0$.

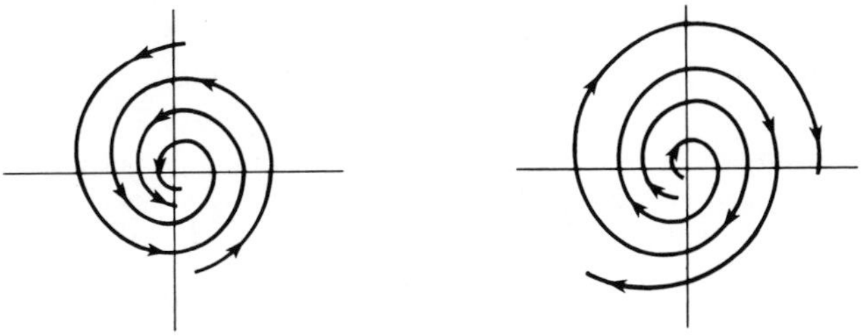

(a) (b)

Figure IV.2 Trajectories near an equilibrium point in $\mathbb{R}^2$, when the $\sigma_J(\mu)$ are complex. (a) Stable focus, $\xi(\mu) < 0$; (b) Unstable focus, $\xi(\mu) > 0$

IV.7 Criticality and Strict Loss of Stability

We shall assume, without loss of generality, that $\mathbf{u} = 0$ is stable ($\xi(\mu) < 0$) when $\mu < 0$ and is unstable ($\xi(\mu) > 0$) when $\mu > 0$. The value $\mu = 0$ is critical. At criticality

$$(\xi(0), \eta(0), \sigma(0)) = (0, \omega_0, i\omega_0). \tag{IV.24}$$

We say that the loss of stability of $\mathbf{u} = 0$ is strict if

$$\xi'(0) \overset{\text{def}}{=} \frac{d\xi(0)}{d\mu} > 0 \tag{IV.25}$$

and recall that in $\mathbb{R}^1$ this condition implies double-point bifurcation. Let $(\cdot)' \overset{\text{def}}{=} d(\cdot)/d\mu$. Then assuming that σ and x are differentiable we find by differentiating (IV.20):

$$\mathbf{A}(\mu) \cdot \mathbf{x}' + \mathbf{A}'(\mu) \cdot \mathbf{x} = \sigma'\mathbf{x} + \sigma\mathbf{x}'. \tag{IV.26}$$

Now we form the scalar product of (IV.26) with the adjoint eigenvector belonging to σ and note that

$$\langle \mathbf{A} \cdot \mathbf{x}', \mathbf{y} \rangle = \langle \mathbf{x}', \mathbf{A}^{\mathrm{T}} \cdot \mathbf{y} \rangle = \langle \sigma\mathbf{x}', \mathbf{y} \rangle$$

so that

$$\sigma'\langle \mathbf{x}, \mathbf{y} \rangle = \langle \mathbf{A}' \cdot \mathbf{x}, \mathbf{y} \rangle \tag{IV.27}$$

where

$$[\mathbf{A}'(\mu)] = \begin{bmatrix} a'(\mu) & b'(\mu) \\ c'(\mu) & d'(\mu) \end{bmatrix}.$$

The formula (IV.27) holds if $\sigma(\mu)$ is simple. In the semi-simple case two different eigenvectors $\mathbf{x}_1$ and $\mathbf{x}_2$ belong to one and the same σ and, in addition to finding $\sigma'(\mu)$ we must determine the linear combinations of $\mathbf{x}_1$ and $\mathbf{x}_2$ for which (IV.26) holds (eigenvectors of $\mathbf{A}'$ for $\mu = 0$). Equation (IV.27) is meaningless when σ is a multiple eigenvalue with Riesz index greater than one; for in this case $\langle \mathbf{x}, \mathbf{y} \rangle = 0$ for all eigenvectors and we *cannot* normalize so that $\langle \mathbf{x}, \mathbf{y} \rangle = 1$.

Strict crossing in the case of a simple eigenvalue means that

$$\sigma'(0) = \langle \mathbf{A}'(0) \cdot \mathbf{x}_1, \mathbf{y}_1 \rangle, \qquad \operatorname{Re} \sigma'(0) > 0 \tag{IV.28}$$

when the eigenvalue with greatest real part satisfies $\xi_1(0) = 0$.

The possible ways in which $\sigma(\mu)$ can cross the line $\xi = 0$ in the simple and semi-simple case are shown in Figure IV.3.

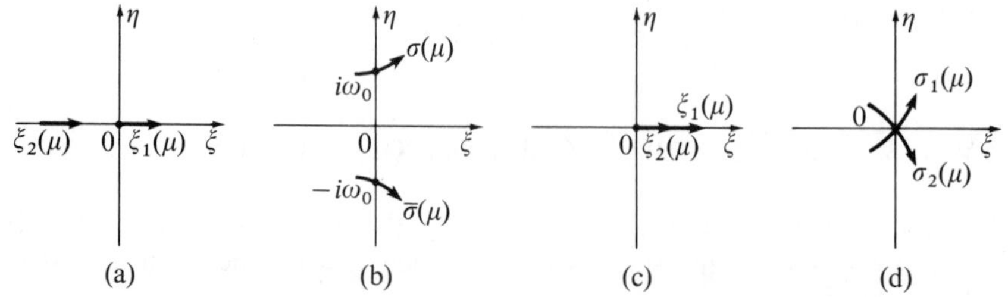

Figure IV.3 Strict crossing $\xi_1(0) = 0$, $\xi_1'(0) > 0$ directions in simple (a, b) and semi-simple (c, d) cases. (a) Two eigenvalues are real and distinct; $\xi_1(0) = 0$. (b) A complex-conjugate pair crosses over. (c) Two real eigenvalues cross together, but they may cross at different rates (or in different directions). (d) Two complex-conjugate eigenvalues cross together

The perturbation of $\sigma(\mu)$ at $\mu = 0$ when $\sigma(0)$ is a double eigenvalue of index two (not semi-simple) is special because $\sigma(\cdot)$ is not differentiable in general at $\mu = 0$. $\sigma(0)$ is an algebraically double eigenvalue of index two when $\Delta = 0$ and $|q| + |b| + |c| \neq 0$. There are three possibilities: (1) $b \neq 0$, $c = q = 0$; (2) $c \neq 0$, $b = q = 0$; (3) c, b, q are not zero and $q^2 + bc = 0$. These three cases are equivalent, since the matrices $\mathbf{A}(\mu)$ corresponding to the three cases differ from one another by a similarity transformation.

We take case (1) as canonical and write

$$[\mathbf{A}(\mu)] = \begin{bmatrix} 0 & 1 \\ 0 & 0 \end{bmatrix} + \mu \begin{bmatrix} a'(\mu) & b'(\mu) \\ c'(\mu) & d'(\mu) \end{bmatrix} \tag{IV.29}$$

where a', b', c', and d' are bounded at $\mu = 0$. The two eigenvalues of $\mathbf{A}(\mu)$ are determined by the quadratic equation

$$\det \begin{bmatrix} -\sigma + \mu a' & 1 + \mu b' \\ \mu c' & -\sigma + \mu d' \end{bmatrix} = \sigma^2 - \sigma\mu(a' + d') + \mu^2 a'd' - \mu c'(1 + \mu b')$$
$$= 0.$$

Hence

$$\begin{aligned} \sigma_{\pm}(\mu) &= \tfrac{1}{2}\mu(a' + d') \pm \sqrt{\mu^2[(a' + d')^2/4 - a'd' + b'c'] + \mu c'} \\ &= \pm \sqrt{\mu c'} + \tfrac{1}{2}\mu(a' + d') + O(\mu^{3/2}) \end{aligned} \tag{IV.30}$$

where we chose one of the two complex determinations for $\sqrt{\mu c'}$. We see that $\sigma_{\pm}(0) = 0$ and note that $|d\sigma_{\pm}(0)/d\mu| = \infty$ but $d\sigma/d\sqrt{\mu} = \pm\sqrt{c'}$ if c' and $\mu > 0$. There are two eigenvalues and they are complex when $\mu c' < 0$. Since there is always a positive eigenvalue when $\mu c' > 0$, we conclude that $\mathbf{u} = 0$ is unstable when $|\mu|$ is small and $\mu c' > 0$. Since $\mu c' > 0$ implies instability we can get stability only when $\mu c' < 0$, so $\sqrt{\mu c'}$ is imaginary and stability is determined by the sign of $\mu(a' + d')$. If $c' > 0$, then $\mathbf{u} = 0$ is unstable for small $\mu > 0$ and is stable for $\mu < 0$ if $a' + d' > 0$. $\mathbf{u} = 0$ can be unstable on both sides of criticality.

Appendix IV.1 Biorthogonality for Generalized Eigenvectors

Let $\mathbf{A}$ be an $n \times n$ matrix, $n \geq 2$ and define

$$\mathbf{T} = \mathbf{A} - \sigma\mathbf{I} \tag{IV.31}$$

where σ is any of the eigenvalues of $\mathbf{A}$. Let

$$N_l = \{\boldsymbol{\psi}: \mathbf{T}^l \cdot \boldsymbol{\psi} = 0\}$$

be the null space of the lth power of matrix $\mathbf{T}$,

$$\mathbf{T}^l = \mathbf{T} \cdot \mathbf{T} \cdots \mathbf{T} \qquad (l \text{ times})$$

and let

$$n_l = \dim N_l$$

be the number of *independent* vectors $\boldsymbol{\psi}$ which are annihilated by $\mathbf{T}^l$ where $l \geq 1$. If $\mathbf{T} \cdot \boldsymbol{\psi} = 0$ then $\mathbf{T}^n \cdot \boldsymbol{\psi} = 0$ for $n \in \mathbb{N}$, $n \geq 2$. So, for example $N_2 \supseteq N_1$ and $N_{n+1} \supseteq N_n$. We saw in § IV.2 that the Riesz index ν is the largest integer for which we have

$$N_1 \subset N_2 \subset \cdots \subset N_\nu = N_{\nu+k} \quad \text{for all } k \in \mathbb{N}$$

where the inclusions are strict. We have already defined

$$n_\nu = \text{algebraic multiplicity of } \sigma$$

$$n_1 = \text{geometric multiplicity of } \sigma,$$

and of course, $n_\nu \geq n_1$. The vectors $\boldsymbol{\psi} \in N_1$ are the proper eigenvectors of σ; they satisfy $\mathbf{T} \cdot \boldsymbol{\psi} = 0$. The vectors $\boldsymbol{\psi} \in N_l$, $l > 1$ are called generalized eigenvectors. There are no generalized eigenvectors when $\nu = 1$. In this case σ is a semi-simple eigenvalue of $\mathbf{A}$, simple if $n_1 = 1$ and of higher multiplicity otherwise. A Riesz index of one means that σ is semi-simple.

We also have generalized null spaces for the adjoint:

$$N_l^* = \{\boldsymbol{\psi}^*: (\mathbf{T}^{\mathrm{T}})^l \cdot \boldsymbol{\psi}^* = 0\}.$$

They have the same dimension as N_l,

$$n_l = \dim N_l^* = \dim N_l \tag{IV.32}$$

for $1 \le l \le \nu$.

We shall now show that the eigenvalues σ of a real symmetric matrix $\mathbf{A}$ are real and semi-simple, so the eigenvectors of $\mathbf{A}$ are proper and not generalized. We have $\mathbf{A} \cdot \mathbf{x} = \sigma \mathbf{x}$ and $\langle \mathbf{A} \cdot \mathbf{x}, \mathbf{x} \rangle = \langle \mathbf{x}, \mathbf{A} \cdot \mathbf{x} \rangle$ so that

$$\sigma \langle \mathbf{x}, \mathbf{x} \rangle = \bar{\sigma} \langle \mathbf{x}, \mathbf{x} \rangle$$

and $\sigma = \bar{\sigma}$ if $\mathbf{x} \neq 0$. Now assume that $\mathbf{T}^2 \cdot \boldsymbol{\psi} = 0$. Then

$$0 = \langle \mathbf{T}^2 \cdot \boldsymbol{\psi}, \boldsymbol{\psi} \rangle = \langle \mathbf{T} \cdot \boldsymbol{\psi}, \mathbf{T} \cdot \boldsymbol{\psi} \rangle$$

where $\overline{\mathbf{T}}^{\mathrm{T}} = \mathbf{T}$ because $\bar{\sigma} = \sigma$. It follows that $\mathbf{T} \cdot \boldsymbol{\psi} = 0$ so that $\boldsymbol{\psi}$ is proper and not generalized.

In the general case it can be shown (in books on linear algebra which treat Jordan bases) that it is possible to choose the generalized eigenvectors of $\mathbf{T}$: $\boldsymbol{\psi}_i^{(1)}, \ldots, \boldsymbol{\psi}_i^{(\nu_i)}$ so that

$$\begin{aligned} \mathbf{T} \cdot \boldsymbol{\psi}_i^{(1)} &= 0, \\ \mathbf{T} \cdot \boldsymbol{\psi}_i^{(2)} &= \boldsymbol{\psi}_i^{(1)}, \\ \mathbf{T} \cdot \boldsymbol{\psi}_i^{(\nu_i)} &= \boldsymbol{\psi}_i^{(\nu_i - 1)}, \qquad i = 1, \ldots, k. \end{aligned} \tag{IV.33}$$

We can also choose the generalized eigenvectors of $\mathbf{T}^{\mathrm{T}}$: $\bar{\boldsymbol{\psi}}_i^{*(1)}, \ldots, \bar{\boldsymbol{\psi}}_i^{*(\nu_i)}$, so that

$$\begin{aligned} \mathbf{T}^{\mathrm{T}} \cdot \bar{\boldsymbol{\psi}}_i^{*(\nu_i)} &= 0, \\ \mathbf{T}^{\mathrm{T}} \cdot \bar{\boldsymbol{\psi}}_{\mathrm{i}}^{(\nu_i - 1)} &= \bar{\boldsymbol{\psi}}_i^{*(\nu_i)} \\ \mathbf{T}^{\mathrm{T}} \cdot \bar{\boldsymbol{\psi}}_i^{*(1)} &= \bar{\boldsymbol{\psi}}_i^{*(2)}, \qquad i = 1, \ldots, k, \end{aligned} \tag{IV.34}$$

where

$$\langle \boldsymbol{\psi}_i^{(l)}, \boldsymbol{\psi}_j^{*(m)} \rangle = \delta_{ij} \delta^{lm}. \tag{IV.35}$$

In fact, some of the relations (IV.35) are automatically realized and the others may be realized by a proper choice of the $\boldsymbol{\psi}_j^{*(k)}$ (see the example which follows). The biorthogonality conditions which are automatically realized may be derived as follows

$$\begin{aligned} \langle \boldsymbol{\psi}_i^{(m)}, \boldsymbol{\psi}_j^{*(n)} \rangle &= \langle \mathbf{T} \cdot \boldsymbol{\psi}_i^{(m+1)}, \boldsymbol{\psi}_j^{*(n)} \rangle \\ &= \langle \boldsymbol{\psi}_i^{(m+1)}, \mathbf{T}^{\mathrm{T}} \cdot \boldsymbol{\psi}_j^{*(n)} \rangle \\ &= \langle \boldsymbol{\psi}_i^{(m+1)}, \boldsymbol{\psi}_j^{*(n+1)} \rangle, \end{aligned}$$

provided that $1 \le m < \nu_i$, $1 < n \le \nu_j$. We deduce that

$$\langle \boldsymbol{\psi}_i^{(m)}, \boldsymbol{\psi}_j^{*(\nu_j)} \rangle = 0, \qquad 1 \le m < \nu_i,$$

and

$$\langle \psi_i^{(1)}, \psi_j^{*(n)} \rangle = 0, \qquad 1 < n \le \nu_j,$$
$$\langle \psi_i^{(m)}, \psi_i^{*(n)} \rangle = 0, \qquad n \ge m + 1.$$

EXAMPLE IV.1. Consider the matrix

$$\mathbf{T} = \begin{bmatrix} 0 & 0 & 1 \\ 0 & 0 & -1 \\ 0 & 0 & 0 \end{bmatrix}.$$

Zero is an eigenvalue of $\mathbf{T}$ of algebraic multiplicity three and

$$\mathbf{x}_1 = \alpha \begin{bmatrix} 1 \\ 0 \\ 0 \end{bmatrix}, \qquad \mathbf{x}_2 = \beta \begin{bmatrix} 0 \\ 1 \\ 0 \end{bmatrix}$$

are the eigenvectors belonging to the eigenvalue zero. It is easy to verify that the only combination $\alpha\mathbf{x}_1 + \beta\mathbf{x}_2$ satisfying $\mathbf{T} \cdot \psi = \alpha\mathbf{x}_1 + \beta\mathbf{x}_2$ is the one for which $\alpha = 1$, $\beta = -1$; that is,

$$\mathbf{T} \cdot \psi = \mathbf{x}_1 - \mathbf{x}_2 = \begin{bmatrix} 1 \\ -1 \\ 0 \end{bmatrix}. \tag{IV.36}$$

It follows that it is not possible to solve the equations

$$\mathbf{T} \cdot \psi = \mathbf{x}_1, \qquad \mathbf{T} \cdot \psi = \mathbf{x}_2$$

for the generalized eigenvector $\psi = \psi^{(2)}$. However, we may choose the proper eigenvector of $\mathbf{T}$ in the form

$$\psi_1^{(1)} = \begin{bmatrix} 1 \\ 0 \\ 0 \end{bmatrix} \quad \text{and} \quad \psi_2^{(1)} = \begin{bmatrix} 1 \\ -1 \\ 0 \end{bmatrix}.$$

The generalized eigenvector $\psi = \psi_2^{(2)}$ satisfying (IV.36) is

$$\psi_2^{(2)} = \begin{bmatrix} 0 \\ 0 \\ 1 \end{bmatrix}.$$

The adjoint eigenvectors satisfying

$$\begin{aligned} \mathbf{T}^{\mathrm{T}} \cdot \bar{\psi}_2^{*(2)} &= 0 \\ \mathbf{T}^{\mathrm{T}} \cdot \bar{\psi}_2^{*(1)} &= \bar{\psi}_2^{*(2)} \\ \mathbf{T}^{\mathrm{T}} \cdot \bar{\psi}_1^{*(1)} &= 0, \end{aligned} \tag{IV.37}$$

where

$$\mathbf{T}^{\mathrm{T}} = \begin{bmatrix} 0 & 0 & 0 \\ 0 & 0 & 0 \\ 1 & -1 & 0 \end{bmatrix}$$

are

$$\boldsymbol{\psi}_1^{*(1)} = \begin{bmatrix} \alpha_1 \\ \alpha_1 \\ \beta_1 \end{bmatrix}, \qquad \boldsymbol{\psi}_2^{*(2)} = \begin{bmatrix} 0 \\ 0 \\ \beta_2 \end{bmatrix}, \qquad \boldsymbol{\psi}_2^{*(1)} = \begin{bmatrix} \alpha_3 + \beta_2 \\ \alpha_3 \\ \beta_3 \end{bmatrix}.$$

These vectors also satisfy (IV.35) provided that

$$\boldsymbol{\psi}_1^{*(1)} = \begin{bmatrix} 1 \\ 1 \\ 0 \end{bmatrix}, \qquad \boldsymbol{\psi}_2^{*(2)} = \begin{bmatrix} 0 \\ 0 \\ 1 \end{bmatrix}, \qquad \boldsymbol{\psi}_2^{*(1)} = \begin{bmatrix} 0 \\ -1 \\ 0 \end{bmatrix}.$$

Appendix IV.2 Projections

Consider the linear operator $\mathbf{P}$ defined in $\mathbb{R}^n$ by

$$\mathbf{P} \cdot \mathbf{x} = \langle \mathbf{x}, \boldsymbol{\psi}^* \rangle \boldsymbol{\psi} \qquad \text{for any } \mathbf{x} \text{ in } \mathbb{R}^n, \tag{IV.38}$$

where $\boldsymbol{\psi}$ and $\boldsymbol{\psi}^*$ are two vectors satisfying $\langle \boldsymbol{\psi}, \boldsymbol{\psi}^* \rangle = 1$. Then $\mathbf{P}^2 \cdot \mathbf{x} = \mathbf{P}(\mathbf{P} \cdot \mathbf{x}) = \langle \mathbf{x}, \boldsymbol{\psi}^* \rangle \mathbf{P} \cdot \boldsymbol{\psi} = \langle \mathbf{x}, \boldsymbol{\psi}^* \rangle \mathbf{I} \cdot \boldsymbol{\psi} = \mathbf{P} \cdot \mathbf{x}$, so we have

$$\mathbf{P}^2 = \mathbf{P}. \tag{IV.39}$$

More generally, any linear operator $\mathbf{P}$ in $\mathbb{R}^n$ satisfying $\mathbf{P}^2 = \mathbf{P}$ is called a projection.

Now consider a family of p vectors $(\boldsymbol{\psi}_1, \ldots, \boldsymbol{\psi}_p)$ of $\mathbb{R}^n$, and a biorthogonal family $(\boldsymbol{\psi}_1^*, \ldots, \boldsymbol{\psi}_p^*)$ satisfying

$$\langle \boldsymbol{\psi}_i, \boldsymbol{\psi}_j^* \rangle = \delta_{ij}, \qquad i, j = 1, \ldots, p. \tag{IV.40}$$

Then we may define a linear operator $\mathbf{P}$ as follows:

$$\mathbf{P} \cdot \mathbf{x} = \sum_{i=1}^{p} \langle \mathbf{x}, \boldsymbol{\psi}_i^* \rangle \boldsymbol{\psi}_i. \tag{IV.41}$$

It is easy to verify that (IV.39) is satisfied, hence (IV.41) defines a projection on a p-dimensional subspace of $\mathbb{R}^n$. The condition (IV.40) is necessary, otherwise we would not have $\mathbf{P}^2 = \mathbf{P}$ for this operator.

EXAMPLE IV.2 (in $\mathbb{R}^3$). The vectors

$$\psi_1 = \begin{bmatrix} 1 \\ -1 \\ 0 \end{bmatrix}, \quad \psi_2 = \begin{bmatrix} 0 \\ 0 \\ 1 \end{bmatrix}, \quad \psi_1^* = \begin{bmatrix} 0 \\ -1 \\ 0 \end{bmatrix}, \quad \psi_2^* = \begin{bmatrix} 0 \\ 0 \\ 1 \end{bmatrix}$$

satisfy $\langle \psi_i, \psi_j^* \rangle = \delta_{ij}$, $i, j = 1, 2$.

The matrix of the projection $\mathbf{P}$ defined by

$$\mathbf{P} \cdot \mathbf{x} = \sum_{i=1}^{2} \langle \mathbf{x}, \psi_i^* \rangle \psi_i \tag{IV.42}$$

may be formed from the columns $\mathbf{P} \cdot \mathbf{e}_j$ of $\mathbf{P}$ where the $\mathbf{e}_j$ are the standard orthonormal basis vectors in $\mathbb{R}^n$, $\langle \mathbf{e}_i, \mathbf{e}_j \rangle = \delta_{ij}$. Hence

$$(\mathbf{P} \cdot \mathbf{x}) \cdot \mathbf{e}_l = (\mathbf{P} \cdot \mathbf{x})_l = \sum_{j=1}^{3} (\mathbf{P} \cdot \mathbf{e}_j)_l x_j.$$

So $P_{lj} = (\mathbf{P} \cdot \mathbf{e}_j)_l$. For example, (IV.42) may be written as

$$(\mathbf{P} \cdot \mathbf{x})_l = \sum_{i=1}^{2} \sum_{j=1}^{3} \langle \mathbf{e}_j, \psi_i^* \rangle \langle \psi_i, \mathbf{e}_l \rangle x_j$$

so that

$$P_{lj} = \sum_{i=1}^{2} \langle \mathbf{e}_j, \psi_i^* \rangle \langle \psi_i, \mathbf{e}_l \rangle.$$

Hence

$$\mathbf{P} = \begin{bmatrix} 0 & -1 & 0 \\ 0 & 1 & 0 \\ 0 & 0 & 1 \end{bmatrix} \quad \text{and} \quad \mathbf{P}^2 = \mathbf{P}.$$

Projections are the mathematical tool we use to reduce the dimension of bifurcation problems. The more frequently used projections are such that they commute with a linear operator $\mathbf{A}$:

$$\mathbf{P} \cdot \mathbf{A} = \mathbf{A} \cdot \mathbf{P} \tag{IV.43}$$

Let $\mathbf{A}$ be a linear operator in $\mathbb{R}^n$ (represented by an $n \times n$ matrix) having an eigenvalue σ of multiplicity n_ν and Riesz index ν. Results stated in Appendix IV.1 guarantee that we may construct a projection $\mathbf{P}$ defined by

$$\mathbf{P} \cdot \mathbf{x} = \sum_m \langle \mathbf{x}, \psi_i^{*(m)} \rangle \psi_i^{(m)}, \qquad 1 \le m \le \nu_i.$$

We may assert and then demonstrate that

$$\mathbf{P} \cdot \mathbf{A} \cdot \mathbf{x} = \mathbf{A} \cdot \mathbf{P} \cdot \mathbf{x} \quad \text{for any } \mathbf{x}.$$

For the demonstration, we define $\mathbf{T} = \mathbf{A} - \sigma\mathbf{I}$ and show that

$$\mathbf{P}\cdot\mathbf{T}\cdot\mathbf{x} = \mathbf{T}\cdot\mathbf{P}\cdot\mathbf{x}$$

is equivalent to (IV.43). We have

$$\begin{aligned}\mathbf{T}\cdot\mathbf{P}\cdot\mathbf{x} &= \sum_m \langle\mathbf{x}, \boldsymbol{\psi}_i^{*(m)}\rangle\mathbf{T}\cdot\boldsymbol{\psi}_i^{(m)} \quad \text{for } 2 \le m \le \nu_i \\ &= \sum_m \langle\mathbf{x}, \boldsymbol{\psi}_i^{*(m+1)}\rangle\boldsymbol{\psi}_i^{(m)} \quad \text{for } 1 \le m \le \nu_i - 1.\end{aligned}$$

In addition,

$$\begin{aligned}\mathbf{P}\cdot\mathbf{T}\cdot\mathbf{x} &= \sum_m \langle\mathbf{x}, \mathbf{T}^{\mathrm{T}}\cdot\boldsymbol{\psi}_i^{*(m)}\rangle\boldsymbol{\psi}_i^{(m)} \quad \text{for } 1 \le m \le \nu_i \\ &= \sum_m \langle\mathbf{x}, \boldsymbol{\psi}_i^{*(m+1)}\rangle\boldsymbol{\psi}_i^{(m)} \quad \text{for } 1 \le m \le \nu_i - 1.\end{aligned}$$

proving (IV.43).

Example IV.3 (in $\mathbb{R}^4$).

$$\mathbf{T} = \begin{bmatrix} 0 & 0 & 1 & 1 \\ 0 & 0 & -1 & 1 \\ 0 & 0 & 0 & 1 \\ 0 & 0 & 0 & 1 \end{bmatrix},$$

the eigenvalue zero has a multiplicity 3 and index 2, and we may choose the following system of generalized eigenvectors:

$$\boldsymbol{\psi}_1^{(1)} = \begin{bmatrix} 1 \\ 0 \\ 0 \\ 0 \end{bmatrix}, \quad \boldsymbol{\psi}_2^{(1)} = \begin{bmatrix} 1 \\ -1 \\ 0 \\ 0 \end{bmatrix}, \quad \boldsymbol{\psi}_2^{(2)} = \begin{bmatrix} 0 \\ 0 \\ 1 \\ 0 \end{bmatrix},$$

$$\boldsymbol{\psi}_1^{*(1)} = \begin{bmatrix} 1 \\ 1 \\ 0 \\ -2 \end{bmatrix}, \quad \boldsymbol{\psi}_2^{*(1)} = \begin{bmatrix} 0 \\ -1 \\ 0 \\ 0 \end{bmatrix}, \quad \boldsymbol{\psi}_2^{*(2)} = \begin{bmatrix} 0 \\ 0 \\ 1 \\ -1 \end{bmatrix}.$$

The projection (which is the sum of two commuting projections of the type just described)

$$\mathbf{P} = \langle\cdot, \boldsymbol{\psi}_1^{*(1)}\rangle\boldsymbol{\psi}_1^{(1)} + \langle\cdot, \boldsymbol{\psi}_2^{*(1)}\rangle\boldsymbol{\psi}_2^{(1)} + \langle\cdot, \boldsymbol{\psi}_2^{*(2)}\rangle\boldsymbol{\psi}_2^{(2)}$$

has the matrix representation.

$$\mathbf{P} = \begin{bmatrix} 1 & 0 & 0 & -2 \\ 0 & 1 & 0 & 0 \\ 0 & 0 & 1 & -1 \\ 0 & 0 & 0 & 0 \end{bmatrix},$$

and we may verify that

$$\mathbf{P} \cdot \mathbf{T} = \mathbf{T} \cdot \mathbf{P} = \begin{bmatrix} 0 & 0 & 1 & -1 \\ 0 & 0 & -1 & 1 \\ 0 & 0 & 0 & 0 \\ 0 & 0 & 0 & 0 \end{bmatrix}$$

is in fact identical with $\mathbf{T}$ on the subspace generated by $\boldsymbol{\psi}_1^{(1)}, \boldsymbol{\psi}_2^{(1)}, \boldsymbol{\psi}_2^{(2)}$; that is, $\mathbf{T} \cdot \mathbf{x} = \mathbf{P} \cdot \mathbf{T} \cdot \mathbf{x}$ when $\mathbf{x} = \alpha\boldsymbol{\psi}_1^{(1)} + \beta\boldsymbol{\psi}_2^{(1)} + \gamma\boldsymbol{\psi}_2^{(2)}$. In the present case this subspace includes all vectors whose fourth component is zero.

CHAPTER V

Bifurcation of Steady Solutions in Two Dimensions and the Stability of the Bifurcating Solutions

We turn now to the analysis of steady bifurcating solutions of the two-dimensional autonomous problem (IV.1). In this chapter we emphasize the fact that the problem is two-dimensional by writing (IV.1) in component form as

$$\frac{du_l}{dt} = f^{(l)}(\mu, u_1, u_2), \qquad l = 1, 2, \tag{V.1}$$

where

$$\begin{aligned}
f^{(1)}(\mu, u_1, u_2) &= a_0 u_1 + b_0 u_2 + \mu(a'(\mu)u_1 + b'(\mu)u_2) \\
&\quad + \alpha_1(\mu)u_1^2 + 2\beta_1(\mu)u_1u_2 + \gamma_1(\mu)u_2^2 + O(\|\mathbf{u}\|^3), \\
f^{(2)}(\mu, u_1, u_2) &= c_0 u_1 + d_0 u_2 + \mu(c'(\mu)u_1 + d'(\mu)u_2) \\
&\quad + \alpha_2(\mu)u_1^2 + 2\beta_2(\mu)u_1u_2 + \gamma_2(\mu)u_2^2 + O(\|\mathbf{u}\|^3).
\end{aligned}$$

The leading terms in $f^{(l)}$ are components of the matrices $(A_{lj}(\mu))$ and $(B_{ljk}(\mu))$ defined by (IV.2), and $\|\mathbf{u}\|^2 = u_1^2 + u_2^2$.

We studied the stability of the solution $\mathbf{u} = 0$ in Chapter IV. We framed our study in terms of the eigenvalues $\sigma(\mu)$ of $\mathbf{A}(\mu)$. Now we shall find the conditions under which new steady solutions of (V.1) can bifurcate and specify the conditions under which they are stable to small disturbances.

V.1 The Form of Steady Bifurcating Solutions and Their Stability

There are many equivalent ways to parameterize one and the same bifurcating solution. We may use the given physical parameter μ and seek the bifurcating solutions in the form $(\mu, u_1(\mu), u_2(\mu))$. Or we may introduce an amplitude ε

defined by some function of u_1 and u_2 and seek solutions in the form $(\varepsilon, \mu(\varepsilon), u_1(\varepsilon), u_2(\varepsilon))$. For example, we can choose $\varepsilon = u_1$ or $\varepsilon = u_2$ or $\varepsilon = f(u_1, u_2)$ for some good function f. In more general cases, like those arising in bifurcation of solutions of partial differential equations, the amplitude of the bifurcating solution can be defined by a suitably chosen functional. Definitions like (VI.7), which are based on projecting a component of the bifurcating solution, are particularly convenient. Different definitions of ε are equivalent if they are related to one another by an invertible transformation.

In this chapter we often parameterize the bifurcating branches by

$$\begin{aligned} u_1 &= \varepsilon, \\ u_2 &= \varepsilon y(\varepsilon), \\ \mu &= \varepsilon \lambda(\varepsilon). \end{aligned} \tag{V.2}$$

Before starting the analysis of the solutions (V.2) it is of value to frame an analysis of the problem in terms of the given parameter μ:

$$(\mu, u_1, u_2) = (\mu, u_1(\mu), u_2(\mu)). \tag{V.3}$$

To obtain solutions in the form (V.3) it is enough that (u_1, u_2) be intersection points of the two curves of the (u_1, u_2) plane of equations

$$f_l(\mu, u_1, u_2) = 0, \qquad l = 1, 2,$$

for some $\mu = \mu_0$ and that at the same μ_0, the criterion

$$\det \mathscr{J} \neq 0,$$

where

$$\mathscr{J} = \begin{bmatrix} \dfrac{\partial f_1}{\partial u_1} & \dfrac{\partial f_1}{\partial u_2} \\[2ex] \dfrac{\partial f_2}{\partial u_1} & \dfrac{\partial f_2}{\partial u_2} \end{bmatrix},$$

of the implicit function theorem in $\mathbb{R}^2$ (see Appendix V.1) is satisfied.

There is an intimate connection between the sufficient condition $\det \mathscr{J} \neq 0$ for the existence of a branch and the stability of the solutions. The stability of any solution to small disturbances v_l of (V.3) or (V.2) is determined from the linearized evolution problem governing v_l:

$$\frac{dv_l}{dt} = \frac{\partial f_l}{\partial u_1} v_1 + \frac{\partial f_l}{\partial u_2} v_2, \qquad l = 1, 2,$$

where, for (V.3),

$$f_l = f_l(\mu_0, u_1(\mu_0), u_2(\mu_0))$$

or, for (V.2)

$$f_l = f_l(\varepsilon_0 \lambda(\varepsilon_0), \varepsilon_0, \varepsilon_0 y(\varepsilon_0)).$$

The evolution problem is solved by exponentials

$$\mathbf{v}(t) = e^{\gamma t}\boldsymbol{\zeta} \qquad (v_l(t) = e^{\gamma t}\zeta_l)$$

if the eigenvalue γ and eigenfunction $\boldsymbol{\zeta}$ satisfy the spectral problem

$$\gamma \quad = \mathscr{J} \cdot \boldsymbol{\zeta}.$$

The eigenvalues of the Jacobian matrix $\mathscr{J}$ are $\gamma_1(\mu_0)$ and $\gamma_2(\mu_0)$ and a solution (V.3) is stable when $\mu = \mu_0$ (or $\varepsilon = \varepsilon_0$ for $\mu_0 = \mu(\varepsilon_0)$) if the real part of both eigenvalues is negative. Obviously, $f_l(\mu_0, u_1, u_2) = 0$ and

$$\det \mathscr{J} = \gamma_1(\mu_0)\gamma_2(\mu_0) \neq 0$$

is a sufficient condition for the existence of a continuous branch of the solution (V.3) for μ in a neighborhood of μ_0 framed in terms of the eigenvalues governing stability.

If $\gamma_1(\mu_0)$ is complex, then $\gamma_2(\mu_0) = \bar{\gamma}_1(\mu_0)$ and $\det \mathscr{J} = |\gamma_1(\mu_0)|^2 > 0$. If $\gamma_1(\mu_0)$ is real-valued then $\gamma_2(\mu_0)$ is also real. The ambiguous case in which the existence of a branch through (μ_0, u_1, u_2) cannot be established by the argument following (V.3) using the implicit function theorem is when one of the two eigenvalues governing stability vanishes. We may describe this ambiguous case in geometric terms as follows: $\det \mathscr{J} \neq 0$ when $\mu = \mu_0$ if and only if the curves relating u_1 and u_2 in the (u_1, u_2) plane intersect transversally when $\mu = \mu_0$; that is, $f_1(\mu_0, u_1, u_2) = 0$ and $f_2(\mu_0, u_1, u_2) = 0$. If these curves have the same tangent at the point where they intersect, then the two equations

$$\frac{\partial f_l}{\partial u_1}\delta u_1 + \frac{\partial f_l}{\partial u_2}\delta u_2 = 0, \qquad l = 1, 2,$$

have a nonzero solution $(\delta u_1, \delta u_2)$; that is, $\det \mathscr{J} = \gamma_1(\mu_0)\gamma_2(\mu_0) = 0$.

After setting $(\mu, u_1, u_2) = (\varepsilon\lambda, \varepsilon, \varepsilon y)$ we get

$$\mathscr{J} = \begin{bmatrix} \dfrac{\partial f_1}{\partial u_1} & \dfrac{\partial f_1}{\partial u_2} \\[2ex] \dfrac{\partial f_2}{\partial u_1} & \dfrac{\partial f_2}{\partial u_2} \end{bmatrix}$$

$$= \begin{bmatrix} a_0 + \varepsilon(\lambda a' + 2\alpha_1 + 2\beta_1 y) & b_0 + \varepsilon(\lambda b' + 2\beta_1 + 2\gamma_1 y) \\ c_0 + \varepsilon(\lambda c' + 2\alpha_2 + 2\beta_2 y) & d_0 + \varepsilon(\lambda d' + 2\beta_2 + 2\gamma_2 y) \end{bmatrix} + O(\varepsilon^2). \tag{V.4}$$

V.2 Classification of the Three Types of Bifurcation of Steady Solutions

Now we shall use the implicit function theorem to prove the existence of unique functions $\lambda(\varepsilon)$ and $y(\varepsilon)$ satisfying

$$f_1(\mu, u_1, u_2) = \varepsilon g_1(\lambda(\varepsilon), \varepsilon, y(\varepsilon)) = 0$$
$$f_2(\mu, u_1, u_2) = \varepsilon g_2(\lambda(\varepsilon), \varepsilon, y(\varepsilon)) = 0,$$

where

$$g_1(\lambda, \varepsilon, y) = a_0 + b_0 y + \varepsilon[\lambda(a' + b'y) + \alpha_1 + 2\beta_1 y + \gamma_1 y^2] + O(\varepsilon^2),$$
$$g_2(\lambda, \varepsilon, y) = c_0 + d_0 y + \varepsilon[\lambda(c' + d'y) + \alpha_2 + 2\beta_2 y + \gamma_2 y^2] + O(\varepsilon^2).$$

We seek steady bifurcating solutions of the equation

$$g_i(\lambda(\varepsilon), \varepsilon, y(\varepsilon)) = 0. \tag{V.5}$$

To solve (V.5) we first require that

$$g_i(\lambda_0, 0, y_0) = 0, \qquad i = 1, 2. \tag{V.6}$$

Equations (V.6) imply that

$$a_0 + b_0 y_0 = 0$$
$$c_0 + d_0 y_0 = 0. \tag{V.7}$$

Equations (V.7) are always satisfied by the eigenvector

$$\mathbf{x}_1 = \begin{bmatrix} 1 \\ y_0 \end{bmatrix}$$

belonging to $\sigma_1(0) = 0$ (see §IV.4.2). There are three ways to satisfy (V.7):

(i) $\Delta = (a_0 + d_0)^2/4 - a_0 d_0 + b_0 c_0 \neq 0$. Then $\sigma_1(0)$ and $\sigma_2(0)$ are distinct. Moreover, (V.7) shows that $a_0 d_0 - b_0 c_0 = 0$ and

$$\Delta = \frac{(a_0 + d_0)^2}{4} > 0,$$

so that the two eigenvalues $\sigma_1(0) = \xi_1(0) = 0$, $\sigma_2(0) = \xi_2(0) < 0$ $(a_0 + d_0 < 0)$ are real-valued. In this case we have bifurcation at the simple eigenvalue $\xi_1(0) = 0$. This problem is reduced to $\mathbb{R}^1$ in projection in Chapter VI.

(ii) $\sigma_1(0) = \sigma_2(0) = 0$ has Riesz index two. We shall study this case, without loss of generality (see §IV.4.4.2) when the parameters are such that

$$a_0 = c_0 = d_0 = b_0 - 1 = y_0 = 0.$$

(iii) $\sigma_1(0) = \sigma_2(0) = 0$ has Riesz index one. This is the case of bifurcation at a double semi-simple eigenvalue.

V.3 Bifurcation at a Simple Eigenvalue

We set

$$y = y_0 + \varepsilon \tilde{y} \tag{V.8}$$

and define

$$\varepsilon h_i(\lambda, \varepsilon, \tilde{y}) = g_i(\lambda, \varepsilon, y_0 + \varepsilon \tilde{y}) = 0 \tag{V.9}$$

where

$$\begin{aligned} h_1(\lambda, \varepsilon, \tilde{y}) &= b_0 \tilde{y} + \lambda(a' + b' y_0) + \alpha_1 + 2\beta_1 y_0 + \gamma_1 y_0^2 + O(\varepsilon) = 0 \\ h_2(\lambda, \varepsilon, \tilde{y}) &= d_0 \tilde{y} + \lambda(c' + d' y_0) + \alpha_2 + 2\beta_2 y_2 + \gamma_2 y_0^2 + O(\varepsilon) = 0. \end{aligned} \tag{V.10}$$

We find that

$$\begin{aligned} h_1(\lambda_0, 0, \tilde{y}_0) &= b_0 \tilde{y}_0 + \lambda_0(a_0' + b_0' y_0) \\ &\quad + \alpha_{10} + 2\beta_{10} y_0 + \gamma_{10} y_0 = 0 \\ h_2(\lambda_0, 0, \tilde{y}_0) &= d_0 \tilde{y}_0 + \lambda_0(c_0' + d_0' y_0) \\ &\quad + \alpha_{20} + 2\beta_{20} y_0 + \gamma_{20} y_0^2 = 0. \end{aligned} \tag{V.11}$$

Recalling that $y_0 = -a_0/b_0 = -c_0/d_0$, we may verify that Equations (V.11) determine unique values of $\tilde{y}_0$ and λ_0 in terms of coefficients evaluated at $\mu = 0$, provided only that the determinant of the coefficients of $\tilde{y}_0$ and λ_0 does not vanish. This determinant is the same as the determinant of the Jacobian matrix

$$\begin{bmatrix} \dfrac{\partial h_1}{\partial \lambda} & \dfrac{\partial h_1}{\partial \tilde{y}} \\[2ex] \dfrac{\partial h_2}{\partial \lambda} & \dfrac{\partial h_2}{\partial \tilde{y}} \end{bmatrix} = \begin{bmatrix} a_0' + b_0' y_0 & b_0 \\ c_0' + d_0' y_0 & d_0 \end{bmatrix}$$

evaluated at $\varepsilon = 0$ and it can be shown to be

$$d_0(a_0' + b_0' y_0) - b_0(c_0' + d_0' y_0) = \xi_1'(0)\xi_2(0) < 0. \tag{V.12}$$

Given (V.12), it follows from the implicit function theorem that (V.10) can be solved for $\lambda(\varepsilon)$ and $\tilde{y}(\varepsilon)$. Then tracing back through (V.8) we get the bifurcating solutions given by (V.2).

To prove (V.12) we note that (V.7) implies

$$\begin{aligned} &d_0(a_0' + b_0' y_0) - b_0(c_0' + d_0' y_0) \\ &\quad = d_0 a_0' - c_0 b_0' - b_0 c_0' + a_0 d_0' \\ &\quad = (ad - bc)_0'. \end{aligned} \tag{V.13}$$

Since

$$\det \mathbf{A} = ad - bc = \xi_1 \xi_2,$$

it follows that

$$(ad - bc)' = \xi_1(0)\xi_2'(0) + \xi_1'(0)\xi_2(0) = \xi_1'(0)\xi_2(0),$$

proving (V.12).

Another method for constructing the same bifurcating solution, which is just as easy to implement, is given in §VI.2.

V.4 Stability of the Steady Solution Bifurcating at a Simple Eigenvalue

We can determine the stability of the bifurcating solution in the linearized theory by studying the eigenvalues of the matrix $\mathscr{J}$. However, we are not really interested in all the eigenvalues but only in the largest one which is zero at criticality (the other eigenvalue is negative). So it is a good idea to project into $\mathbb{R}^1$; then we can study the interesting eigenvalue, the one controlling stability. This projection is carried out in §VI.4, where we show that supercritical steady solutions which bifurcate at a simple eigenvalue are stable and subcritical solutions are unstable. This is exactly the same result which we already proved in the analysis of bifurcation problems in $\mathbb{R}^1$ (see Figure II.3).

V.5 Bifurcation at a Double Eigenvalue of Index Two*

In case (ii) of §V.2 we seek the steady solutions which bifurcate when $\sigma(0) = 0$ is a double eigenvalue of index two with $a_0 = c_0 = d_0 = b_0 - 1 = y_0 = 0$. The steady solutions are in the form (V.2) with $\lambda(\varepsilon)$ and $y(\varepsilon)$ to be determined from the two nonlinear equations

$$g_1(\lambda, \varepsilon, y) = \tilde{g}_2(\lambda, \varepsilon, y) \stackrel{\text{def}}{=} \frac{g_2(\lambda, \varepsilon, y)}{\varepsilon} = 0 \qquad (\text{V.14})_1$$

where

$$\begin{aligned} g_1 &= y + \varepsilon\{\lambda(a' + b'y) + \alpha_1 + 2\beta_1 y + \gamma_1 y^2\} + O(\varepsilon^2) \\ \tilde{g}_2 &= \lambda(c' + d'y) + \alpha_2 + 2\beta_2 y + \gamma_2 y^2 + O(\varepsilon). \end{aligned} \qquad (\text{V.14})_2$$

* This problem has been studied in a more general context by K. A. Landman and S. Rosenblat, Bifurcation from a multiple eigenvalue and stability of solutions, *SIAM J. Appl. Math.*, **34**, 743 (1978).

We shall use the implicit function theorem to solve (V.14). First, we note that (V.14) is satisfied with

$$\varepsilon = y_0 = 0 \quad \text{only if} \quad \lambda_0 = -\frac{\alpha_{20}}{c_0'}; \quad \text{that is,}$$

$$g_1\left(-\frac{\alpha_{20}}{c_0'}, 0, 0\right) = 0 \tag{V.15}$$

$$\tilde{g}_2\left(-\frac{\alpha_{20}}{c_0'}, 0, 0\right) = 0.$$

We can solve (V.14) for $(\lambda(\varepsilon), y(\varepsilon))$ when ε is small and $(\lambda(\varepsilon), y(\varepsilon))$ is close to $(\lambda(0), y(0)) = (-\alpha_{20}/c_0', 0)$ provided that the determinant of

$$\tilde{\mathbf{J}} = \begin{bmatrix} \dfrac{\partial g_1}{\partial \lambda} & \dfrac{\partial g_1}{\partial y} \\ \dfrac{\partial \tilde{g}_2}{\partial \lambda} & \dfrac{\partial \tilde{g}_2}{\partial \tilde{y}} \end{bmatrix} \tag{V.16}$$

does not vanish when $(\lambda, y, \varepsilon) = (-\alpha_{20}/c_0', 0, 0)$; that is when

$$\det\ \tilde{\mathbf{J}} = \det \begin{bmatrix} 0 & 1 \\ c_0' & \lambda_0 d_0' + 2\beta_2 \end{bmatrix} = -c_0' \neq 0. \tag{V.17}$$

If $c_0' = 0$ and $\alpha_{20} \neq 0$ we can find the steady bifurcating solution in the form $(u_1(\mu), u_2(\mu))$. The construction of this solution is given as Exercise V.1 at the end of this chapter.

Supposing now $c_0' \neq 0$ we can easily find the bifurcating solution as a power series in ε. The coefficients of the power series can be computed by repeated differentiation of (V.14) with respect to ε at $\varepsilon = 0$. For the first derivatives we get

$$\left.\frac{dg_1(\lambda(\varepsilon), \varepsilon, y(\varepsilon))}{d\varepsilon}\right|_{\varepsilon=0} = \frac{dy}{d\varepsilon} + \lambda_0(a_0' + \alpha_{10}) = 0 \tag{V.18}$$

and

$$\left.\frac{d\tilde{g}_2(\lambda(\varepsilon), \varepsilon, y(\varepsilon))}{d\varepsilon}\right|_{\varepsilon=0} = c_0' \frac{d\lambda}{d\varepsilon} + \lambda_0\left(\lambda_0 \frac{dc'(0)}{d\mu} + d_0' \frac{dy}{d\varepsilon}\right) + 2\beta_{20} \frac{dy}{d\varepsilon} = 0, \tag{V.19}$$

Equation (V.18) gives $dy/d\varepsilon$, and (V.19) gives $d\lambda/d\varepsilon$.

Another method for constructing the solution bifurcating from a double eigenvalue of index two will be given in §VI.11. The method given there projects problems of infinite dimensions into one dimension. The dimension of the important projection for infinite-dimensional problems is the geometric multiplicity n_1, irrespective of the algebraic multiplicity μ_1 (see §§IV.2 and

VI.11). In the present case $n_1 = 1$ even though $\mu_1 = 2$; we have one eigenvector and a double eigenvalue.

V.6 Stability of the Steady Solution Bifurcating at a Double Eigenvalue of Index Two

The linearized stability of the bifurcating solution computed in (V.5) may be determined from the sign of the real part of the Jacobian matrix $\mathscr{J}$ defined by (V.4). In the present case with $a_0 = c_0 = d_0 = y_0 = b_0 - 1 = \lambda_0 c_0' + \alpha_{20} = 0$,

$$\mathscr{J} = \begin{bmatrix} \varepsilon(\lambda_0 a_0' + 2\alpha_{10}) & 1 + \varepsilon(\lambda_0 b_0' + 2\beta_{10}) \\ \varepsilon(\lambda_0 c_0' + 2\alpha_{20}) & \varepsilon(\lambda_0 d_0' + 2\beta_{20}) \end{bmatrix} + O(\varepsilon^2).$$

The eigenvalues of $\mathscr{J}$ are $\gamma_1(\varepsilon)$ and $\gamma_2(\varepsilon)$ and are given by det $[\mathscr{J} - \gamma\mathbf{1}] = 0$; that is,

$$\begin{bmatrix} \gamma_1(\varepsilon) \\ \gamma_2(\varepsilon) \end{bmatrix} = [-\varepsilon\lambda_0 c_0']^{1/2}\begin{bmatrix} 1 \\ -1 \end{bmatrix} + \frac{\varepsilon}{2}\{\lambda_0(a_0' + d_0')$$
$$+ 2\alpha_{10} + 2\beta_{20}\}\begin{bmatrix} 1 \\ 1 \end{bmatrix} + O(\varepsilon^{3/2}).$$

Choose one determination of $\sqrt{\varepsilon}$, say $\sqrt{\varepsilon}$ is real and positive if $\varepsilon > 0$, then $\gamma_1(\varepsilon)$ and $\gamma_2(\varepsilon)$ are analytic in $\sqrt{\varepsilon}$. If $\lambda_0 \neq 0$, we have

$$\begin{bmatrix} \gamma_1(\varepsilon) \\ \gamma_2(\varepsilon) \end{bmatrix} = (-\mu c_0')^{1/2}\begin{bmatrix} 1 \\ -1 \end{bmatrix} + O[\mu(\varepsilon)].$$

Compare this with the formula (IV.30)

$$\begin{bmatrix} \sigma_1(\mu) \\ \sigma_2(\mu) \end{bmatrix} = (\mu c_0')^{1/2}\begin{bmatrix} 1 \\ -1 \end{bmatrix} + O(\mu)$$

giving the eigenvalues for the stability of solution $\mathbf{u} = 0$. When $\mu c_0' > 0$, zero is unstable and the stability of the bifurcated solution is determined by the eigenvalues $\gamma_i(\varepsilon)$ at order $\mu(\varepsilon)$; that is, at order $O(\varepsilon)$, since $(-\mu c_0')^{1/2}$ is imaginary. When $\mu c_0' < 0$, the bifurcated solution is unstable while the zero solution has its stability determined by the terms of order μ in $\sigma_2(\mu)$. Let us assume that $(a_0' + d_0')c_0' < 0$ and $c_0' > 0$. This implies that the zero solution is stable for $\mu < 0$ and is unstable for $\mu > 0$. (The opposite is true if $c_0' < 0$.) In this case the bifurcated solution is unstable for $\mu < 0$ and may be stable or unstable for $\mu > 0$ depending on the coefficient of ε in $\gamma_i(\varepsilon)$ (see Figure V.1). We draw the readers attention to the fact that neither the zero solution nor the bifurcated solution can be stable on both sides of criticality.

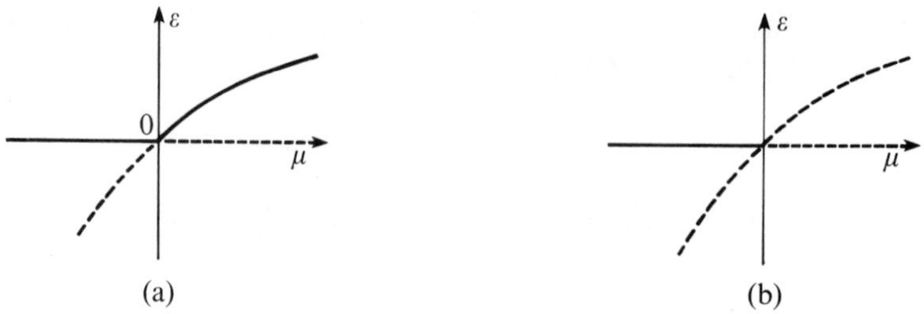

Figure V.1 Possible distributions of stability for steady solutions bifurcating from $\varepsilon = 0$ at a double eigenvalue of index 2 in the case $c_0' > 0$, where it is assumed that the zero solution loses stability strictly as μ is increased past zero (see (IV.30))

If $\lambda_0 = 0$ and $\mu_{\varepsilon\varepsilon}(0) \neq 0$, the bifurcation is one-sided and the determination of stability depends on details of the specific problem.

V.7 Bifurcation and Stability of Steady Solutions in the Form (V.2) at a Double Eigenvalue of Index One (Semi-Simple)*

Here and in §V.8 we are concerned with bifurcation and stability of steady solutions under the hypothesis laid down under (iii) of §V.2. The hypothesis is that $\sigma(0) = 0$ is a double eigenvalue of index one (semi-simple). It follows from this hypothesis that $a_0 = b_0 = c_0 = d_0 = 0$. We may find the bifurcating solutions in the form (V.2) or (V.3). When both solutions exist they are equivalent and we may invert $(\mu, u_1(\mu), u_2(\mu)) \leftrightarrows (\varepsilon\lambda(\varepsilon), \varepsilon, \varepsilon y(\varepsilon))$. Here we study (V.2) and in §V.8 we study (V.3).

The steady solutions (V.2) are determined as roots $\lambda(\varepsilon)$, $y(\varepsilon)$ of the two nonlinear equations

$$\tilde{g}_l(\lambda, \varepsilon, y) \overset{\text{def}}{=} \frac{g_l(\lambda, \varepsilon, y)}{\varepsilon} = 0, \qquad l = 1, 2, \tag{V.20}$$

where

$$\begin{aligned} \tilde{g}_1(\lambda, \varepsilon, y) &= \lambda(a' + b'y) + \alpha_1 + 2\beta_1 y + \gamma_1 y^2 + O(\varepsilon) \\ \tilde{g}_2(\lambda, \varepsilon, y) &= \lambda(c' + d'y) + \alpha_2 + 2\beta_2 y + \gamma_2 y^2 + O(\varepsilon). \end{aligned} \tag{V.21}$$

* This problem has been studied by J. B. McLeod and D. H. Sattinger, Loss of stability and bifurcation at a double eigenvalue, *J. Functional Analysis*, **14**, 62, (1973). Problems involving multiple semi-simple eigenvalues arise in bifurcation problems which break spatial symmetry (see the book by Sattinger listed in the references at the end of Chapter I). It is frequently the case that the eigenvalues depend on more than one parameter and are semi-simple only for special relationships between the parameters. It is then useful to see what happens when the parameters are moved in such a way as to split the multiple eigenvalues. It is even possible to find secondary bifurcations in this way (see Example V.6 and Exercise V.6 at the end of this chapter).

We seek solutions of (V.20) which bifurcate at $(\varepsilon, \lambda, y) = (0, \lambda_0, y_0)$. It follows that

$$\begin{aligned} \tilde{g}_1(\lambda_0, 0, y_0) &= \lambda_0(a_0' + b_0' y_0) + \alpha_{10} + 2\beta_{10} y_0 + \gamma_{10} y_0^2 = 0 \\ \tilde{g}_2(\lambda_0, 0, y_0) &= \lambda_0(c_0' + d_0' y_0) + \alpha_{20} + 2\beta_{20} y_0 + \gamma_{20} y_0^2 = 0. \end{aligned} \tag{V.22}$$

We can guarantee bifurcation, using the implicit function theorem in $\mathbb{R}^2$ (see Appendix V.1), when

$$\det \tilde{\tilde{\mathbf{J}}} \neq 0 \tag{V.23}$$

where the Jacobian $\tilde{\tilde{\mathbf{J}}}$ is given by

$$\tilde{\tilde{\mathbf{J}}} = \begin{bmatrix} \dfrac{\partial \tilde{g}_1}{\partial \lambda} & \dfrac{\partial \tilde{g}_1}{\partial y} \\ \dfrac{\partial \tilde{g}_2}{\partial \lambda} & \dfrac{\partial \tilde{g}_2}{\partial y} \end{bmatrix} = \begin{bmatrix} a_0' + b_0' y_0 & 2\beta_{10} + 2\gamma_{10} y_0 + \lambda_0 b_0' \\ c_0' + d_0' y_0 & 2\beta_{20} + 2\gamma_{20} y_0 + \lambda_0 d_0' \end{bmatrix} \tag{V.24}$$

at $(\varepsilon, \lambda, y) = (0, \lambda_0, y_0)$.

The stability of $(\mu(\varepsilon), u_1(\varepsilon), u_2(\varepsilon)) = (\varepsilon\lambda(\varepsilon), \varepsilon, \varepsilon y(\varepsilon))$ may be determined from the eigenvalues of

$$\begin{aligned} \mathbf{J} &= \varepsilon \begin{bmatrix} \lambda_0 a_0' + 2\alpha_{10} + 2\beta_{10} y_0 & \lambda_0 b_0' + 2\beta_{10} + 2\gamma_{10} y_0 \\ \lambda_0 c_0' + 2\alpha_{20} + 2\beta_{20} y_0 & \lambda_0 d_0' + 2\beta_{20} + 2\gamma_{20} y_0 \end{bmatrix} + O(\varepsilon^2) \\ &\stackrel{\text{def}}{=} \varepsilon\, \mathbf{J}_0 + O(\varepsilon^2). \end{aligned} \tag{V.25}$$

When ε is small we may determine stability from the sign of the real parts of the two eigenvalues belonging to $\mathbf{J}_0$.

Now we want to call the readers' attention to a big difference between the problem of bifurcation in the semi-simple case (iii) being treated now and the previously treated cases (i) and (ii) of bifurcation at a simple eigenvalue and at a double eigenvalue of index two. In the other two cases $f_l = \varepsilon g_l$, where g_1 and g_2 do not vanish identically when $\varepsilon = 0$. The lowest-order balance occurs then at the linear order ε and leads to linear equations (V.7) or (V.15) determining a unique y_0 or a unique λ_0. In the present, semi-simple case, $f_l = \varepsilon^2 \tilde{g}_l$ where $\tilde{g}_1$ and $\tilde{g}_2$ are finite at $\varepsilon = 0$. The lowest-order balance occurs at order ε^2 and is nonlinear. From this nonlinear balance we can sometimes get more than one bifurcating solution.

To see how the semi-simple problem can lead to multiple solutions we need only to note that (V.22) is equivalent to a cubic equation for y_0, $\mathscr{C}(y_0) = 0$ if $d_0'\gamma_{10} - b_0'\gamma_{20} \neq 0$, where

$$\begin{aligned} \mathscr{C}(y_0) = &(c_0' + d_0' y_0)(\alpha_{10} + 2\beta_{10} y_0 + \gamma_{10} y_0^2) \\ &- (a_0' + b_0' y_0)(\alpha_{20} + 2\beta_{20} y_0 + \gamma_{20} y_0^2). \end{aligned} \tag{V.26}$$

This equation has three roots, two real roots one of which is double, or one real root which may be triple (see Figure V.2).

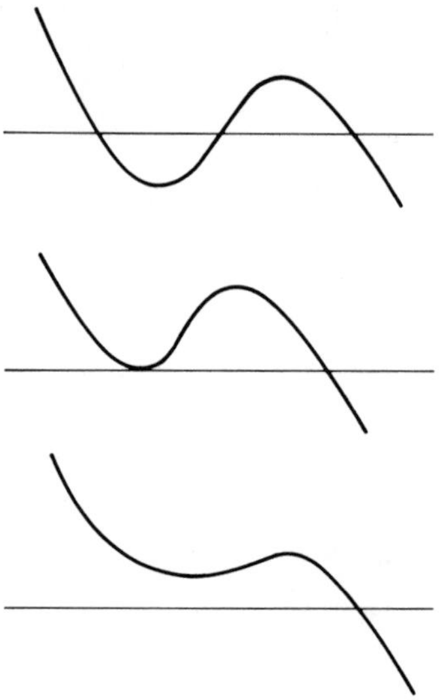

Figure V.2 The number of solutions which can bifurcate from $\varepsilon = 0$ at a double semi-simple eigenvalue correspond to the roots of a cubic $\mathscr{C}(y_0) = 0$ if $d_0'\gamma_{10} - b_0'\gamma_{20} \neq 0$

We may construct one bifurcating solution $(\lambda^{[k]}(\varepsilon), y^{[k]}(\varepsilon))$ for each simple real root $y_0^{[k]}$ of the cubic $\mathscr{C}(y_0) = 0$ as a power series in powers of ε. Since $1 \leq k \leq 3$ there can be 1, 2, or 3 solutions, each with a different bifurcation curve $\lambda^{[k]}(\varepsilon)$ as in Figure V.3. To obtain the coefficients of the power series it is only necessary to differentiate $\tilde{g}_l(\lambda^{[k]}(\varepsilon), \varepsilon, y^{[k]}(\varepsilon))$ repeatedly with respect to ε at $\varepsilon = 0$. For example, we may compute $d\lambda^{[k]}(0)/d\varepsilon$ and $dy^{[k]}(0)/d\varepsilon$ from the two equations $(l = 1, 2)$

$$\frac{\partial \tilde{g}_l(\lambda^{[k]}, 0, y^{[k]})}{\partial \varepsilon} + \frac{d\lambda^{[k]}}{d\varepsilon}(0)\frac{\partial \tilde{g}_l(\lambda^{[k]}, 0, y^{[k]})}{\partial \lambda} + \frac{\partial \tilde{g}_l}{dy}\frac{dy^{[k]}}{d\varepsilon}(0) = 0.$$

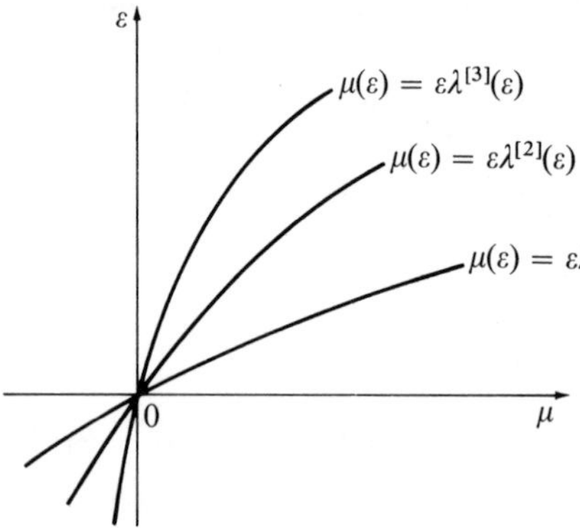

Figure V.3 Bifurcation curves when there are three roots of $\mathscr{C}(y_0) = 0$ and (V.23) holds for each pair $\lambda^{[k]}(0)$, $y^{[k]}(0)$

Second derivatives $(dy^{[k]}/d\varepsilon)(0)$ and $(dy^{[k]}/d\varepsilon)(0)$ may be computed by differentiating $\tilde{g}$ twice, and so on.

The stability of the solutions which bifurcate in the semi-simple case is complicated. Almost anything can happen (see §V.9).

V.8 Bifurcation and Stability of Steady Solutions (V.3) at a Semi-Simple Double Eigenvalue

The bifurcating solution (V.3) may be obtained directly from (V.2) or they may be constructed by the procedure suggested under (V.3). In this procedure we use the fact that $a_0 = b_0 = c_0 = d_0 = 0$ to define

$$\hat{u}_l \overset{\text{def}}{=} \frac{u_l}{\mu}$$
$$\hat{f}_l(\mu, \hat{u}_1, \hat{u}_2) \overset{\text{def}}{=} \frac{f_l(\mu, u_1, u_2)}{\mu^2} = 0, \tag{V.27}$$

where $l = 1, 2$ and

$$\begin{aligned}\hat{f}_1(\mu, \hat{u}_1, \hat{u}_2) &= a'\hat{u}_1 + b'\hat{u}_2 + \alpha_1\hat{u}_1^2 + 2\beta_1\hat{u}_1\hat{u}_2 + \gamma_1\hat{u}_2^2 + O(|\mu|),\\ \hat{f}_2(\mu, \hat{u}_1, \hat{u}_2) &= c'\hat{u}_1 + d'\hat{u}_2 + \alpha_2\hat{u}_1^2 + 2\beta_2\hat{u}_1\hat{u}_2 + \gamma_2\hat{u}_2^2 + O(|\mu|)\end{aligned} \tag{V.28}$$

and a', b', c', d', α_l, β_l, γ_l are functions of μ. When $\mu = 0$ the equations $\hat{f}_l(0, \hat{u}_{10}, \hat{u}_{20}) = 0$ are conic sections:

$$\begin{aligned}\hat{f}_{10} &\overset{\text{def}}{=} \hat{f}_1(0, \hat{u}_{10}, \hat{u}_{20})\\ &= a'_0\hat{u}_{10} + b'_0\hat{u}_{20} + \alpha_{10}\hat{u}_{10}^2 + 2\beta_{10}\hat{u}_{10}\hat{u}_{20} + \gamma_{10}\hat{u}_{20}^2 = 0,\\ \hat{f}_{20} &\overset{\text{def}}{=} \hat{f}_2(0, \hat{u}_{10}, \hat{u}_{20})\\ &= c'_0\hat{u}_{10} + d'_0\hat{u}_{20} + \alpha_{20}\hat{u}_{10}^2 + 2\beta_{20}\hat{u}_{10}\hat{u}_{20} + \gamma_{20}\hat{u}_{20}^2 = 0.\end{aligned} \tag{V.29}$$

Bifurcating solutions are obtained from the points of intersection on the two conics. The conics intersect at the origin $(\hat{u}_1, \hat{u}_2) = (0, 0)$ for any μ. Apart from the origin there are, at most, three other solutions (see Figure V.3). The intersection points of $(V.29)_1$ and $(V.29)_2$, other than $(0, 0)$, correspond to the roots of the cubic equation $\mathscr{C}(y_0) = 0$ given by (V.26). There are three solutions, or two solutions, or one solution, plus the solution $(0, 0)$ (see Figure V.4). The equations given in (V.7) are related to those given here by the transformation $(\mu, \hat{u}_1, \hat{u}_2) = (\varepsilon\lambda, 1/\lambda, y/\lambda)$.

The connection between the existence and stability of bifurcating solutions which was mentioned in §V.1 has an especially nice form relative to the parameterization (V.27). We first note that $\mathscr{J}(\mu) = \mu\mathscr{J}_0 + O(\mu^2)$ and find that

$$\det \mathscr{J} = \mu^2 \det \mathscr{J}_0 + O(|\mu|^3)$$

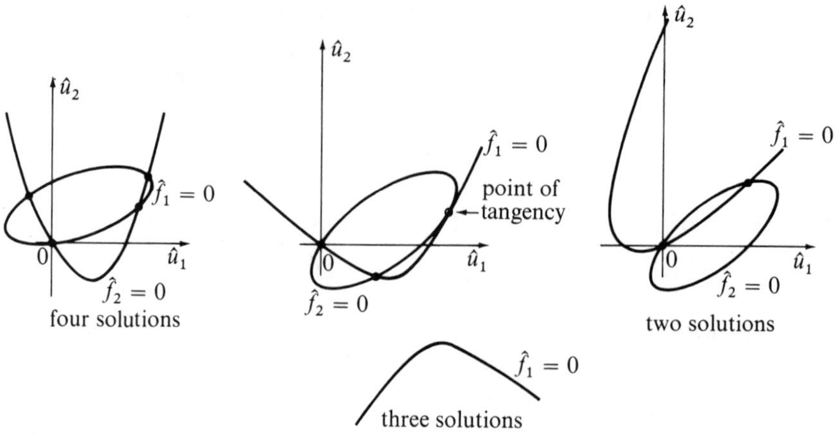

Figure V.4 Bifurcating solutions of the form (V.4) are intersection points of the two conic sections (V.28). The implicit function theorem guarantees bifurcation when the points intersect transversally and not at points of tangency

where

$$\mathscr{J}_0 = \begin{bmatrix} \dfrac{\partial \hat{f}_{10}}{\partial \hat{u}_{10}} & \dfrac{\partial \hat{f}_{10}}{\partial \hat{u}_{10}} \\[2ex] \dfrac{\partial \hat{f}_{20}}{\partial \hat{u}_{20}} & \dfrac{\partial \hat{f}_{20}}{\partial \hat{u}_{20}} \end{bmatrix}.$$

The stability of the solutions (V.3) to small disturbances is determined by the sign of the real part of the eigenvalues $\gamma_1(\mu)$, $\gamma_2(\mu)$ of $\mathscr{J}(\mu)$. When μ is small,

$$(\gamma_1(\mu), \gamma_2(\mu)) = \mu(\gamma_{10}, \gamma_{20}) + o(\mu)$$

where $(\gamma_{10}, \gamma_{20})$ are the eigenvalues of $\mathscr{J}_0$. The solutions (V.3) of $f_l(\mu, u_1, u_2) = 0$ are stable when $\mu > 0$ is small if Re $\gamma_{10} < 0$ and Re $\gamma_{20} < 0$. If Re $\gamma_{10} > 0$ or Re $\gamma_{20} > 0$ then the solutions (V.3) are unstable.

The implicit function theorem guarantees the existence of bifurcation for μ in a neighborhood of zero when (V.29) holds and

$$\det \mathscr{J}_0 = \gamma_{10}\gamma_{20} \neq 0. \tag{V.30}$$

This criterion fails at a point of tangency (see Figure V.4). If det $\mathscr{J}_0 < 0$, then either $\gamma_{10} > 0$ or $\gamma_{20} > 0$ and one of the two eigenvalues

$$\gamma_1(\mu) = \gamma_{10}\mu + O(|\mu|^2) \qquad \text{(V.31)}_1$$

or

$$\gamma_2(\mu) = \gamma_{20}\mu + O(|\mu|^2) \qquad \text{(V.31)}_2$$

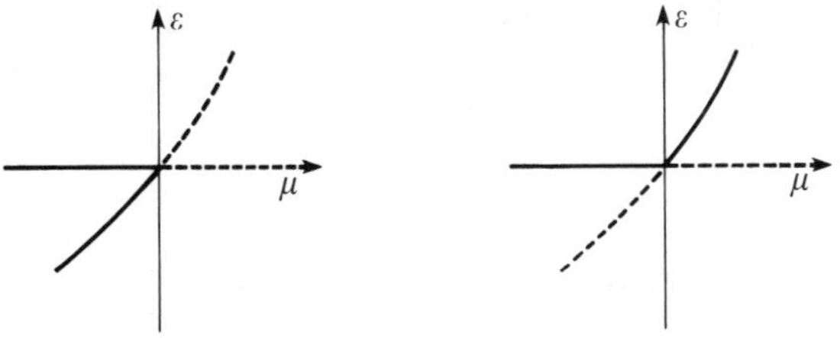

(a) (b)

Figure V.5 Distribution of stability when det $\mathscr{J}_0 > 0$ and the eigenvalues are real. Here $\varepsilon = u_1$ or $\varepsilon = u_2$. In (a) the supercritical solution is unstable and subcritical one is stable

will be positive when $|\mu|$ is small. It follows then that the bifurcating solution (V.3) is unstable on both sides of criticality whenever det $\mathscr{J}_0 < 0$ (as in Figure V.1(b)).

If γ_{10} is real-valued and det $\mathscr{J}_0 > 0$, then γ_{10} and γ_{20} are both positive or both negative. Equation (V.31) then shows that the bifurcating solution is stable on one side of criticality and is unstable on the other side (see Figure V.5).

The case in which $\gamma_{20} = \bar{\gamma}_{10}$ is complex is not so obviously related to $\mathscr{J}_0$. This case is one in which a double eigenvalue splits into a conjugate pair, as in Fig. IV.3(d) (with γ replacing σ). In this case, if Re γ_{10} = Re $\gamma_{20} \neq 0$ (trace of $\mathscr{J} \neq 0$) we have the same type of stability analysis as at Figure V.5.

V.9 Examples of Stability Analysis at a Double Semi-Simple (Index-One) Eigenvalue

We now give some restricted examples in which the two conics intersect either in four distinct points (including the origin) or in two distinct simple points, not at infinity (see exercises for other cases). We also restrict ourselves to the case when the zero solution is stable for $\mu < 0$ and loses it stability for $\mu > 0$.

Example V.1

$$\frac{du_1}{dt} = \mu u_1 - \mu u_2 - u_1^2 + u_2^2$$

$$\frac{du_2}{dt} = \mu u_2 + u_1 u_2.$$

The conics of §V.8 intersect at the points (0, 0), $A = (1, 0)$, $B = (-1, 2)$, $C = (-1, -1)$. The linear theory of stability may be applied at each point.

We find that (0, 0) is a node (stable for $\mu > 0$, unstable for $\mu > 0$), and μA, μB, μC are all saddle points so that the bifurcating solutions are unstable on both sides of $\mu = 0$.

Example V.2

$$\frac{du_1}{dt} = 3\mu u_1 - 5\mu u_2 - u_1^2 + u_2^2$$

$$\frac{du_2}{dt} = 2\mu u_1 - u_1 u_2.$$

The conics of §V.8 intersect at the points (0, 0) and $A = (0, 5)$. The origin and the bifurcating solution μA are foci with the same stability. (Stable for $\mu < 0$, unstable for $\mu > 0$.)

Example V.3

$$\frac{du_1}{dt} = 3\mu u_1 - 3\mu u_2 - u_1^2 + u_2^2$$

$$\frac{du_2}{dt} = \mu u_1 - u_1 u_2.$$

The conics of §V.8 intersect at the points (0, 0), $A = (2, 1)$, $B = (0, 3)$, $C = (1, 1)$. The bifurcated solution μB and the origin are foci with the same stability (stable for $\mu < 0$, unstable for $\mu > 0$). The bifurcated solution μA is a node, stable for $\mu > 0$, unstable for $\mu < 0$, and the solution μC is a saddle point, so it is unstable for $\mu > 0$ and $\mu < 0$.

Example V.4

$$\frac{du_1}{dt} = \mu u_1 - u_1^2 - u_1 u_2$$

$$\frac{du_2}{dt} = -2\mu u_1 + 2\mu u_2 + u_1 u_2 - u_2^2.$$

The conics of §V.8 intersect at the points (0, 0), $A = (-1, 2)$, $B = (0, 2)$, $C = (\frac{1}{2}, \frac{1}{2})$. The bifurcated solution μB is a node like the origin but with the opposite stability (stable for $\mu > 0$, unstable for $\mu < 0$). The two other bifurcated solutions are saddles (unstable for $\mu > 0$ and $\mu < 0$).

Example V.5

$$\frac{du_1}{dt} = \mu u_2 + u_1 u_2$$

$$\frac{du_2}{dt} = -\mu u_1 + \mu u_2 + u_1^2 + u_2^2.$$

The conics of §V.8 intersect at the point (0, 0) and $A = (1, 0)$. The origin is a focus (stable for $\mu < 0$, unstable for $\mu > 0$) and the bifurcated solution μA is a saddle (unstable for $\mu < 0$ and $\mu > 0$).

Comments on Examples. V.1–5. Some general properties of stability of bifurcating solutions may be determined from the following geometric interpretation of the eigenvalue problem for $\mathscr{J}_0$. Consider the intersections of the two conics

$$\hat{f}_{10}(\hat{u}_{10}, \hat{u}_{20}) = 0, \qquad \hat{f}_{20}(\hat{u}_{10}, \hat{u}_{20}) = 0$$

and assume all such intersections to be nontangential. Then

$$\begin{aligned}\det \mathscr{J}_0 &= \frac{\partial \hat{f}_{10}}{\partial \hat{u}_{10}} \frac{\partial \hat{f}_{20}}{\partial \hat{u}_{20}} - \frac{\partial \hat{f}_{10}}{\partial \hat{u}_{20}} \frac{\partial \hat{f}_{20}}{\partial \hat{u}_{10}} \\ &= (\nabla \hat{f}_{10} \wedge \nabla \hat{f}_{20}) \cdot \hat{k} \\ &= \gamma_{10} \gamma_{20},\end{aligned}$$

where $\hat{k} = \hat{\imath} \wedge \hat{\jmath}$ and $\hat{\imath}$ and $\hat{\jmath}$ are the orthonormal base vectors along the coordinate axis and

$$\nabla \hat{f}_{10} = \left(\frac{\partial \hat{f}_{10}}{\partial \hat{u}_{10}}, \frac{\partial \hat{f}_{10}}{\partial \hat{u}_{20}} \right)$$

is the gradient vector of $\hat{f}_{10}$ at an intersection point, hence orthogonal to the corresponding conic ($\hat{f}_{10} = 0$).

Each conic has a one-signed curvature (zero curvature if it is a product of lines), so if we consider two *consecutive* (nondegenerate) intersection points on one regular arc of one of these conics (see Figure V.6), the vectors $\nabla \hat{f}_{10} \wedge \nabla \hat{f}_{20}$ have opposite directions at these two points.

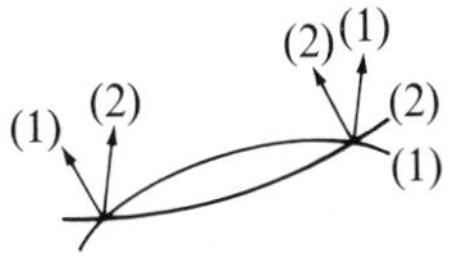

Figure V.6

In the case of intersecting hyperbolas, we have to distinguish between the case when the asymptotes of the hyperbolas separate one another as in Figure V.7 and the opposite case as in Figure V.8. We observe from these figures that the signs of det $\mathscr{J}_0$ at the intersection points (including the origin) take the following values:

i. $(+, +, +, -)$ or $(-, -, -, +)$, or $(+, +)$ or $(-, -)$ in the case of two hyperbolas whose asymptotes alternate in direction, and
ii. $(+, +, -, -)$ or $(+, -)$ in the other case.

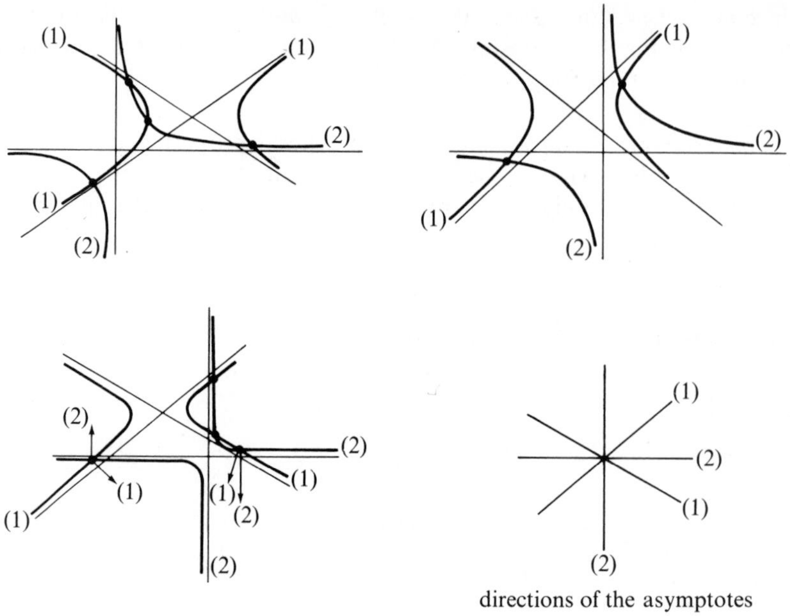

Figure V.7

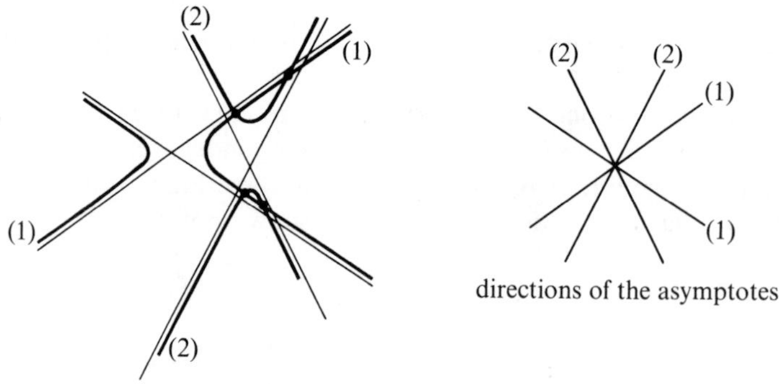

Figure V.8

This result was rigorously proved in the paper by McLeod and Sattinger (1973) cited in §V.7.

In Example V.1 the signs of the determinant are $(+, -, -, -)$; in Example V.2 the signs are $(+, +)$; in Example V.3 they are $(+, +, +, -)$; and in all we impose "+" for the origin.

In Example V.4 the signs are $(+, -)$ while in Example V.5 they are $(+, +, -, -)$, where the origin corresponds to a "+".

The stability of a solution whose Jacobian determinant is "+" is indeterminate, and there is no relation given by these results between the stabilities of different solutions with "+" Jacobian determinants.

EXAMPLE V.6. Consider the differential system in $\mathbb{R}^2$

$$\frac{dx_1}{dt} = \mu x_1 - x_1^2 - 2x_1x_2,$$

$$\frac{dx_2}{dt} = (\mu - \sigma)x_2 + x_1x_2 + x_2^2$$

where σ is a fixed parameter and μ the bifurcation parameter. If $\sigma = 0$, then zero is a double semi-simple eigenvalue of the linearized operator for $\mu = 0$. If we assume that $\sigma > 0$, we may represent the solutions as in Figure V.9. They are given by

(1) $x_1 = 0, \quad x_2 = 0;$

(2) $x_1 = 0, \quad x_2 = \sigma - \mu,$

(3) $x_1 = \mu, \quad x_2 = 0,$

(4) $x_1 = 2\sigma - 3\mu, \quad x_2 = 2\mu - \sigma.$

The points $x_1 = 0, x_2 = \sigma/3$ for $\mu = 2\sigma/3$ and $x_1 = \sigma/2, x_2 = 0$ for $\mu = \sigma/2$ are points of secondary bifurcation.

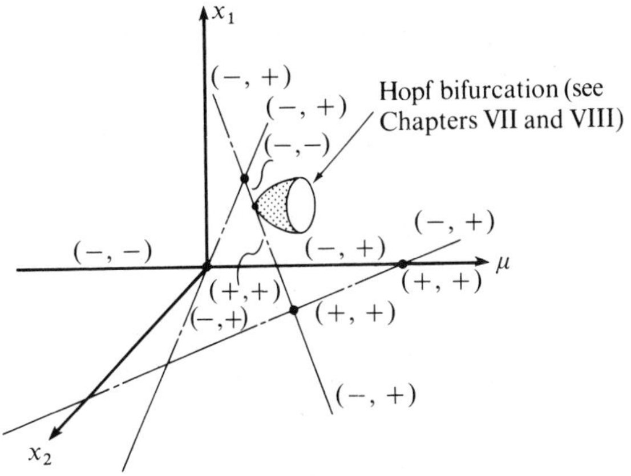

Figure V.9

To determine the stability of a steady solution we find the signs of the real parts of the two eigenvalues for that solution. These signs are shown in Figure V.9. The solution (4) is stable (−, −) for supercritical points $\mu > \sigma/2$ near the point $(x_1, x_2) = (\sigma/2, 0)$ of secondary bifurcation and is unstable (+, +) for subcritical points $\mu < 2\sigma/3$ close the point $(x_1, x_2) = (0, \sigma/3)$ of secondary bifurcation. Between these two points two conjugate complex eigenvalues cross the imaginary axis. To see this, set

$$(x_1, x_2) = (2\sigma - 3\mu + x_1', 2\mu - \sigma + x_2')$$

where (x_1', x_2') satisfy

$$\frac{dx_1'}{dt} = (3\mu - 2\sigma)(x_1' + 2x_2') - x_1'^2 - 2x_1'x_2'$$

$$\frac{dx_2'}{dt} = (2\mu - \sigma)(x_1' + x_2') + x_1'x_2' + x_2'^2.$$

The eigenvalues γ_1 and γ_2 of the linearized system satisfy

$$\gamma_1\gamma_2 = (2\mu - \sigma)(2\sigma - 3\mu) > 0$$

and

$$\gamma_1 + \gamma_2 = 5\mu - 3\sigma.$$

When $\mu = 3\sigma/5$, the two eigenvalues are on the imaginary axis.

In Chapters VII and VIII we show that when a complex-conjugate pair crosses the imaginary axis a time-periodic solution of the original problem will bifurcate. This is known as Hopf bifurcation. In this example the tertiary bifurcation is degenerate: it holds for a single value $\mu = 3\sigma/5$. If we add higher order terms in the differential system, we obtain a non-degenerate Hopf bifurcation as shown in Fig. V.9.

Appendix V.1 Implicit Function Theorem for a System of Two Equations in Two Unknown Functions of One Variable

Consider the following system of two equations:

$$\begin{aligned} f_1(x_1, x_2, \varepsilon) &= 0 \\ f_2(x_1, x_2, \varepsilon) &= 0, \end{aligned} \tag{V.32}$$

where f_1 and f_2 are continuously differentiable in the open "cube": $A_1 < x_1 < B_1$, $A_2 < x_2 < B_2$, $\varepsilon_1 < \varepsilon < \varepsilon_2$. Assume that

$$f_1(x_{10}, x_{20}, \varepsilon_0) = f_2(x_{10}, x_{20}, \varepsilon_0) = 0, \tag{V.33}$$

and $A_1 < x_{10} < B_1$, $A_2 < x_{20} < B_2$, $\varepsilon_1 < \varepsilon_0 < \varepsilon_2$, and that the Jacobian matrix

$$\mathscr{J} = \begin{bmatrix} \dfrac{\partial f_1}{\partial x_1} & \dfrac{\partial f_1}{\partial x_2} \\[2ex] \dfrac{\partial f_2}{\partial x_1} & \dfrac{\partial f_2}{\partial x_2} \end{bmatrix} \tag{V.34}$$

computed at the point $(x_{10}, x_{20}, \varepsilon_0)$, has a nonzero determinant: $\det \mathscr{J} \neq 0$. Then, there exists $\alpha > 0$ and $\beta > 0$ such that the following assertions hold:

(i) There is a unique continuous pair of functions x_1 and x_2 defined for $\varepsilon_0 - \alpha < \varepsilon < \varepsilon_0 + \alpha$, satisfying $x_{i0} - \beta < x_i(\varepsilon) < x_{i0} + \beta$, $i = 1, 2$, and $f_i(x_1(\varepsilon), x_2(\varepsilon), \varepsilon) = 0$, $i = 1, 2$.

(ii) Moreover, x_1 and x_2 are continuously differentiable for $\varepsilon_0 - \alpha < \varepsilon < \varepsilon_0 + \alpha$ and

$$\begin{bmatrix} x_1'(\varepsilon) \\ x_2'(\varepsilon) \end{bmatrix} = -\mathscr{J}^{-1}(x_1(\varepsilon), x_2(\varepsilon), \varepsilon) \cdot \begin{bmatrix} \dfrac{\partial f_1(x_1(\varepsilon), x_2(\varepsilon), \varepsilon)}{\partial \varepsilon} \\ \dfrac{\partial f_2(x_1(\varepsilon), x_2(\varepsilon), \varepsilon)}{\partial \varepsilon} \end{bmatrix}. \tag{V.35}$$

If f_1 and f_2 are analytic functions of all variables, then $x_1(\varepsilon)$ and $x_2(\varepsilon)$ are analytic near $\varepsilon = \varepsilon_0$.

Remark. This theorem is sufficient for our needs of the moment. Its proof in a more general frame may be found in any book on advanced calculus.

The condition that $\det \mathscr{J} \neq 0$ also arises from Cramer's rule for solving for the higher-order derivatives of $x_1(\varepsilon)$ and $x_2(\varepsilon)$. If all derivatives of $f_i(x_1, x_2, \varepsilon)$ through order n are known at $(x_{10}, x_{20}, \varepsilon_0)$ and if $\partial^k x_j(\varepsilon_0)/\partial \varepsilon^k$, $j = 1, 2$, $k = 1, \ldots, n - 1$, are also known, then the nth derivative of $f(x_1(\varepsilon), x_2(\varepsilon), \varepsilon)$ vanishes and is of the form

$$\frac{\partial f_1}{\partial x_1}\frac{\partial^n x_1}{\partial \varepsilon^n} + \frac{\partial f_1}{\partial x_2}\frac{\partial^n x_2}{\partial \varepsilon^n} + g_1 = 0$$

$$\frac{\partial f_2}{\partial x_1}\frac{\partial^n x_1}{\partial \varepsilon^n} + \frac{\partial f_2}{\partial x_2}\frac{\partial^n x_2}{\partial \varepsilon^n} + g_2 = 0,$$

where g_1 and g_2 contain only known terms of lower order. Cramer's rule says that these linear equations can be solved if $\det \mathscr{J} \neq 0$.

The functions $x_1(\varepsilon)$ and $x_2(\varepsilon)$ can be constructed as a power series in ε up to the order allowed by their differentiability. As an exercise, the reader should show that the construction can be carried out provided that $\det \mathscr{J} \neq 0$.

Exercises

V.1 (see §V.5). Consider the system

$$\frac{du_1}{dt} = u_2 + \mu(a_0' u_1 + b_0' u_2) + \alpha_{10} u_1^2 + 2\beta_{10} u_1 u_2 + \gamma_{10} u_2^2,$$

$$\frac{du_2}{dt} = \mu d_0' u_2 + \alpha_{20} u_1^2 + 2\beta_{20} u_1 u_2 + \gamma_{20} u_2^2$$

where $\alpha_{20} \neq 0$.

(i) Construct a steady bifurcating solution $(u_1(\mu), u_2(\mu))$ in the form

$$u_i(\mu) = \sum_{n=1}^{\infty} u_{in}\mu^n, \qquad i = 1, 2.$$

Note that we are dealing with a case in which zero is a double eigenvalue of the linearized operator for $\mu = 0$, of index 2, and that $c_0' = 0$ as in §V.5.

Hint. First show that $u_{21} = u_{11} = u_{22} = 0$, $u_{12} = a_0' d_0'/\alpha_{20}$, etc.

(ii) Assume the zero solution is stable for $\mu < 0$ and loses stability strictly as μ increases past zero (with $a_0' > 0$ and $d_0' > 0$). Then show that the bifurcating solution is *unstable* for $\mu < 0$ and for $\mu > 0$, when $|\mu|$ is small (see Figure V.10)

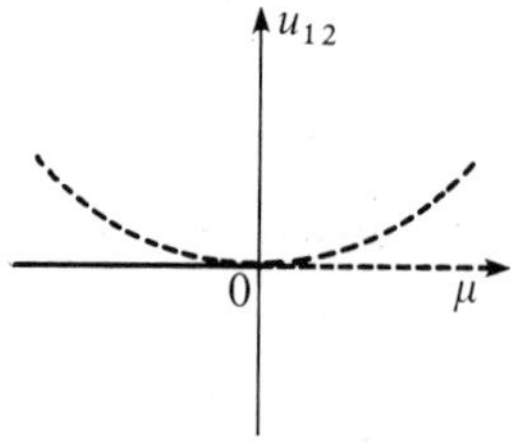

Figure V.10

V.2. Consider the system

$$\frac{du_1}{dt} = \mu u_1 + 2u_1u_2 + u_1^2 + O(|\mu|\,\|u\|^2 + |\mu|^2\|u\| + \|u\|^3),$$

$$\frac{du_2}{dt} = \mu u_2 - u_1u_2 + u_2^2 + O(|\mu|\|u\|^2 + |\mu|^2\,\|u\| + \|u\|^3),$$

which enters into the frame of §V.7 and §V.8.

(i) Show that if you look for bifurcating solutions in the form

$$u_1 = \varepsilon,\ u_2 = \varepsilon y(\varepsilon),\ \mu = \varepsilon\lambda(\varepsilon),$$

you obtain *only two* steady solutions bifurcating from zero:

(1) $$u_1 = \varepsilon, \qquad u_2 = O(\varepsilon^2), \qquad \mu = -\varepsilon + O(\varepsilon^2),$$

(2) $$u_1 = \varepsilon, \qquad u_2 = -2\varepsilon + O(\varepsilon^2), \qquad \mu = 3\varepsilon + O(\varepsilon^2).$$

(ii) Show that if you look for bifurcating solutions in the form

$$u_2 = \varepsilon, \qquad u_1 = \varepsilon x(\varepsilon), \qquad \mu = \varepsilon\lambda(\varepsilon),$$

you also obtain two bifurcating solutions. One solution is the same as (2). The other is different: (3) $u_1 = O(\varepsilon^2)$, $u_2 = \varepsilon$, $\mu = -\varepsilon + O(\varepsilon^2)$.

(iii) Show that if you look for bifurcating solutions in the form

$$u_1 = u_1(\mu), \qquad u_2 = u_2(\mu)$$

then you will find the three solutions (1), (2), (3) at once.

Remark. Here we have the relationships $d_0'\gamma_{10} - b_0'\gamma_{20} = 0$ and $c_0'\alpha_{10} - a_0'\alpha_{20} = 0$ which say that the "cubic" (V.26) is only quadratic in both cases (i) and (ii).

V.3. Consider the system

$$\frac{du_1}{dt} = \mu u_1 + \mu u_2 + u_1^2 + u_1 u_2 + u_2^3$$

$$\frac{du_2}{dt} = \mu u_1 - \mu u_2 + 2u_1^2 - 2u_1 u_2,$$

which enters into the frame of §V.7 and §V.8.

(i) Show by the method of §V.8 that you obtain only two nonzero bifurcating solutions: (1) $u_1 = -\mu + O(\mu^2)$, $u_2 = -\mu + O(\mu^2)$, (2) $u_1 = -\frac{1}{2}\mu + O(\mu^2)$, $u_2 = \frac{1}{2}\mu + O(\mu^2)$.

(ii) Show that the method of §V.7 gives a third bifurcating solution of the form (3) $u_2 = \varepsilon$, $u_1 = \varepsilon^2 + O(\varepsilon^3)$, $\mu = -2\varepsilon^2 + O(\varepsilon^3)$.

Remark. This situation is due to the fact that the two conics

$$u_1 + u_2 + u_1^2 + u_1 u_2 = 0$$

$$u_1 - u_2 + 2u_1^2 - 2u_1 u_2 = 0$$

have a common asymptote. This common asymptote corresponds to the 3rd solution computed under (ii) by the method of §V.7 with $\lambda_0 = 0$.

V.4. Consider the system $(S_\pm)$

$$\frac{du_1}{dt} = \mu u_1 - \mu u_2 - u_1^2 + u_2^2 \pm \mu u_1 u_2^2$$

$$\frac{du_2}{dt} = -\mu u_2 + u_1^2 + u_2^2, \qquad (S_\pm)$$

which enters into the frame of §V.7 and §V.8. Compute the steady solutions

(1) $$u_1 = u_2 = 0$$

(2) $$u_1 = 0, \qquad u_2 = \mu$$

and show that (3) there are no other solutions for (S_+), and *two other solutions* for (S_-).

V.5. Consider the system

$$\frac{du_1}{dt} = \mu u_2 + u_1(u_1^2 + u_2^2)$$

$$\frac{du_2}{dt} = -\mu u_1 + u_2(u_1^2 + u_2^2),$$

which enters into the frame of §V.7 and §V.8. Show that there is *no bifurcation at all.*

Remark. In this case the "conics" have disappeared, so the method fails.

V.6 (*Secondary bifurcation* obtained by splitting a double semi-simple eigenvalue, saving symmetry). Consider the system

$$\frac{du_1}{dt} = \mu u_1 - u_1^2 + u_2^2$$

$$\frac{du_2}{dt} = \mu u_2 - c u_1 u_2, \qquad c \neq 0, 1, \qquad (1)$$

which is invariant under the transformation $u_2 \to -u_2$. This system enters into the frame of §V.8.

(i) Show that the two conics are hyperbolas which intersect at 2 points (including (0, 0)) if $c > 1$, or 4 points if $c < 1$. Show that the directions of their asymptotes alternate (as in Examples V.1–3 in §V.9). Show that the steady bifurcating solutions are

$$(u_1, u_2) = (\mu, 0), \qquad \left(\frac{\mu}{c}, \frac{\mu}{c}\sqrt{1-c}\right), \qquad \left(\frac{\mu}{c}, -\frac{\mu}{c}\sqrt{1-c}\right).$$

(ii) Study the stability of the 0 solution and of the bifurcated solutions ($c \neq 0, 1$). Suppose $c > 1$, and show that the origin and the bifurcated solutions are both nodes but with different stabilities. Suppose $c < 1$ and show that the origin is a node (stable for $\mu < 0$, unstable for $\mu > 0$); $(\mu, 0)$ is a saddle; and $(\mu/c, \pm(\mu/c)\sqrt{1-c})$ are saddles if $c < 0$ and nodes if $0 < c < 1$ (stable for $\mu > 0$, unstable for $\mu < 0$).

(iii) Consider now the "imperfect" system

$$\begin{aligned} \frac{du_1}{dt} &= \mu u_1 - u_1^2 + u_2^2 + \alpha \\ \frac{du_2}{dt} &= \mu u_2 - c u_1 u_2 + \beta u_2 \end{aligned} \tag{2}$$

obtained by perturbing (1) by adding perturbations which keep the invariance under the transformation $u_2 \to -u_2$. The problem is now to see how the bifurcation described under (i) behaves under perturbation.

Show that the steady solutions of (2) are given in the (u_1, u_2, μ) space by two conics defined by $u_2 = 0$, $\mu u_1 - u_1^2 + \alpha = 0$ (hyperbola centered at 0, in the plane $u_2 = 0$) and

$$\mu = c u_1 - \beta, \qquad (c-1)u_1^2 + u_2^2 - \beta u_1 + \alpha = 0$$

which is an ellipse if $c > 1$, a hyperbola if $c < 1$, in a plane parallel to the u_2 axis (see Figure V.11). Note that when $\mu^2 > -4\alpha$ and

$$\frac{1}{c^2}(\mu + \beta)^2 - \frac{\mu}{c}(\mu + \beta) - \alpha > 0,$$

there are four steady solutions (u_1, u_2) of (2). Deduce that if $\beta^2 + 4\alpha(1 - c) > 0$ there are two bifurcations ("secondary bifurcations"), and that for $\beta^2 + 4\alpha(1 - c) < 0$ there are 3 or 4 isolated branches depending on whether $c > 1$ or $c < 1$, with no bifurcation at all.

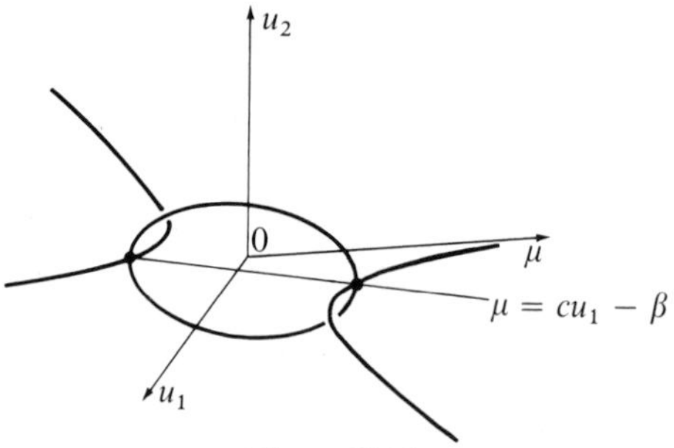

Figure V.11

Remark. In general, an imperfection in a system giving a bifurcation at a double eigenvalue breaks the bifurcation, as in one dimension. The imperfection parameter in the problem is α. When $\alpha = 0$, we get secondary bifurcation for all $\beta \neq 0$. β is a parameter which splits the double eigenvalue $\sigma = 0$ at $\mu = 0$ of the spectral problem for the stability of $(u_1, u_2) = 0$ into two simple eigenvalues, $\sigma = \mu$ and $\sigma = \mu + \beta$. We get secondary bifurcation when we split the double eigenvalue with β and retain the symmetry $u_2 \to -u_2$ of (2). The first persons to note that the splitting of multiple eigenvalues could lead to secondary bifurcation were L. Bauer, H. Keller, and E. Reiss, Multiple eigenvalues lead to secondary bifurcation, *SIAM Review*, **17**, 101 (1975). The first persons to recognize the importance of symmetry in the creation of secondary bifurcation by splitting perturbations were M. Golubitsky and D. Schaeffer, Imperfect bifurcation in the presence of symmetry, *Com. Math. Phys.*, **67**, 205–232 (1979) and Michael Shearer, Secondary bifurcation near a double eigenvalue, *SIAM J. Math. Anal.*, **11**, No. 2, 365–389 (1980).

V.7. (Periodic orbits bifurcating from the origin at a double eigenvalue of index 2). Consider the system

$$\begin{aligned} \frac{du_1}{dt} &= u_2 \\ \frac{du_2}{dt} &= \mu c_0' u_1 + \alpha_{20} u_1^2, \qquad c_0' > 0 \end{aligned} \tag{1}$$

which enters into the frame of §V.5, V.6.

(i) Compute and study the stability of the steady bifurcating solution (V.2). (If $\mu < 0, (u_1, u_2) = (0, 0)$ is a *center* while (V.2) is a saddle, if $\mu > 0$ the situation is reversed.)

(ii) Integrate the second-order nonlinear equation equivalent to (1) once and show that for each μ there is an infinite number of periodic solutions of (1). (See Figure V.12.)

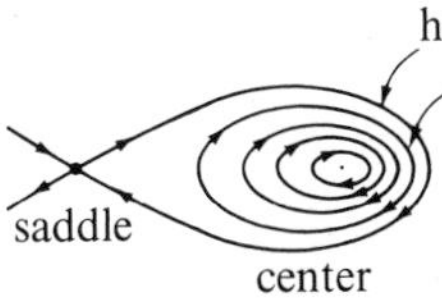

Figure V.12 Phase-plane portraits of solutions of (1)

Remark. The system (1) has the form $\dot{u}_l = f_l(u_1, u_2, \mu)$, $l = 1, 2$ with the special property $(\partial f_1/\partial u_1) + (\partial f_2/\partial u_2) = 0$. This property implies that (1) is a conservative and not a dissipative system. Conservative systems do not exhibit asymptotic stability and they possess other special properties which will not be studied in this book.

V.8. Suppose that the quadratic terms in $f_1(\mu, u_1, u_2) = 0$ and $f_2(\mu, u_1, u_2) = 0$ vanish but that cubic terms do not. Show that, in general, 0, 2, or 4 branches bifurcate.

Hint. Recall that two cubics intersect in 1, 3, 5, 7, or 9 points, and use symmetry.

CHAPTER VI

Methods of Projection for General Problems of Bifurcation into Steady Solutions

We wish now to make precise the sense in which one- and two-dimensional problems arise out of higher-dimensional problems, partial differential equations, and integro-differential equations by methods of projection.

It is best to start with a problem which we have already treated in Chapter V, using different notation, namely, the problem of bifurcation into steady solutions in $\mathbb{R}^2$ when the eigenvalues of $\mathbf{f}_u(0|\cdot) = \mathbf{A}(0)(\cdot)$ are real and distinct. This problem is essentially a one-dimensional problem after a projection associated with eigenvalue $\xi_1(0) = 0$ at criticality and with eigenvector $\mathbf{x}_1$ and adjoint eigenvector $\mathbf{y}_1$. For maximum efficiency it is best to write this demonstration of projection of the bifurcation problem into $\mathbb{R}^1$ in a notation which may directly be generalized to the problem of bifurcation at a real-valued simple eigenvalue for infinite-dimensional problems such as those which arise in the study of partial differential equations.

VI.1 The Evolution Equation and the Spectral Problem

We first write the governing problem again using the functional notation introduced in (I.21):

$$\frac{d\mathbf{u}}{dt} = \mathbf{f}(\mu, \mathbf{u}) = \mathbf{f}_u(\mu|\mathbf{u}) + \tfrac{1}{2}\mathbf{f}_{uu}(\mu|\mathbf{u}|\mathbf{u}) + O(\|\mathbf{u}\|^3) \tag{VI.1}$$

where $\mathbf{f}_u(\mu|\mathbf{u}) = \mathbf{A}(\mu)\cdot\mathbf{u}$ etc., as in (I.22). For the moment we think of (VI.1) as the two-dimensional problem treated under case (i) in Chapter V. The spectral problem for the stability of $\mathbf{u} = 0$ was already derived in Chapter IV. A small disturbance $\mathbf{v} = e^{\sigma(\mu)t}\mathbf{x}$ satisfies $\dot{\mathbf{v}} = \mathbf{f}_u(\mu|\mathbf{v})$ and

$$\sigma(\mu)\mathbf{x} = \mathbf{f}_u(\mu|\mathbf{x}). \tag{VI.2}$$

In $\mathbb{R}^2$ we imagine that $\mathbf{A}(\mu) = \mathbf{f}_u(\mu|\cdot)$ has two distinct real eigenvalues $\xi_1(\mu)$ and $\xi_2(\mu)$ and two eigenvectors $\mathbf{x}_1(\mu)$ and $\mathbf{x}_2(\mu)$ (see §IV.2). So

$$\xi_j(\mu)\mathbf{x}_j = \mathbf{f}_u(\mu|\mathbf{x}_j), \qquad j = 1, 2 \tag{VI.3}$$

for μ in an interval around zero. The problem adjoint to (VI.2) in the scalar product (IV.7) is

$$\bar{\sigma}(\mu)\mathbf{y} = \mathbf{f}_u^*(\mu|\mathbf{y}) \tag{VI.4}$$

where $\mathbf{f}_u^*(\mu|\cdot) \stackrel{\text{def}}{=} [\mathbf{f}_u(\mu|\cdot)]^*$ is the linear operator adjoint to $\mathbf{f}_u$ in the scalar product $\langle \mathbf{a}, \mathbf{b}\rangle = \overline{\langle \mathbf{b}, \mathbf{a}\rangle}$ whose form in $\mathbb{C}^n$ is discussed in §IV.3. We define $\mathbf{f}_u^*$ by

$$\langle \mathbf{a}, \mathbf{f}_u(\mu|\mathbf{b})\rangle = \langle [\mathbf{f}_u(\mu|\mathbf{a})]^*, \mathbf{b}\rangle = \langle \mathbf{f}_u^*(\mu|\mathbf{a}), \mathbf{b}\rangle$$

for all $\mathbf{a}$ and $\mathbf{b}$ in a suitable space ($\mathbf{f}_u^*$ should *not* be interpreted as the linearization of some $\mathbf{f}^*$). In $\mathbb{R}^n$

$$\mathbf{f}_u^*(\mu|\cdot) = \mathbf{A}^{\mathrm{T}}(\mu)$$

may be represented by a matrix.

Under our assumptions about the eigenvalues of $\mathbf{f}_u(\mu|\cdot)$, (VI.4) reduces to

$$\xi_i(\mu)\mathbf{y}_i = \mathbf{f}_u^*(\mu|\mathbf{y}_i), \qquad i = 1, 2. \tag{VI.5}$$

When two eigenvalues are real and distinct

$$\langle \mathbf{x}_j, \mathbf{y}_i\rangle = \delta_{ij}.$$

At criticality, $\xi_1(0) = 0$, $\xi_2(0) < 0$, and $\xi_1'(0) > 0$ and

$$\xi_1'(0) = \langle \mathbf{f}_{u\mu}(0|\mathbf{x}_1), \mathbf{y}_1\rangle \tag{VI.6}$$

(see Figure IV.3(a)).

VI.2 Construction of Steady Bifurcating Solutions as Power Series in the Amplitude

We first define the amplitude by a projection on the eigensubspace associated with the adjoint eigenvector $\mathbf{y}_1 = \mathbf{y}_1(0)$ belonging to the eigenvalue $\xi_1(0) = 0$:

$$\varepsilon \stackrel{\text{def}}{=} \langle \mathbf{u}, \mathbf{y}_1\rangle. \tag{VI.7}$$

We then seek solutions as a power series in ε:

$$\begin{bmatrix} \mathbf{u}(\varepsilon) \\ \mu(\varepsilon) \end{bmatrix} = \sum_{n=1}^{\infty} \frac{\varepsilon^n}{n!} \begin{bmatrix} \mathbf{u}_n \\ \mu_n \end{bmatrix}. \tag{VI.8}$$

We are assuming that $\mathbf{f}(\mu, \mathbf{u})$ is analytic* in $(\mu, \mathbf{u})$ in a neighborhood of (0, 0). After inserting (VI.7) into (VI.1) we find by identification that

$$\begin{aligned} &\mathbf{f}_u(0|\mathbf{u}_1) = 0 \\ &\mathbf{f}_u(0|\mathbf{u}_2) + 2\mu_1\mathbf{f}_{u\mu}(0|\mathbf{u}_1) + \mathbf{f}_{uu}(0|\mathbf{u}_1|\mathbf{u}_1) = 0 \\ &\mathbf{f}_u(0|\mathbf{u}_3) + 3\mu_1\mathbf{f}_{uu\mu}(0|\mathbf{u}_1|\mathbf{u}_1) + 3\mu_1^2\mathbf{f}_{u\mu\mu}(0|\mathbf{u}_1) \\ &\quad + 3\mu_1\mathbf{f}_{u\mu}(0|\mathbf{u}_2) + 3\mathbf{f}_{uu}(0|\mathbf{u}_1|\mathbf{u}_2) \\ &\quad + 3\mu_2\mathbf{f}_{u\mu}(0|\mathbf{u}_1) + \mathbf{f}_{uuu}(0|\mathbf{u}_1|\mathbf{u}_1|\mathbf{u}_1) = 0 \end{aligned} \tag{VI.9}$$

and, in general,

$$\mathbf{f}_u(0|\mathbf{u}_n) + n\mu_{n-1}\mathbf{f}_{u\mu}(0|\mathbf{u}_1) + \mathbf{k}_n = 0, \qquad \text{(VI.9)}_4$$

where $\mathbf{k}_n$ depends on lower-order terms. Equations (VI.9) are to be solved subject to the normalization (VI.7) which implies that

$$\langle \mathbf{u}_1, \mathbf{y}_1 \rangle = 1, \qquad \langle \mathbf{u}_n, \mathbf{y}_1 \rangle = 0 \quad \text{for } n > 1. \tag{VI.10}$$

The solution of $(\text{VI.9})_1$ and $(\text{VI.10})_1$ is immediate, since the eigenvalue problem $\mathbf{f}_u(0|\mathbf{u}_1) = \mathbf{A}(0) \cdot \mathbf{u}_1 = 0$ with $\langle \mathbf{u}_1, \mathbf{y}_1 \rangle = 1$ has only one solution:

$$\mathbf{u}_1 = \mathbf{x}_1. \tag{VI.11}$$

In this way we eliminate the solution $\mathbf{u} = 0$ of (VI.1).

The other problems are not generally solvable. But they can be made solvable by choosing the derivatives of $\mu(\varepsilon)$ properly. The method of selection is described below.

Solvability Theorem (The Fredholm alternative). *Given* $\mathbf{g} \in \mathbb{R}^2$, *the equation*

$$\mathbf{f}_u(0|\mathbf{u}) = \mathbf{g} \tag{VI.12}$$

is solvable for $\mathbf{u} \in \mathbb{R}^2$ *if and only if*

$$\langle \mathbf{g}, \mathbf{y}_1 \rangle = 0. \tag{VI.13}$$

PROOF. Equation (VI.13) is necessary because

$$\langle \mathbf{f}_u(0|\mathbf{u}), \mathbf{y}_1 \rangle = \langle \mathbf{u}, \mathbf{f}_u^*(0|\mathbf{y}_1) \rangle = 0.$$

For sufficiency, we note that $\mathbf{f}_u(0|\mathbf{u}) = \mathbf{A}(0) \cdot \mathbf{u}$, so that

$$\begin{aligned} a_0 u_1 + b_0 u_2 - g_1 &= 0 \\ c_0 u_1 + d_0 u_2 - g_2 &= 0. \end{aligned} \tag{VI.14}$$

Writing out $\mathbf{f}_u^*(0|\mathbf{y}_1) = \mathbf{A}^T(0) \cdot \mathbf{y}_1 = 0$, we have

$$\begin{aligned} a_0 y_{11} + c_0 y_{12} &= 0 \\ b_0 y_{11} + d_0 y_{12} &= 0 \end{aligned} \tag{VI.15}$$

* When $\mathbf{f}(\mu, \mathbf{u})$ is sufficiently smooth but not analytic, our construction gives the derivatives of $\mathbf{u}(\varepsilon)$ and $\mu(\varepsilon)$ up to a certain order, and the truncated Taylor series (VI.8) gives the asymptotic expansion of the solution.

and

$$\langle \mathbf{g}, \mathbf{y}_1 \rangle = g_1 y_{11} + g_2 y_{12} = 0. \tag{VI.16}$$

Equations (VI.15) and (VI.16) show that the two equations (VI.14) are dependent; in essence there is just one equation and it may be solved, for example, in u_1 for any given value of u_2. To *make the solution unique it is necessary to add any normalizing condition*, say

$$\langle \mathbf{u}, \mathbf{y}_1 \rangle = u_1 y_{11} + u_2 y_{12} = k, \tag{VI.17}$$

for any k, say 1 or 0 as in (VI.10), or ε as in (VI.7). We get a unique solution (u_1, u_2) by solving (VI.14) and (VI.17). Only one equation of (VI.14) is useful in fact, the other being automatically satisfied.

Applying (VI.13) to $(\text{VI.9})_2$, using (VI.11), we find that

$$2\mu_1 \langle \mathbf{f}_{u\mu}(0|\mathbf{x}_1), \mathbf{y}_1 \rangle + \langle \mathbf{f}_{uu}(0|\mathbf{x}_1|\mathbf{x}_1), \mathbf{y}_1 \rangle = 0. \tag{VI.18}$$

Moreover, from (IV.27), we have

$$\langle \mathbf{f}_{u\mu}(0|\mathbf{x}_1), \mathbf{y}_1 \rangle = \langle \mathbf{A}'(0) \cdot \mathbf{x}_1, \mathbf{y}_1 \rangle = \xi_1'(0) > 0. \tag{VI.19}$$

Hence

$$2\mu_1 \xi_1'(0) + \langle \mathbf{f}_{uu}(0|\mathbf{x}_1|\mathbf{x}_1), \mathbf{y}_1 \rangle = 0. \tag*{(VI.20)$_1$}$$

Since $\xi_1'(0) \neq 0$ we may solve $(\text{VI.20})_1$ for μ_1. We may then find a unique $\mathbf{u}_2$ satisfying $(\text{VI.9})_2$ and (VI.10).

In the same way, we get

$$\begin{aligned} 3\mu_2 \xi_1'(0) &+ 3\langle \mathbf{f}_{uu}(0|\mathbf{x}_1|\mathbf{u}_2), \mathbf{y}_1 \rangle + \langle \mathbf{f}_{uuu}(0|\mathbf{x}_1|\mathbf{x}_1|\mathbf{x}_1), \mathbf{y}_1 \rangle \\ &+ 3\mu_1 \langle \mathbf{f}_{uu\mu}(0|\mathbf{x}_1|\mathbf{x}_1), \mathbf{y}_1 \rangle + 3\mu_1^2 \langle \mathbf{f}_{u\mu\mu}(0|\mathbf{x}_1), \mathbf{y}_1 \rangle \\ &+ 3\mu_1 \langle \mathbf{f}_{u\mu}(0|\mathbf{u}_2), \mathbf{y}_1 \rangle = 0 \end{aligned} \tag*{(VI.20)$_2$}$$

which leads to the determination of μ_2 and $\mathbf{u}_3$ solving $(\text{VI.9})_3$ and (VI.10). More generally, the equation

$$n\mu_n \xi_1'(0) + \langle \mathbf{k}_n, \mathbf{y}_1 \rangle = 0 \tag*{(VI.20)$_3$}$$

determines μ_n and $\mathbf{u}_{n-1}$, solving $(\text{VI.9})_4$ and (VI.10) as a function of the lower-order coefficients.

VI.3 $\mathbb{R}^1$ and $\mathbb{R}^1$ in Projection

Equations (VI.19) and (VI.20) are essentially equations in $\mathbb{R}^1$ arising in the projection on $\mathbf{x}_1$. It is instructive to compare these equations with those that arise in $\mathbb{R}^1$ directly. To make this comparison we set $\mathbf{F}(\mu, \varepsilon) = \mathbf{f}(\mu, \varepsilon)$, where $\mathbf{f}(\mu, 0)$ is reduced to local form. Then (II.52) is the $\mathbb{R}^1$ analogue of (VI.19). And if we replace $\sigma_\mu^{(1)}(0)$ with $\xi_1'(0)$, (II.52) becomes

$$\xi_1'(0) = f_{\mu\varepsilon}(0, 0) > 0, \tag{VI.21}$$

and we may obtain μ_n, $n \geq 1$ by expanding $\mathbf{f}(\mu(\varepsilon), \varepsilon)$ in powers of ε at $\varepsilon = 0$. Identifying independent powers of ε, using (VI.21), we get

$$2\mu_1\xi_1'(0) + f_{\varepsilon\varepsilon}(0, 0) = 0$$

$$3\mu_2\xi_1'(0) + f_{\varepsilon\varepsilon\varepsilon}(0, 0) + 3\mu_1 f_{\varepsilon\varepsilon\mu} + 3\mu_1^2 f_{\varepsilon\mu\mu} = 0 \tag{VI.22}$$

$$n\mu_n\xi_1'(0) + k_n = 0,$$

where k_n depends on lower-order coefficients. So (VI.21) and (VI.22), which arise in $\mathbb{R}^1$, are nearly identical to (VI.19) and (VI.20), which arise in $\mathbb{R}^1$ in projection.

VI.4 Stability of the Bifurcating Solution

We now turn to the study of the stability of the steady bifurcating solutions in $\mathbb{R}^2$. For any steady bifurcating solution $(\mu(\varepsilon), \mathbf{u}(\varepsilon))$, we have $\mathbf{f}(\mu(\varepsilon), \mathbf{u}(\varepsilon)) = 0$. An infinitesimal disturbance $\mathbf{v}$ of $\mathbf{u}(\varepsilon)$ satisfies

$$\frac{d\mathbf{v}}{dt} = \mathbf{f}_u(\mu(\varepsilon), \mathbf{u}(\varepsilon) \,|\, \mathbf{v}) \tag{VI.23}$$

where in $\mathbb{R}^2$

$$\mathbf{f}_u(\mu(\varepsilon), \mathbf{u}(\varepsilon) \,|\, \mathbf{v}) = \mathscr{A}(\varepsilon) \cdot \mathbf{v}$$

and $\mathscr{A}(\varepsilon)$ is a 2×2 matrix. Setting $\mathbf{v} = e^{\gamma t}\boldsymbol{\zeta}$ we find, using (VI.23), the spectral equation

$$\gamma\boldsymbol{\zeta} = \mathbf{f}_u(\mu(\varepsilon), \mathbf{u}(\varepsilon) \,|\, \boldsymbol{\zeta}). \tag*{(VI.24)$_1$}$$

There are two eigenvalues γ of $\mathscr{A}(\varepsilon)$ in $\mathbb{R}^2$. One is close to $\xi_1(0) = 0$ and the other to $\xi_2(0)$. When ε is close to zero the two eigenvalues are necessarily real-valued if $\xi_2(0) \neq 0$ because complex eigenvalues must occur in conjugate pairs. The eigenvalue problem adjoint to (VI.24)$_1$ is

$$\bar{\gamma}\boldsymbol{\zeta}^* = \mathbf{f}_u^*(\mu(\varepsilon), \mathbf{u}(\varepsilon) \,|\, \boldsymbol{\zeta}^*) \tag*{(VI.24)$_2$}$$

where in $\mathbb{R}^2$

$$\mathbf{f}_u^*(\mu(\varepsilon), \mathbf{u}(\varepsilon) \,|\, \boldsymbol{\zeta}^*) = \mathscr{A}^{\mathrm{T}}(\varepsilon) \cdot \boldsymbol{\zeta}^*$$

and $\mathscr{A}^{\mathrm{T}}(\varepsilon)$ is the transpose of $\mathscr{A}(\varepsilon)$.

Factorization Theorem. *Let $\boldsymbol{\zeta}(\varepsilon)$ and $\bar{\boldsymbol{\zeta}}^*(\varepsilon)$ be the eigenvectors belonging to $\gamma(\varepsilon)$ and suppose $\gamma(\varepsilon)$ is a simple eigenvalue of $\mathbf{f}_u(\mu(\varepsilon), \mathbf{u}(\varepsilon) \,|\, \cdot)$ and $\langle \mathbf{u}_\varepsilon(\varepsilon), \boldsymbol{\zeta}^*(\varepsilon)\rangle \neq 0$. Then*

$$\gamma(\varepsilon) = \frac{-\mu_\varepsilon(\varepsilon)\langle \mathbf{f}_\mu(\mu(\varepsilon), \mathbf{u}(\varepsilon)), \boldsymbol{\zeta}^*(\varepsilon)\rangle}{\langle \mathbf{u}_\varepsilon(\varepsilon), \boldsymbol{\zeta}^*(\varepsilon)\rangle}, \tag{VI.25}$$

$$\boldsymbol{\zeta}(\varepsilon) = \frac{1}{\langle \mathbf{u}_\varepsilon(\varepsilon), \boldsymbol{\zeta}^*(\varepsilon)\rangle}\{\mathbf{u}_\varepsilon(\varepsilon) + \mu_\varepsilon(\varepsilon)\mathbf{q}(\varepsilon)\}, \tag{VI.26}$$

where $\mathbf{q}(\varepsilon)$ *satisfies*

$$-\frac{\langle \mathbf{f}_\mu, \boldsymbol{\zeta}^* \rangle}{\langle \mathbf{u}_\varepsilon, \boldsymbol{\zeta}^* \rangle}\mathbf{u}_\varepsilon + \mathbf{f}_\mu(\mu(\varepsilon), \mathbf{u}(\varepsilon)) + (\gamma\mathbf{q} - \mathbf{f}_u(\mu(\varepsilon), \mathbf{u}(\varepsilon)|\mathbf{q}) = 0 \quad \text{(VI.27)}$$

and

$$\langle \mathbf{q}(\varepsilon), \boldsymbol{\zeta}^*(\varepsilon) \rangle = 0.$$

Moreover, as $\varepsilon \to 0$, $\mu(\varepsilon) \to 0$, $\boldsymbol{\zeta}^*(\varepsilon) \to \mathbf{y}_1$, $\mathbf{u}_\varepsilon(\varepsilon) \to \mathbf{x}_1$, $\langle \mathbf{f}_\mu(\mu(\varepsilon), \mathbf{u}(\varepsilon)), \boldsymbol{\zeta}^* \rangle \to \langle \mathbf{f}_{\mu u}(0, 0|\mathbf{x}_1), \mathbf{y}_1 \rangle\varepsilon = \xi_1'(0)\varepsilon$ *and*

$$\gamma_\varepsilon(\varepsilon) = -\mu_\varepsilon(\varepsilon)\{\xi_1'(0)\varepsilon + O(\varepsilon^2)\}. \quad \text{(VI.28)}$$

PROOF. Since $\mathbf{f}(\mu(\varepsilon), \mathbf{u}(\varepsilon)) = 0$, we find by differentiating once with respect to ε that

$$\mu_\varepsilon(\varepsilon)\mathbf{f}_\mu(\mu(\varepsilon), \mathbf{u}(\varepsilon)) + \mathbf{f}_u(\mu(\varepsilon), \mathbf{u}(\varepsilon)|\mathbf{u}_\varepsilon(\varepsilon)) = 0. \quad \text{(VI.29)}$$

Equation (VI.25) arises from the scalar product $\langle$(VI.29), $\boldsymbol{\zeta}^*\rangle$, using $\langle \mathbf{f}_u(\mu, \mathbf{u}|\mathbf{u}_\varepsilon), \boldsymbol{\zeta}^* \rangle = \langle \mathbf{u}_\varepsilon, \mathbf{f}_u^*(\mu, \mathbf{u}|\boldsymbol{\zeta}^*) \rangle = \gamma\langle \mathbf{u}_\varepsilon, \boldsymbol{\zeta}^* \rangle$. To find (VI.28), substitute (VI.25) and (VI.26) into (VI.24) and (VI.29) to eliminate $\mathbf{f}_u(\mu, \mathbf{u}|\mathbf{u}_\varepsilon)$. In this way you will find that $\mu_\varepsilon(\varepsilon)$ times the left side of (VI.27) vanishes. Now it may be shown that $\mathbf{q}(\varepsilon)$ solving (VI.27) is smooth in ε, so that all terms on the left of (VI.27) are smooth in ε. Hence $\mu_\varepsilon(\varepsilon)$ is a true factor and the left side of (VI.27) vanishes. To compute $\mathbf{q}$ solving (VI.27) it is first necessary to compute (VI.24).

The factorization (VI.25) for $\mathbb{R}^1$ in projection is the analogue of the factorization (II.44) for $\mathbb{R}^1$. And (VI.28) for $\mathbb{R}^1$ in projection is the same as (II.53) for $\mathbb{R}^1$. All the conclusions about the local properties of the bifurcating solutions are the same in $\mathbb{R}^1$ and $\mathbb{R}^1$ in projection.

VI.5 The Extra Little Part for $\mathbb{R}^1$ in Projection

We draw your attention to fact that the terms $3\langle \mathbf{f}_{uu}(0|\mathbf{x}_1|\mathbf{u}_2), \mathbf{y}_1 \rangle$ and $3\mu_1\langle \mathbf{f}_{u\mu}(0|\mathbf{u}_2), \mathbf{y}_1 \rangle$ in $(\text{VI.20})_2$ for $\mathbb{R}^1$ in projection have no counterpart in $(\text{VI.22})_2$ for $\mathbb{R}^1$. The extra term arises from the fact that there is a passive part of the solution in $\mathbb{R}^2$ which arises at higher orders as a result of nonlinear coupling and has a zero projection into $\mathbb{R}^1$.

The extra little part is the term $\mathbf{w}$ in the decomposition

$$\mathbf{u} = a(t)\mathbf{x}_1 + \mathbf{w}, \qquad \mathbf{f}_u(\mu|\mathbf{x}_1(\mu)) = \xi_1(\mu)\mathbf{x}_1(\mu) \quad \text{(VI.30)}$$

into the projection $a(t)\mathbf{x}_1$ where $a(t) = \langle \mathbf{u}, \mathbf{y}_1 \rangle$, and the complementary part $\langle \mathbf{w}, \mathbf{y}_1 \rangle = 0$ with zero projection. It is of interest to derive the equation satisfied by $a(t)$ and by $\mathbf{w}$. We write (VI.I) as

$$\frac{d\mathbf{u}}{dt} = \mathbf{f}_u(\mu|\mathbf{u}) + \mathbf{N}(\mu, \mathbf{u}) \quad \text{(VI.31)}$$

where

$$N(\mu, \mathbf{u}) = \tfrac{1}{2}\mathbf{f}_{uu}(\mu|\mathbf{u}|\mathbf{u}) + O(\|u\|^3).$$

Substituting (VI.30) in (VI.31) we find

$$(\dot{a} - \xi_1(\mu)a)\mathbf{x}_1 + \frac{d\mathbf{w}}{dt} = \mathbf{f}_u(\mu|\mathbf{w}) + \mathbf{N}(\mu, \mathbf{u}). \tag{VI.32}$$

Since

$$\left\langle \frac{d\mathbf{w}}{dt}, \mathbf{y}_1 \right\rangle = \frac{d}{dt}\langle \mathbf{w}, \mathbf{y}_1 \rangle = 0$$

and

$$\langle \mathbf{f}_u(\mu|\mathbf{w}), \mathbf{y}_1 \rangle = \langle \mathbf{w}, \mathbf{f}_u^*(\mu|\mathbf{y}_1)\rangle = \xi_1\langle \mathbf{w}, y_1 \rangle = 0,$$

we find that the projected part of the solution satisfies

$$\dot{a} - \xi_1(\mu)a = \langle \mathbf{N}(\mu, \mathbf{u}), \mathbf{y}_1 \rangle, \tag{VI.33}$$

while the complementary part with zero projection on $\mathbf{x}_1$ satisfies

$$\frac{d\mathbf{w}}{dt} = \mathbf{f}_u(\mu|\mathbf{w}) + \{\mathbf{N}(\mu, \mathbf{u}) - \langle \mathbf{N}(\mu, \mathbf{u}), \mathbf{y}_1 \rangle \mathbf{x}_1\}. \tag{VI.34}$$

Equation (VI.34) shows that if all the eigenvalues of $\mathbf{f}_u(\mu|\cdot)$ except $\xi_1(\mu)$ have a negative real part (in $\mathbb{R}^2$ there is only one other eigenvalue, $\xi_2(\mu) < 0$) then, for t large enough, $\mathbf{w} = O(a^2)$ because

$$\begin{aligned}\mathbf{N}(\mu, \mathbf{u}) = {} & a^2\mathbf{f}_{uu}(\mu|\mathbf{x}_1|\mathbf{x}_1) + 2a\mathbf{f}_{uu}(\mu|\mathbf{x}_1|\mathbf{w}) \\ & + \mathbf{f}_{uu}(\mu|\mathbf{w}|\mathbf{w}) + O(\|\mathbf{u}\|^3).\end{aligned}$$

Hence we may write (VI.33) as

$$\dot{a} - \xi_1(\mu)a = \alpha_1(\mu)a^2 + O(a^3) \tag{VI.35}$$

where $\alpha_1(\mu) = \langle \mathbf{f}_{uu}(\mu|\mathbf{x}_1|\mathbf{x}_1), \mathbf{y}_1 \rangle$. In $\mathbb{R}^2$ we may set

$$\mathbf{w} = b(t)\mathbf{x}_2, \qquad \mathbf{f}_u(\mu|\mathbf{x}_2) = \xi_2(\mu)\mathbf{x}_2$$

into (VI.34), and after projecting the result with $\mathbf{y}_2$ find that

$$\begin{aligned}\dot{b} - \xi_2(\mu)b = {} & \{\alpha_2(\mu)a^2 + 2\beta_2(\mu)ab + \gamma_2(\mu)b^2\} \\ & + O(|a| + |b|)^3,\end{aligned} \tag{VI.36}$$

where

$$\alpha_2(\mu) = \langle \mathbf{f}_{uu}(\mu|\mathbf{x}_1|\mathbf{x}_1), \mathbf{y}_2 \rangle$$
$$\beta_2(\mu) = \langle \mathbf{f}_{uu}(\mu|\mathbf{x}_1|\mathbf{x}_2), \mathbf{y}_2 \rangle$$
$$\gamma_2(\mu) = \langle \mathbf{f}_{uu}(\mu|\mathbf{x}_2|\mathbf{x}_2), \mathbf{y}_2 \rangle.$$

The amplitude $b(t)$ of $\mathbf{w}$ enters (VI.35) first at

$$O(a^3) = 2ab\beta_1(\mu) + b^2\gamma_1(\mu) + O(\|\mathbf{u}\|^3), \tag{VI.37}$$

where

$$\begin{aligned}\beta_1(\mu) &= \langle \mathbf{f}_{uu}(\mu|\mathbf{x}_1|\mathbf{x}_2|), \mathbf{y}_1\rangle \\ \gamma_1(\mu) &= \langle \mathbf{f}_{uu}(\mu|\mathbf{x}_2|\mathbf{x}_2), \mathbf{y}_1\rangle.\end{aligned} \tag{VI.38}$$

We may obtain the bifurcation results of § VI.2 from (VI.35), (VI.36), and (VI.37). For example, we can find a steady bifurcating solution in the form

$$a = \varepsilon$$

$$b = \sum_{n=2}^{\infty} \frac{1}{n!} b_n \varepsilon^n$$

$$\mu = \sum_{n=1}^{\infty} \frac{1}{n!} \mu_n \varepsilon^n.$$

VI.6 Projections of Higher-Dimensional Problems

We are going to treat infinite-dimensional problems, in particular, problems involving partial differential equations and other problems which can be framed as evolution equations in Hilbert space, as if they were in $\mathbb{R}^n$. The idea is to show that bifurcation of high-dimensional problems takes place in low-dimensional spaces $\mathbb{R}^1$ or $\mathbb{R}^2$ for a wide class, so we may "project" the problems into the low-dimensional space. In $\mathbb{R}^n$ we use an operator notation which is formally identical to what we would use when treating an evolution equation in a Hilbert space, and all the formal operations are the same. So the results we compute in $\mathbb{R}^n$ are exactly the same for a wide class of problems in Hilbert spaces.* The only question then is to show that some boundary value problem can be set up as an evolution equation in a suitable Hilbert space. This type of showing is fairly routine for many of the common problems of continuum physics, but in any event is outside the scope of this elementary book.

* Essentially similar results follow for problems which can be framed more generally as evolution equations in Banach spaces. The really important property which we cannot drop is the Fredholm alternative for the solvability of the perturbation equations. In our very first example in Appendix VI.1 we consider an integro-differential equation which does not enter into the frame of Hilbert spaces for which the Fredholm alternative works well. In fact, it is the Fredholm alternative which is in back of the statement that in problems for which the alternative holds, and the action is going on in the null space of a certain linear operator, the part of the problem having a zero projection into the null space is like the little tail of a lively dog. So our use of Hilbert spaces here is in no way essential; it is just one way for us to show that the analysis we give in $\mathbb{R}^n$ holds more generally—in Hilbert space, for example.

In Appendix VI.1 we give some examples of formal computation which show how this theory works for partial differential equations and integral equations.

In what follows we define H as the set of $\mathbf{u}$ for which the evolution problem is well defined. For instance, for a system in $\mathbb{R}^n$, $H = \mathbb{R}^n$; for a partial differential equation of evolution, H will be the set of all functions satisfying the boundary conditions and possessing smoothness sufficient to assure that spatial derivatives and other defining properties of $\mathbf{f}(\mu, \cdot)$ are satisfied.

From now on we assume that H is a *Hilbert space*. This means, among other things, that it is a vector space with a scalar product $\langle \mathbf{u}_1, \mathbf{u}_2 \rangle$ for $\mathbf{u}_1$ and $\mathbf{u}_2$ in H which can be extended to complex elements, $\langle \overline{\mathbf{u}_1, \mathbf{u}_2} \rangle = \langle \mathbf{u}_2, \mathbf{u}_1 \rangle$, and generally possesses the properties of the scalar product (see §IV.3) in $\mathbb{C}^n$.

Consider a linear operator $\mathbf{A}$ with values in H, whose domain is some dense subset of H, as in the case of partial differential equations. (When $H = \mathbb{R}^n$, $\mathbf{A}$ is defined in all of H and may be expressed by an $n \times n$ matrix.) In the general case the spectral problem

$$\mathbf{A} \cdot \boldsymbol{\zeta} = \sigma \boldsymbol{\zeta} \quad \text{for} \quad \boldsymbol{\zeta} \text{ in } H \tag{VI.39}$$

has a meaning and defines eigenvalues σ and eigenvectors $\boldsymbol{\zeta}$. In the simplest of cases, all of the values of σ for which (VI.39) possesses nonzero solutions are isolated, algebraically simple eigenvalues. The difficulty is that in H there may be an infinite number of eigenvalues and even a continuum of them. Even in the case when the continuum of eigenvalues is excluded, it may happen that all of the solutions of the linear evolution problem

$$\frac{d\mathbf{v}}{dt} = \mathbf{A} \cdot \mathbf{v} \quad \text{in } H \tag{VI.40}$$

cannot be expressed as linear combinations (even infinite combinations) of terms proportional to $e^{\sigma_n t}\boldsymbol{\zeta}_n$. In the most simple case when the spectrum is of simple eigenvalues and $\mathbf{A}$ satisfies other requirements which occur frequently in applications (for example, in $\mathbb{R}^n$ and for fields defined over bounded regions of space which are governed by equations of the reaction–diffusion type or of the Navier–Stokes type) the solutions may be expressed in such linear combinations, as in (VI.57).

In fact, our theory holds more generally; we do not require that solutions be expressible in such linear combinations. It is enough that the evolution problem should possess one solution of the form $e^{\sigma_1 t}\boldsymbol{\zeta}$ where σ_1 is an *isolated eigenvalue* possessing the following properties.

(i) No other value of σ for which (VI.39) possesses nontrivial solutions should have a larger real part than σ_1.

(ii) σ_1 is of finite multiplicity and finite Riesz index (the definition of index and multiplicity are given in Chapter IV).

(iii) The same properties (i) and (ii) are required of the eigenvalues $\bar{\sigma}$ of the adjoint operator $\mathbf{A}^*$ satisfying

$$\langle \mathbf{A} \cdot \mathbf{u}, \mathbf{v} \rangle = \langle \mathbf{u}, \mathbf{A}^* \cdot \mathbf{v} \rangle \tag{VI.41}$$

for all $\mathbf{u}$ in the domain of $\mathbf{A}$ and all $\mathbf{v}$ in the domain of $\mathbf{A}^*$.

Then, as in §IV.1, we can find elements $\{\boldsymbol{\psi}_i\}_{i=1,\ldots,n}$ $\{\boldsymbol{\psi}_i^*\}_{i=1,\ldots,n}$ in H satisfying

$$(\sigma \mathbf{1} - \mathbf{A}) \cdot \boldsymbol{\psi}_i = 0$$
$$(\bar{\sigma} \mathbf{1} - \mathbf{A}^*) \cdot \boldsymbol{\psi}_i^* = 0,$$

and

$$\langle \boldsymbol{\psi}_i, \boldsymbol{\psi}_j^* \rangle = \delta_{ij}, \tag{VI.42}$$

where n is the multiplicity of the eigenvalue σ, and v its Riesz index. Moreover, here, as in Appendix IV.2, we may define a projection $\mathbf{P}$:

$$\mathbf{P} \cdot \mathbf{u} = \sum_{i=1}^{n} \langle \mathbf{u}, \boldsymbol{\psi}_i^* \rangle \boldsymbol{\psi}_i \tag{VI.43}$$

and

$$\mathbf{P} \cdot \mathbf{P} \cdot \mathbf{u} = \mathbf{P} \cdot \mathbf{u}, \qquad \mathbf{P} \cdot \mathbf{A} \cdot \mathbf{u} = \mathbf{A} \cdot \mathbf{P} \cdot \mathbf{u} \tag{VI.44}$$

for any $\mathbf{u}$ in H*. In fact, (VI.43) and (VI.44) are fundamental properties of projections in a general setting.

In all of our problems the multiplicity and the index will not be larger than two, so that the range of the projection $\mathbf{P}$ has one or two dimensions. In these problems it is frequently more convenient to work with the components $\langle \mathbf{u}, \boldsymbol{\psi}_i^* \rangle$ of the projection, rather than with the projection itself.

VI.7 The Spectral Problem for the Stability of $\mathbf{u} = 0$

We shall now consider the evolution equation

$$\frac{d\mathbf{u}}{dt} = \mathbf{f}(\mu, \mathbf{u}), \qquad \mathbf{f}(\mu, 0) = 0, \qquad \mathbf{u} \in H. \tag{VI.45}$$

A small disturbance $\mathbf{v} = e^{\sigma t}\boldsymbol{\zeta}$ of $\mathbf{u} = 0$ satisfies

$$\frac{d\mathbf{v}}{dt} = \mathbf{f}_u(\mu | \mathbf{v}), \qquad \sigma\boldsymbol{\zeta} = \mathbf{f}_u(\mu | \boldsymbol{\zeta}) \tag{VI.46}$$

where

$$\mathbf{f}_u(\mu | \cdot) \stackrel{\text{def}}{=} \mathbf{f}_u(\mu, 0 | \cdot) \stackrel{\text{def}}{=} \mathbf{A}(\mu)(\cdot)$$

* In the domain of $\mathbf{A}$ for $\mathbf{P} \cdot \mathbf{A} \cdot \mathbf{u}$.

is the linearized part of $\mathbf{f}(\mu, \mathbf{u})$ evaluated at the point $\mathbf{u} = 0$ and $\mathbf{f}_u(\mu|\cdot)$ is the linearized operator introduced in §VI.6. The eigenvalues

$$\sigma(\mu) = \xi(\mu) + i\eta(\mu) \tag{VI.47}$$

of $\mathbf{f}_u(\mu|\cdot)$ have to satisfy $(\text{VI.46})_2$ for a nonzero $\boldsymbol{\zeta}$. They constitute part of what is called the *spectrum* of $\mathbf{f}_u(\mu|\cdot)$. If $H = \mathbb{R}^n$, the spectrum is entirely of eigenvalues. In many problems of partial differential equations (for example, in evolution problems for parabolic partial differential equations, or in Navier–Stokes problems on bounded domains) the spectrum is entirely of eigenvalues.

We shall be interested in the case when the part of the spectrum which controls the stability of $\mathbf{u} = 0$ is a finite set of isolated eigenvalues. This is true generally in $\mathbb{R}^n$ and in most of the applications involving fields which arise in continuum physics and mechanics.

In general $\xi(\mu)$ and $\eta(\mu)$, hence $\sigma(\mu)$, are continuous in μ. If the eigenvalue σ is *algebraically simple* for some value μ_0 of μ, then in a neighborhood of μ_0, ξ, η, and σ are as smooth as $\mathbf{f}_u(\mu|\cdot)$. The same smoothness holds for *semi-simple* eigenvalues, for example, if the perturbation of the multiple eigenvalue with $\mu - \mu_0$ splits the eigenvalue at first order in $\mu - \mu_0$ into a number of eigenvalues equal to the multiplicity of $\sigma(\mu)$ at $\mu = \mu_0$. We have seen already in §IV.7 that if σ is multiple eigenvalue of Riesz index greater than one for some value μ_0 of μ, then in general ξ, η, σ need not be differentiable with respect to μ at μ_0, even if $\mathbf{f}_u(\mu|\cdot)$ is analytic.

Associated with the operator $\mathbf{f}_u(\mu|\cdot) = \mathbf{A}(\mu)$ is the operator

$$\mathbf{f}_u^*(\mu|\cdot) \stackrel{\text{def}}{=} \mathbf{A}^*(\mu) \tag{VI.48}$$

satisfying (VI.41) and the adjoint spectral equation is

$$\bar{\sigma}\boldsymbol{\zeta}^* = \mathbf{f}_u^*(\mu|\boldsymbol{\zeta}^*). \tag{VI.49}$$

The stability of the solution $\mathbf{u} = 0$ of (VI.45) may be determined from the knowledge of the spectrum of $\mathbf{f}_u(\mu|\cdot)$. Let us assume that the spectrum is exclusively of eigenvalues. Then $\mathbf{u} = 0$ is *conditionally stable* if all eigenvalues have a negative real part ($\xi < 0$), and is unstable if some eigenvalue has a positive real part. Conditionally stable means that $\mathbf{u} = 0$ is stable to small disturbances. A conditionally stable solution may be unstable to large disturbances whose evolution is not governed by the linearized theory.

Assume that for $\mu < 0$ the null solution $\mathbf{u} = 0$ is conditionally stable, and that this solution loses its stability when μ crosses 0. We shall consider two cases.

(a) For $\mu = 0$, a pair of simple eigenvalues $\pm i\omega_0$, $\omega_0 \neq 0$, are on the imaginary axis, the other eigenvalues being of negative real part. At criticality

$$\sigma(0) = i\eta(0) = i\omega_0, \qquad \xi(0) = 0,$$

and we assume that $\xi_\mu(0) > 0$. In this case we get bifurcation into time-periodic solutions (see Chapters VII and VIII). For now it will suffice to remark that the linearized problem

$$\frac{d\mathbf{v}}{dt} = \mathbf{f}_u(0|\mathbf{v})$$

has only two independent time-periodic solutions:

$$\mathbf{v}(t) = e^{i\omega_0 t}\boldsymbol{\zeta} \quad \text{and} \quad \bar{\mathbf{v}}(t).$$

The other solutions decay exponentially in time. In this case the construction of a bifurcating solution requires that we project into a two-dimensional space.

(b) For $\mu = 0$, a single eigenvalue crosses the origin. If $\sigma = 0$ is simple, the problem may be reduced to $\mathbb{R}^1$ by methods which are identical to those used to reduce to $\mathbb{R}^1$ the problem of bifurcation of solutions in $\mathbb{R}^2$ when the eigenvalues of the 2×2 matrix $\mathbf{A}(0)$ are real and distinct (see Figure VI.1).

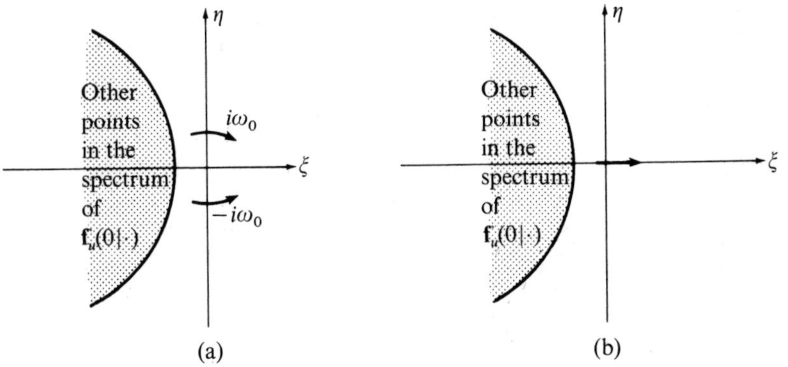

Figure VI.1 Eigenvalues $\sigma(\mu) = \xi(\mu) + i\eta(\mu)$ of $\mathbf{f}_u(\mu|\cdot)$ at criticality $\mu = 0$. In case (a) a pair of conjugate simple eigenvalues cross over. In case (b) a single real-valued eigenvalue passes through the origin

VI.8 The Spectral Problem and the Laplace Transform

A better understanding of the evolution of the linearized initial-value problem

$$\frac{d\mathbf{v}}{dt} = \mathbf{f}_u(\mu|\mathbf{v}), \ \mathbf{v}(0) = \mathbf{v}_0 \tag{VI.50}$$

may be obtained by the method of Laplace transforms. We first define the transform

$$\mathbf{V}(\lambda) = \int_0^\infty \mathbf{v}(t)e^{-\lambda t}\,dt \tag{VI.51}$$

and the Mellin inversion formula

$$\mathbf{v}(t) = \frac{1}{2\pi i}\int_{\hat{\xi}-i\infty}^{\hat{\xi}+i\infty} \mathbf{V}(\lambda)e^{\lambda t}\,d\lambda \tag{VI.52}$$

where $\lambda = \hat{\xi} + i\hat{\eta}$. We next suppose that $\hat{\xi}$ is large enough so that when $t \to \infty$,

$$\mathbf{v}(t)e^{-\hat{\xi}t} \to 0, \tag{VI.53}$$

where $\mathbf{v}(t)$ satisfies (VI.50). We will show that (VI.53) holds if $\hat{\xi} > \xi_1(\mu)$ where $\xi_1(\mu)$ is the largest of the real part of the eigenvalues of $\mathbf{f}_u(\mu|\cdot)$.

Applying the Laplace transform to (VI.50), we find that

$$\mathbf{v}_0 = \lambda\mathbf{V} - \mathbf{f}_u(\mu|\mathbf{V}). \tag{VI.54}$$

The spectrum of $\mathbf{f}_u(\mu|\cdot)$ may now be defined as the set of values of λ for which (VI.54) cannot be solved for $\mathbf{V}$. All of the eigenvalues belong to this singular set. The values of λ not in the spectrum are said to be in the resolvent set. For these values we may invert (VI.54):

$$\mathbf{V}(\lambda, \mu) = \mathbf{R}(\lambda, \mu)\cdot\mathbf{v}_0 \tag{VI.55}$$

where $\mathbf{R}(\lambda, \mu)$ is the resolvent operator, the inverse of the linear operator $\lambda\mathbf{1} - \mathbf{f}_u(\mu|\cdot)$. For many partial differential equations, $\mathbf{R}(\lambda, \mu)$ takes the form of a Green integral operator.

Combining (VI.55) and (VI.52) we find that

$$\mathbf{v}(t, \mu) = \frac{1}{2\pi i}\int_{\hat{\xi}-i\infty}^{\hat{\xi}+i\infty} e^{\lambda t}\mathbf{R}(\lambda, \mu)\cdot\mathbf{v}_0\,d\lambda. \tag{VI.56}$$

We take $\hat{\xi}$ large enough to place all the singularities of $R(\cdot, \mu)$ on the left of the line $\hat{\xi}$ = constant (see Figure VI.2). These singularities are nothing other than the spectrum of $\mathbf{f}_u(\mu|\cdot)$; the eigenvalues σ of $\mathbf{f}_u(\mu|\cdot)$ are poles of $\mathbf{R}(\lambda, \mu)$. Simple eigenvalues are simple poles of $\mathbf{R}(\lambda, \mu)$.

It is instructive to consider the case when a conjugate pair of simple eigenvalues σ_1 and $\bar{\sigma}_1$ have a real part larger than the real part of any of other values in the spectrum of $\mathbf{f}_u(\mu|\cdot)$.

It is not hard to show that

$$\langle \mathbf{v}(t), \zeta_1^* \rangle = \langle \mathbf{v}_0, \zeta_1^* \rangle e^{\sigma_1 t}.$$

To show it, we first project (VI.52) and find that

$$\langle \mathbf{v}(t), \zeta_1^* \rangle = \frac{1}{2\pi i}\int_{-\hat{\xi}-i\infty}^{\hat{\xi}+i\infty} \langle \mathbf{V}(\lambda, \mu), \zeta_1^* \rangle e^{\lambda t}\,d\lambda.$$

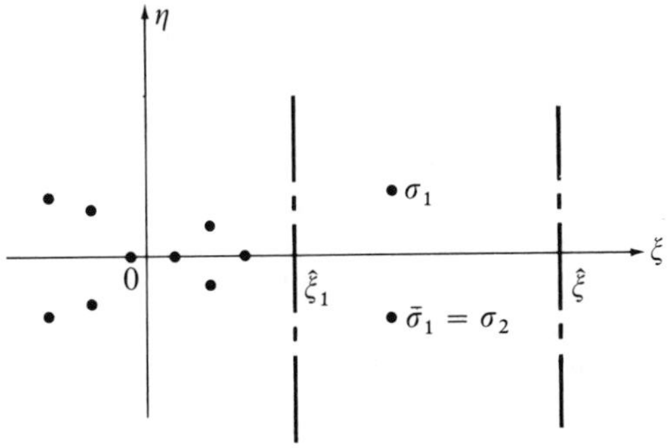

Figure VI.2 Poles of $\mathbf{R}(\lambda, \mu)$ in the complex λ plane are eigenvalues σ of $\mathbf{f}_u(\mu|\cdot)$. Poles plus other singularities of $\mathbf{R}(\cdot|\mu)$ define the entire spectrum of $\mathbf{f}_u(\mu|\cdot)$

Then projecting (VI.54) we calculate

$$\begin{aligned}\langle \mathbf{v}_0, \boldsymbol{\zeta}_1^* \rangle &= \lambda\langle \mathbf{V}, \boldsymbol{\zeta}_1^* \rangle - \langle \mathbf{f}_u(\mu|\mathbf{V}), \boldsymbol{\zeta}_1^* \rangle \\ &= \lambda\langle \mathbf{V}, \boldsymbol{\zeta}_1^* \rangle - \langle \mathbf{V}, \mathbf{f}_u^*(\mu|\boldsymbol{\zeta}_1^*)\rangle \\ &= (\lambda - \sigma_1)\langle \mathbf{V}, \boldsymbol{\zeta}_1^* \rangle.\end{aligned}$$

Hence,

$$\begin{aligned}\langle \mathbf{v}(t), \boldsymbol{\zeta}_1^* \rangle &= \frac{\langle \mathbf{v}_0, \boldsymbol{\zeta}_1^* \rangle}{2\pi i} \int_{\hat{\xi} - i\infty}^{\hat{\xi} + i\infty} \frac{e^{\lambda t}}{\lambda - \sigma_1}\, d\lambda \\ &= \langle \mathbf{v}_0, \boldsymbol{\zeta}_1^* \rangle e^{\sigma_1 t}.\end{aligned}$$

So the method of residues (VI.56) applied in the present case (and suppressing the dependence of $\mathbf{v}$ on μ) leads us to a residue integral representation

$$\mathbf{v}(t) - \langle \mathbf{v}_0, \boldsymbol{\zeta}_1^* \rangle e^{\sigma_1 t}\boldsymbol{\zeta}_1 - \langle \mathbf{v}_0, \bar{\boldsymbol{\zeta}}_1^* \rangle e^{\bar{\sigma}_1 t}\bar{\boldsymbol{\zeta}}_1 = \frac{1}{2\pi i} \int_{\hat{\xi}_1 - i\infty}^{\hat{\xi}_1 + i\infty} e^{\lambda t}\mathbf{V}(\lambda)\, d\lambda \qquad \text{(VI.57)}$$

where $\hat{\xi}_1 < \operatorname{Re} \sigma_1 < \hat{\xi}$ (see Figure VI.2) and the integral at the end of (VI.57) is orthogonal to $\boldsymbol{\zeta}_1^*$ and $\bar{\boldsymbol{\zeta}}_1^*$. When t is large (VI.57) shows that

$$\mathbf{v}(t) \sim e^{\xi_1 t}(e^{i\eta_1 t}\langle \mathbf{v}_0, \boldsymbol{\zeta}_1^* \rangle \boldsymbol{\zeta}_1 + e^{-i\eta_1 t}\langle \mathbf{v}_0, \bar{\boldsymbol{\zeta}}_1^* \rangle \bar{\boldsymbol{\zeta}}_1).$$

More generally, if the eigenvalue σ_1 is multiple, $\bar{\sigma}_1$ is also multiple because $\mathbf{f}_u(\mu|\cdot)$ is real and the method of residues may be used to derive a new form of (VI.56):

$$\mathbf{v}(t) = e^{\sigma_1 t}\mathbf{P}(t) + e^{\bar{\sigma}_1 t}\bar{\mathbf{P}}(t) + \frac{1}{2\pi i} \int_{\hat{\xi}_1 - i\infty}^{\hat{\xi}_1 + i\infty} e^{\lambda t}\mathbf{R}(\lambda, \mu) \cdot \mathbf{v}_0\, d\lambda, \qquad \text{(VI.58)}$$

where $\mathbf{R}(\lambda, \mu) = (\lambda\mathbf{1} - \mathbf{f}_u(\mu|\cdot))^{-1}$ and $\mathbf{P}(t)$ is a polynomial of degree $\nu - 1$ in t, ν being the Riesz index of σ_1, whose coefficients depend on $\mathbf{v}_0$. It is possible to obtain an estimate of the integral in (VI.58) of the following type

$$\|\mathbf{v}(t) - 2\operatorname{Re}(e^{\sigma_1 t}\mathbf{P}(t))\| \le k e^{\hat{\xi}_1 t}\|\mathbf{v}_0\|.$$

Since $\xi_1 = \operatorname{Re} \sigma_1 > \hat{\xi}_1$, the first two terms on the right-hand side of (VI.58) dominate the behavior of $\mathbf{v}(t)$ when $t \to \infty$.

In concluding this section we remark that the formula (VI.56) is a generalization of the exponential of a matrix. When $H = \mathbb{R}^n$

$$\mathbf{v}(t, \mu) = e^{\mathbf{A}(\mu)t} \cdot \mathbf{v}_0, \tag{VI.59}$$

where the exponential is defined for $t \geq 0$ and $t < 0$. In more general cases, for example, in evolution problems where $H \neq \mathbb{R}^n$, we may define (VI.56) only for $t > 0$, so instead of a group property

$$e^{\mathbf{A}(t_1 + t_2)} = e^{\mathbf{A}t_1} e^{\mathbf{A}t_2},$$

where $-\infty < t_j < \infty$, $j = 1, 2$, we have *semi-group* property, because we require that t_1 and $t_2 \geq 0$.

VI.9 Projections into $\mathbb{R}^1$

When a single, real-valued, isolated, simple eigenvalue $\sigma(\mu) = \xi(\mu)$ of $\mathbf{f}_u(\mu|\cdot)$ crosses strictly through the origin of the complex σ-plane when $\mu = 0$, we get bifurcation of $\mathbf{u} = 0$ into a steady solution $\mathbf{u} \neq 0$ for $\mu \neq 0$. The solution may be decomposed into a part on the null space of $\mathbf{f}_u(0|\cdot)$ and a little part which is orthogonal to the eigenvector $\boldsymbol{\zeta}^*$ of $\mathbf{f}_u^*(0|\cdot)$ belonging to $\sigma = 0$. The analysis is *identical* to the one given when $\mathbf{u}$ is in $\mathbb{R}^2$, on the understanding that the extra little part $\mathbf{w}$ defined in (VI.30) is some kind of "superposition of all other modes," eigenfunctions of $\mathbf{f}_u(0|\cdot)$. In $\mathbb{R}^n$, there are $n - 1$ other modes. While for partial differential equations such superpositions, even infinite ones, are not always possible, it is always possible to define $\mathbf{w}$ merely by saying it is orthogonal to $\boldsymbol{\zeta}^*$.

The main thing to check in problems of a general type in H is that the Fredholm alternative applies. The Fredholm alternative is said to apply if a necessary and sufficient condition for the solvability of

$$\mathbf{f}_u(0|\mathbf{v}) = \tilde{\mathbf{w}}, \qquad \mathbf{v} \in H \tag{VI.60}$$

is that

$$\langle \tilde{\mathbf{w}}, \boldsymbol{\zeta}^* \rangle = 0 \tag{VI.61}$$

for all $\boldsymbol{\zeta}^*$ such that $\mathbf{f}_u^*(0|\boldsymbol{\zeta}^*) = 0$. Then the solution $\mathbf{v}$ is determined up to elements $\boldsymbol{\zeta}$ satisfying $\mathbf{f}_u(0|\boldsymbol{\zeta}) = 0$. In the cases treated by us the dimensions of the null spaces of $\mathbf{f}_u(0|\cdot)$ and $\mathbf{f}_u^*(0|\cdot)$ are the same. In this chapter we have one- or two-dimensional null spaces associated with eigenvector $\boldsymbol{\zeta}$ and a one- or two-dimensional adjoint null space associated with $\boldsymbol{\zeta}^*$. So it is possible to find a *unique solution* of $\mathbf{v}$ of (VI.60) satisfying

$$\langle \mathbf{v}, \boldsymbol{\zeta}^* \rangle = 0. \tag{VI.62}$$

for all $\boldsymbol{\zeta}^*$ on the null space of $\mathbf{f}_u^*(0|\cdot)$.

With this understanding we may view the theory of §VI.2 as appropriate to the method of projection in the case of bifurcation at an algebraically simple eigenvalue in $\mathbb{R}^n$ or in a Hilbert space. Following the order already used in treating problems in $\mathbb{R}^1$ and $\mathbb{R}^2$, in §VI.10 we shall discuss the method of projection for isolated solutions which perturb bifurcation at a simple eigenvalue (double-point bifurcation defined for $\mathbb{R}^1$ in Chapter II); in §VI.11 we apply the method or projection for bifurcation at a double eigenvalue of index two; and in §VI.12 we apply the same method to bifurcation at a double semi-simple eigenvalue of index one.

VI.10 The Method of Projection for Isolated Solutions Which Perturb Bifurcation at a Simple Eigenvalue (Imperfection Theory)

Consider an evolution equation of the form

$$\frac{d\mathbf{u}}{dt} = \tilde{\mathscr{F}}(\mu, \mathbf{u}, \delta) \quad \text{for } \mathbf{u} \text{ in } H \tag{VI.63}$$

where H is $\mathbb{R}^n$ or some other function space which is defined for infinite-dimensional problems such as partial differential equations according to conventions adopted in §§VI.6–8. We further specify that $\mathbf{u} = 0$ is a solution of

$$\tilde{\mathscr{F}}(\mu, \mathbf{u}, 0) = 0 \tag{VI.64}$$

which loses stability strictly as μ is increased past zero. It then follows that (VI.64) undergoes double-point bifurcation of $(\mu, \mathbf{u}) = (0, 0)$. The spectral problem for the stability of $\mathbf{u} = 0$ is

$$\sigma(\mu)\zeta(\mu) = \tilde{\mathscr{F}}_u(\mu, 0, 0 \mid \zeta(\mu)).$$

At criticality, $\sigma(0) = 0$ is an isolated simple eigenvalue of $\mathscr{F}_u(\cdot)$ and the adjoint operator $\mathscr{F}_u^*(\cdot)$; i.e.,

$$\mathscr{F}_u(\zeta) \overset{\text{def}}{=} \tilde{\mathscr{F}}_u(0, 0, 0 \mid \zeta) = 0, \qquad \zeta = \zeta(0), \tag{VI.65}$$

and

$$\mathscr{F}_u^*(\zeta^*) = 0. \tag{VI.66}$$

All the other eigenvalues of $\mathscr{F}_u(\cdot)$ have negative real parts. The operator $\mathscr{F}_u(\cdot)$ is such that the equation

$$\mathscr{F}_u(\mathbf{v}) = \mathbf{f} \tag{VI.67}$$

is uniquely solvable to within an additive multiple of ζ satisfying (VI.65), provided only that

$$\langle \mathbf{f}, \zeta^* \rangle = 0, \tag{VI.68}$$

where ζ^* satisfies (VI.66), and

$$\mathbf{v} = \zeta + \mathbf{w}, \qquad \langle \mathbf{w}, \zeta^* \rangle = 0, \tag{VI.69}$$

where $\mathbf{w} = \mathscr{F}_u^{-1}(\mathbf{f})$ and $\mathscr{F}_u^{-1}(\cdot)$ is a well-defined operator with the property that $\langle \mathscr{F}_u^{-1}(\mathbf{f}), \zeta^* \rangle = 0$ for any $\mathbf{f}$ satisfying (VI.68). The strict loss of stability of $\mathbf{u} = 0$ solving (VI.64) as μ is increased past zero implies that

$$\sigma_\mu(0) = \langle \mathscr{F}_{\mu u}(\zeta), \zeta^* \rangle > 0 \tag{VI.70}$$

where

$$\mathscr{F}_{\mu u}(\cdot) \overset{\text{def}}{=} \tilde{\mathscr{F}}_{\mu u}(0, 0, 0 \,|\, \cdot).$$

Returning now to (VI.63) we imagine δ is small and seek isolated steady solutions which break bifurcation at (0, 0), under the condition that

$$\langle \mathscr{F}_\delta, \zeta^* \rangle \neq 0. \tag{VI.71}$$

Recall that $\mathscr{F}_\delta \overset{\text{def}}{=} \tilde{\mathscr{F}}_\delta(0, 0, 0)$. It is also convenient to introduce an amplitude

$$\varepsilon \overset{\text{def}}{=} \langle \mathbf{u}, \zeta^* \rangle. \tag{VI.72}$$

The condition (VI.71) and the implicit function theorem guarantee the existence of a smooth solution $\mathbf{u}(\mu, \varepsilon)$ and $\delta = \Delta(\mu, \varepsilon)$ of

$$\tilde{\mathscr{F}}(\mu, \mathbf{u}(\mu, \varepsilon), \Delta(\mu, \varepsilon)) = 0, \qquad \varepsilon = \langle \mathbf{u}(\mu, \varepsilon), \zeta^* \rangle. \tag{VI.73}$$

Setting $\varepsilon = 0$ in (VI.73) we find that

$$\mathbf{u}(\mu, 0) = 0 \quad \text{and} \quad \Delta(\mu, 0) = 0. \tag{VI.74}$$

The first derivative of (VI.73) with respect to ε at $\varepsilon = 0$ is in the form

$$\mathscr{F}_u(\mathbf{u}_\varepsilon) + \mathscr{F}_\delta \Delta_\varepsilon = 0, \qquad 1 = \langle \mathbf{u}_\varepsilon, \zeta^* \rangle \tag{VI.75}$$

The solvability condition (VI.68) implies that $\Delta_\varepsilon = 0$ and

$$\mathbf{u}_\varepsilon = \zeta. \tag{VI.76}$$

The second partial derivatives of (VI.73) are in the form

$$\mathscr{F}_u(\mathbf{u}_{\varepsilon\varepsilon}) + \mathscr{F}_{uu}(\zeta \,|\, \zeta) + \mathscr{F}_\delta \Delta_{\varepsilon\varepsilon} = 0 \tag{VI.77}$$

$$\mathscr{F}_u(\mathbf{u}_{\mu\varepsilon}) + \mathscr{F}_{\mu u}(\zeta) + \mathscr{F}_\delta \Delta_{\mu\varepsilon} = 0, \tag{VI.78}$$

where $\mathbf{u}_{\varepsilon\varepsilon}$ and $\mathbf{u}_{\mu\varepsilon}$ are orthogonal to ζ^*. The solvability condition (VI.68) implies that

$$\Delta_{\varepsilon\varepsilon} = -\frac{\langle \mathscr{F}_{uu}(\zeta \,|\, \zeta), \zeta^* \rangle}{\langle \mathscr{F}_\delta, \zeta^* \rangle} \tag{VI.79}$$

and

$$\begin{aligned} \Delta_{\mu\varepsilon} &= -\frac{\langle \mathscr{F}_{\mu u}(\zeta), \zeta^* \rangle}{\langle \mathscr{F}_\delta, \zeta^* \rangle} \\ &= -\frac{\sigma_\mu(0)}{\langle \mathscr{F}_\delta, \zeta^* \rangle} \neq 0. \end{aligned} \tag{VI.80}$$

When $\Delta_{\varepsilon\varepsilon}$ and $\Delta_{\mu\varepsilon}$ satisfy (VI.79) and (VI.80), (VI.77) and (VI.78) may be solved uniquely. Turning next to third-order derivatives we compute

$$\mathscr{F}_u(\mathbf{u}_{\varepsilon\varepsilon\varepsilon}) + \Delta_{\varepsilon\varepsilon\varepsilon}\mathscr{F}_\delta + 3\Delta_{\varepsilon\varepsilon}\mathscr{F}_{u\delta}(\zeta) + 3\mathscr{F}_{uu}(\zeta|\mathbf{u}_{\varepsilon\varepsilon}) + \mathscr{F}_{uuu}(\zeta|\zeta|\zeta) = 0 \tag{VI.81}$$

$$\begin{aligned}\mathscr{F}_u(\mathbf{u}_{\mu\varepsilon\varepsilon}) + \Delta_{\mu\varepsilon\varepsilon}\mathscr{F}_\delta + \mathscr{F}_{\delta\mu}\Delta_{\varepsilon\varepsilon} + \mathscr{F}_{\mu u}(\mathbf{u}_{\varepsilon\varepsilon}) + 2\mathscr{F}_{u\delta}(\zeta)\Delta_{\mu\varepsilon}\\ + 2\mathscr{F}_{uu}(\zeta|\mathbf{u}_{\mu\varepsilon}) + \mathscr{F}_{\mu uu}(\zeta, \zeta) = 0\end{aligned} \tag{VI.82}$$

$$\mathscr{F}_u(\mathbf{u}_{\mu\mu\varepsilon}) + \Delta_{\mu\mu\varepsilon}\mathscr{F}_\delta + \mathscr{F}_{\mu\mu u}(\zeta) + 2\mathscr{F}_{\mu u}(u_{\mu\varepsilon}) + 2\mathscr{F}_{\mu\delta}\Delta_{\mu\varepsilon} = 0. \tag{VI.83}$$

These equations are uniquely solvable among vectors orthogonal to ζ^* when $\Delta_{\varepsilon\varepsilon\varepsilon}$, $\Delta_{\mu\varepsilon\varepsilon}$, and $\Delta_{\mu\mu\varepsilon}$ are selected so as to satisfy (VI.68).

Proceeding to higher orders in the same fashion we generate a Taylor series

$$\begin{aligned}\Delta(\mu, \varepsilon) = \frac{1}{2}[\Delta_{\varepsilon\varepsilon}\varepsilon^2 + 2\Delta_{\varepsilon\mu}\varepsilon\mu] + \frac{1}{3!}[\Delta_{\varepsilon\varepsilon\varepsilon}\varepsilon^3 + 3\Delta_{\mu\varepsilon\varepsilon}\mu\varepsilon^2\\ + 3\Delta_{\mu\mu\varepsilon}\mu^2\varepsilon] + O[\varepsilon(|\mu| + |\varepsilon|)^3].\end{aligned} \tag{VI.84}$$

Equation (VI.84) is in the form of (III.26). We now proceed to obtain the function $\mu(\varepsilon, \delta)$ by successive approximations in $\mathbb{R}^1$. So the problem of finding the isolated solutions which break double-point bifurcation has been reduced to one dimension. We leave as an exercise for the reader the demonstration that the stability problem for the isolated solutions can be reduced to one dimension as in (III.32).

VI.11 The Method of Projection at a Double Eigenvalue of Index Two

We are interested in the steady solutions which bifurcate from $\mathbf{u} = 0$ where $\mathbf{u}$ in H satisfies the evolution problem

$$\frac{d\mathbf{u}}{dt} = \mathbf{f}(\mu, \mathbf{u}) = \mathbf{f}_u(\mu|\mathbf{u}) + \tfrac{1}{2}\mathbf{f}_{uu}(\mu|\mathbf{u}|\mathbf{u}) + O(\|\mathbf{u}\|^3). \tag{VI.85}$$

The stability of $\mathbf{u} = 0$ is determined by the sign of the real part of $\sigma(\mu)$, where $\sigma(\mu)$ is the eigenvalue of $\mathbf{f}_u(\mu|\cdot)$ of largest real part

$$\sigma(\mu)\zeta = \mathbf{f}_u(\mu|\zeta) \quad \text{for } \zeta \text{ in } H. \tag{VI.86}$$

We assume that at criticality $\sigma(0)$ is a double eigenvalue of index two of the linear operator

$$\mathbf{f}_u(0|\cdot) \overset{\text{def}}{=} \mathbf{f}_u(\cdot). \tag{VI.87}$$

Since $\sigma(0) = 0$ is a double eigenvalue of index 2, we have the following Jordan chain equations for eigenvectors and generalized eigenvectors (see Appendices IV.1 and IV.2).

$$\begin{aligned} \mathbf{f}_u(\boldsymbol{\zeta}_1) &= 0, \qquad & \mathbf{f}_u^*(\boldsymbol{\zeta}_2^*) &= 0 \\ \mathbf{f}_u(\boldsymbol{\zeta}_2) &= \boldsymbol{\zeta}_1, \qquad & \mathbf{f}_u^*(\boldsymbol{\zeta}_1^*) &= \boldsymbol{\zeta}_2^*, \end{aligned} \tag{VI.88}$$

where $\mathbf{f}_u^*(\cdot) \overset{\text{def}}{=} [\mathbf{f}_u(0|\cdot)]^*$ is the linear adjoint operator defined relative to the scalar product $\langle\cdot,\cdot\rangle$ in H:

$$\langle\boldsymbol{\zeta}_1, \boldsymbol{\zeta}_2^*\rangle = \langle\mathbf{f}_u(\boldsymbol{\zeta}_2), \boldsymbol{\zeta}_2^*\rangle = \langle\boldsymbol{\zeta}_2, \mathbf{f}_u^*(\boldsymbol{\zeta}_2^*)\rangle = 0$$

$$\langle\boldsymbol{\zeta}_1, \boldsymbol{\zeta}_1^*\rangle = \langle\mathbf{f}_u(\boldsymbol{\zeta}_2), \boldsymbol{\zeta}_1^*\rangle = \langle\boldsymbol{\zeta}_2, \mathbf{f}_u^*(\boldsymbol{\zeta}_1^*)\rangle = \langle\boldsymbol{\zeta}_2, \boldsymbol{\zeta}_2^*\rangle$$

We choose $\boldsymbol{\zeta}_2$ and $\boldsymbol{\zeta}_1^*$ so that $\langle\boldsymbol{\zeta}_1, \boldsymbol{\zeta}_1^*\rangle = \langle\boldsymbol{\zeta}_2, \boldsymbol{\zeta}_2^*\rangle = 1$, $\langle\boldsymbol{\zeta}_2, \boldsymbol{\zeta}_1^*\rangle = 0$. Now we seek the solutions which bifurcate from $\mathbf{u} = 0$ as a power series in the amplitude in projection,

$$\varepsilon = \langle\mathbf{u}, \boldsymbol{\zeta}_1^*\rangle, \tag{VI.89}$$

that is,

$$\begin{bmatrix}\mathbf{u}(\varepsilon)\\ \mu(\varepsilon)\end{bmatrix} = \sum_{n=1} \frac{\varepsilon^n}{n!}\begin{bmatrix}\mathbf{u}_n\\ \mu_n\end{bmatrix}. \tag{VI.90}$$

We find that

$$\begin{gathered} \mathbf{f}_u(\mathbf{u}_1) = 0, \qquad \langle\mathbf{u}_1, \boldsymbol{\zeta}_1^*\rangle = 1 \\ \mathbf{f}_u(\mathbf{u}_2) + 2\mu_1\mathbf{f}_{u\mu}(\mathbf{u}_1) + \mathbf{f}_{uu}(0|\mathbf{u}_1|\mathbf{u}_1) = 0, \\ \langle\mathbf{u}_2, \boldsymbol{\zeta}_1^*\rangle = 0, \end{gathered} \tag{VI.91}$$

and

$$\mathbf{f}_u(\mathbf{u}_n) + n\mu_{n-1}\mathbf{f}_{u\mu}(\mathbf{u}_1) + \not{f}_n = 0, \qquad \langle\mathbf{u}_n, \boldsymbol{\zeta}_1^*\rangle = 0 \tag{VI.92}$$

where $\not{f}_n$ depends on the derivatives of μ and $\mathbf{u}$ of lower order. It is always possible to satisfy the orthogonality condition $\langle\mathbf{u}_n, \boldsymbol{\zeta}_1^*\rangle = 0$ because if $\tilde{\mathbf{u}}_n$ is a solution of the equation which does not satisfy the orthogonality condition then $\mathbf{u}_n = \tilde{\mathbf{u}}_n - \langle\tilde{\mathbf{u}}_n, \boldsymbol{\zeta}_1^*\rangle\boldsymbol{\zeta}_1$ satisfies the equation and the orthogonality condition.

Equation $(\text{VI.91})_1$ shows that

$$\mathbf{u}_1 = \boldsymbol{\zeta}_1. \tag{VI.93}$$

To complete the solution we need the following result.

Lemma (Fredholm alternative when zero is a double eigenvalue of index two of $\mathbf{f}_u(\cdot)$). *The equation*

$$\mathbf{f}_u(\boldsymbol{\phi}) = \boldsymbol{\psi}, \qquad \boldsymbol{\phi} \in H \tag{VI.94}$$

is solvable if and only if

$$\langle\boldsymbol{\psi}, \boldsymbol{\zeta}_2^*\rangle = 0. \tag{VI.95}$$

The criterion (VI.95) is no different from (VI.61); the inhomogeneous term in (VI.99) must be orthogonal to all of the independent null vectors of $\mathbf{f}_u^*$. There is only one null vector at a double eigenvalue of Riesz index two. For solvability it is the *geometric* multiplicity which counts. It is obvious that (VI.95) is a necessary condition. The proof that (VI.95) is also sufficient for solvability in $\mathbb{R}^n$ follows from linear algebra. For more general problems the requirement (VI.95) is exactly the form taken by the Fredholm alternative defined in §VI.9 at an eigenvalue of index two.

Applying (VI.95) to $(\text{VI.91})_2$ we find that

$$2\mu_1 C_0' + \langle \mathbf{f}_{uu}(0|\zeta_1|\zeta_1), \zeta_2^* \rangle = 0,$$

and, at order n,

$$n\mu_{n-1} C_0' + \langle \not{f}_n, \zeta_2^* \rangle = 0 \tag{VI.96}$$

where

$$C_0' \stackrel{\text{def}}{=} \langle \mathbf{f}_{u\mu}(\zeta_1), \zeta_2^* \rangle. \tag{VI.97}$$

Our *bifurcation assumption*

$$C_0' \neq 0 \tag{VI.98}$$

is equivalent to the assumption (V.17) which we made in $\mathbb{R}^2$. So if $C_0' \neq 0$ we can compute the series (VI.90) (each step determining $\mu_{n-1}, \mathbf{u}_n$).

Turning next to the stability of the bifurcating solution (VI.90) we derive the spectral problem

$$\gamma\boldsymbol{\psi} = \mathbf{f}_u(\mu(\varepsilon), \mathbf{u}(\varepsilon)|\boldsymbol{\psi}) \tag{VI.99}$$

for small disturbances $e^{\gamma t}\boldsymbol{\psi}$ of $\mathbf{u}(\varepsilon)$. Inserting the expansion (VI.90) into (VI.99) we find that

$$\gamma\boldsymbol{\psi} = \mathbf{f}_u(\boldsymbol{\psi}) + \varepsilon[\mu_1 \mathbf{f}_{u\mu}(\boldsymbol{\psi}) + \mathbf{f}_{uu}(0|\zeta_1|\boldsymbol{\psi})] + O(\varepsilon^2)\cdot\boldsymbol{\psi}. \tag{VI.100}$$

Let us decompose $\boldsymbol{\psi}$ as

$$\boldsymbol{\psi} = \alpha_1\zeta_1 + \alpha_2\zeta_2 + \mathbf{W}, \tag{VI.101}$$

where

$$\alpha_i = \langle \boldsymbol{\psi}, \zeta_i^* \rangle, \qquad \langle \mathbf{W}, \zeta_i^* \rangle = 0, \qquad i = 1, 2.$$

Noting that

$$\langle \mathbf{f}_u(\boldsymbol{\psi}), \zeta_2^* \rangle = 0, \qquad \langle \mathbf{f}_u(\psi), \zeta_1^* \rangle = \langle \boldsymbol{\psi}, \zeta_2^* \rangle = \alpha_2, \tag{VI.102}$$

and using (VI.97), the equation (VI.100) takes the form of a system:

$$\begin{aligned} \gamma\alpha_1 &= \alpha_2 + \varepsilon(a_1\alpha_1 + b_1\alpha_2) + O(\varepsilon)\cdot\mathbf{W} + O(\varepsilon^2)\cdot\boldsymbol{\psi} \\ \gamma\alpha_2 &= -\varepsilon\mu_1 C_0'\alpha_1 + \varepsilon b_2\alpha_2 + O(\varepsilon)\cdot\mathbf{W} + O(\varepsilon^2)\cdot\boldsymbol{\psi} \\ \gamma\mathbf{W} &= \mathbf{f}_u(\mathbf{W}) + O(\varepsilon)\boldsymbol{\psi}. \end{aligned} \tag{VI.103}$$

Now, because γ is close to zero ($O(\sqrt{\varepsilon})$) and $\mathbf{f}_u$ is invertible on the space orthogonal to $\{\zeta_1^*, \zeta_2^*\}$, $(\text{VI.103})_3$ leads to $\mathbf{W} = O(\varepsilon)\cdot\boldsymbol{\psi}$. Replacing $\mathbf{W}$ in $(\text{VI.103})_1$ and $(\text{VI.103})_2$ we may justify* that γ is an eigenvalue of a 2×2 matrix:

$$\begin{bmatrix} \varepsilon a_1 + O(\varepsilon^2) & 1 + \varepsilon b_1 + O(\varepsilon^2) \\ -\varepsilon\mu_1 C_0' + O(\varepsilon^2) & \varepsilon b_2 + O(\varepsilon^2) \end{bmatrix}. \tag{VI.104}$$

So we have

$$\gamma_\pm = \pm\sqrt{-\varepsilon\mu_1 C_0'} + \tfrac{1}{2}\varepsilon(a_1 + b_2) + O(|\varepsilon|^{3/2}) \tag{VI.105}$$

and we derive exactly the same stability results for the Riesz-index-two case in $\mathbb{R}^2$ in projection as we derived in §V.6 for the same case in $\mathbb{R}^2$. The stability results given in §V.6 and Figure V.1 fully describe the implications of (VI.105).

VI.12 The Method of Projection at a Double Semi-Simple Eigenvalue

We want to show that bifurcation of $\mathbf{u} = 0$ at a double semi-simple eigenvalue associated with the higher-dimensional problem (VI.45) is the same as in $\mathbb{R}^2$ (see §§V.7–9) except for a small passive part which is orthogonal to the projected part and is of higher order in the amplitude.

The stability of the steady solution $\mathbf{u} = 0$ of (VI.45) is governed by the spectral problem (VI.46) and the associated adjoint spectral problem (VI.49). We assume that $\sigma(0) = 0$ is a double semi-simple eigenvalue of $\mathbf{f}_u(0|\cdot) \overset{\text{def}}{=} \mathbf{f}_u(\cdot)$. Unlike the matrix $\mathbf{A}_0$ in $\mathbb{R}^2$, the operator $\mathbf{f}_u(\cdot)$ does not vanish identically. For example, if $\mathbf{f}_u(\cdot) = \mathbf{A}_0(\cdot)$ in $\mathbb{R}^3$ we have

$$\mathbf{A}_0 \cdot \zeta = \sigma\zeta$$

where

$$\mathbf{A}_0 = \begin{bmatrix} 0 & 0 & 0 \\ 0 & 0 & 0 \\ 0 & 0 & -\lambda \end{bmatrix} \neq \mathbf{0}$$

and $\sigma = 0$ is a double semi-simple eigenvalue of $\mathbf{A}_0$. Two independent vectors ζ_1 and ζ_2 which are annihilated by $\mathbf{A}_0$,

$$\mathbf{A}_0 \cdot \zeta_1 = 0, \qquad \mathbf{A}_0 \cdot \zeta_2 = 0,$$

* It is possible to justify this using the theory developed in T. Kato, *Perturbation Theory for Linear Operators* (New York–Heidelberg–Berlin: Springer-Verlag, 1966).

are said to lie in the null space of $\mathbf{A}_0$ and we may choose these vectors so that

$$\zeta_1 = \begin{bmatrix} 1 \\ 0 \\ 0 \end{bmatrix}, \qquad \zeta_2 = \begin{bmatrix} 0 \\ 1 \\ 0 \end{bmatrix}.$$

The third eigenvector

$$\zeta_3 = \begin{bmatrix} 0 \\ 0 \\ 1 \end{bmatrix}$$

is determined uniquely by $\mathbf{A}_0 \cdot \zeta_3 = -\lambda\zeta_3$ and a normalizing condition.

In the general problem, $\mathbf{f}_u^*(\cdot)$ has a two-dimensional null space (with eigenvector ζ_1^* and ζ_2^* belonging to $\sigma = 0$) when $\mathbf{f}_u(\cdot)$ does and the equation

$$\mathbf{f}_u(\boldsymbol{\phi}) = \boldsymbol{\psi} \tag{VI.106}$$

is solvable for $\boldsymbol{\phi}$ in H if and only if $\langle \boldsymbol{\psi}, \zeta_1^* \rangle = \langle \boldsymbol{\psi}, \zeta_2^* \rangle = 0$.

Now we consider the problem of bifurcation and show that it reduces to the one already considered in $\mathbb{R}^2$. We first look for a solution in the form (V.3) of §V.8:

$$\mathbf{u}(\mu) = \mathbf{u}_1\mu + \frac{1}{2}\mathbf{u}_2\mu^2 + \frac{1}{3!}\mathbf{u}_3\mu^3 + O(\mu^4). \tag{VI.107}$$

(Similar methods can be used to obtain solutions in the form (V.2) of §V.7.) Substituting (VI.107) into (VI.45) we find, after identification, that

$$\frac{d\mathbf{u}_1}{dt} = f_u(\mathbf{u}_1).$$

It follows that $\langle \mathbf{u}_1, \zeta_1^* \rangle$ and $\langle \mathbf{u}_1, \zeta_2^* \rangle$ are independent of time. Let us seek steady bifurcating solutions of (VI.45). These satisfy

$$\begin{aligned} &\mathbf{f}_u(\mathbf{u}_1) = 0 \\ &2\mathbf{f}_{\mu u}(\mathbf{u}_1) + \mathbf{f}_{uu}(\mathbf{u}_1|\mathbf{u}_1) + \mathbf{f}_u(\mathbf{u}_2) = 0 \\ &\mathbf{f}_{uuu}(\mathbf{u}_1|\mathbf{u}_1|\mathbf{u}_1) + 3\mathbf{f}_{uu\mu}(\mathbf{u}_1, \mathbf{u}_1) + 3\mathbf{f}_{u\mu\mu}(\mathbf{u}_1) \\ &\qquad + 3\mathbf{f}_{u\mu}(\mathbf{u}_2) + 3\mathbf{f}_{uu}(\mathbf{u}_1|\mathbf{u}_2) + \mathbf{f}_u(\mathbf{u}_3) = 0, \end{aligned} \tag{VI.108}$$

where, in a simplified notation, we have evaluated the derivatives of $\mathbf{f}$ at $(\mu, \mathbf{u}) = (0, 0)$. Equations $(\text{VI.108})_2$ and $(\text{VI.108})_3$ are solvable if and only if, for $l = 1$ and $l = 2$,

$$2\langle \mathbf{f}_{\mu u}(\mathbf{u}_1), \zeta_l^* \rangle + \langle \mathbf{f}_{uu}(\mathbf{u}_1|\mathbf{u}_1), \zeta_l^* \rangle = 0 \tag{VI.109}$$

and

$$\langle \mathbf{M}(\mathbf{u}_1), \zeta_l^* \rangle + 3\langle \mathbf{f}_{u\mu}(\mathbf{u}_2), \zeta_l^* \rangle + 3\langle \mathbf{f}_{uu}(\mathbf{u}_1|\mathbf{u}_2), \zeta_l^* \rangle = 0 \tag{VI.110}$$

where

$$\mathbf{M}(\mathbf{u}_1) = \mathbf{f}_{uuu}(\mathbf{u}_1 | \mathbf{u}_1 | \mathbf{u}_1) + 3\mathbf{f}_{uu\mu}(\mathbf{u}_1 | \mathbf{u}_1) + 3\mathbf{f}_{u\mu\mu}(\mathbf{u}_1).$$

The solution $\mathbf{u}(\mu)$ may always be decomposed into a part on the two-dimensional null space of $\mathbf{f}_u(\cdot)$ and a part which is orthogonal to ζ_1^* and ζ_2^*:

$$\begin{aligned} \mathbf{u}(\mu) &= \mu\{\chi(\mu)\zeta_1 + \theta(\mu)\zeta_2\} + \mu\mathbf{W}(\mu) \\ \langle \mathbf{u}(\mu), \zeta_1^* \rangle &= \mu\chi(\mu) \\ \langle \mathbf{u}(\mu), \zeta_2^* \rangle &= \mu\theta(\mu) \\ \langle \mathbf{W}, \zeta_1^* \rangle &= \langle \mathbf{W}, \zeta_2^* \rangle = 0. \end{aligned} \tag{VI.111}$$

Since all solutions of $\mathbf{f}_u(u_1) = 0$ can be composed of the two independent ones, we have

$$\mathbf{u}_1 = \chi_0\zeta_1 + \theta_0\zeta_2 \tag{VI.112}$$

and $\mathbf{W}_0 = 0$. Combining (VI.112) and (VI.109) we find that

$$\begin{aligned} f_1(\chi_0, \theta_0) &= a_1\chi_0 + b_1\theta_0 + \alpha_1\chi_0^2 + 2\beta_1\chi_0\theta_0 + \gamma_1\theta_0^2 = 0 \\ f_2(\chi_0, \theta_0) &= a_2\chi_0 + b_2\theta_0 + \alpha_2\chi_0^2 + 2\beta_2\chi_0\theta_0 + \gamma_2\theta_0^2 = 0, \end{aligned} \tag{VI.113}$$

where, for $j = 1, 2$,

$$\begin{aligned} a_j &= \langle \mathbf{f}_{u\mu}(\zeta_1), \zeta_j^* \rangle, \\ b_j &= \langle \mathbf{f}_{u\mu}(\zeta_2), \zeta_j^* \rangle, \\ \alpha_j &= \frac{\langle \mathbf{f}_{uu}(\zeta_1 | \zeta_1), \zeta_j^* \rangle}{2}, \\ \beta_j &= \frac{\langle \mathbf{f}_{uu}(\zeta_1 | \zeta_2), \zeta_j^* \rangle}{2}, \\ \gamma_j &= \frac{\langle \mathbf{f}_{uu}(\zeta_2 | \zeta_2), \zeta_j^* \rangle}{2}. \end{aligned}$$

Equations (VI.113) are the intersecting conics (V.29) which we derived in $\mathbb{R}^2$; $(\chi_0, \theta_0) = (0, 0)$ is always a point of intersection. Besides this point of intersection there may be as many as three other points. If, as in $\mathbb{R}^2$, the condition

$$\det \mathscr{J}_0 \neq 0 \quad \text{where} \quad \mathscr{J}_0 = \begin{bmatrix} \dfrac{\partial f_1}{\partial \chi_0} & \dfrac{\partial f_1}{\partial \theta_1} \\[2ex] \dfrac{\partial f_2}{\partial \chi_0} & \dfrac{\partial f_2}{\partial \theta_0} \end{bmatrix} \tag{VI.114}$$

is satisfied at a point of intersection, then a solution bifurcates there. In general, we may, besides $\mathbf{u}(\mu) = 0$, have as few as one other solution or as many as three

$$\mathbf{u}^{[k]}(\mu) = \mathbf{u}_1^{[k]}\mu + \frac{1}{2}\mathbf{u}_2^{[k]}\mu^2 + \frac{1}{3!}\mathbf{u}_3^{[k]}\mu^3 + O(\mu^4). \quad \text{(VI.115)}$$

To show that (VI.113) and (VI.114) are sufficient for the existence of bifurcation we first note that

$$\mathbf{u}_2 = \chi_1\zeta_1 + \theta_1\zeta_2 + \mathbf{W}_1, \quad \text{(VI.116)}$$

where $\mathbf{f}_u(\mathbf{W}_1) = 2\mathbf{f}_{\mu u}(\mathbf{u}_1) + \mathbf{f}_{uu}(\mathbf{u}_1|\mathbf{u}_1)$, which is made solvable by (VI.113). There is a $\mathbf{W}_1^{[k]}$ for each solution $\mathbf{u}_1^{[k]}$. Combining (VI.116) and (VI.110) we find that for $l = 1$ and $l = 2$

$$\begin{aligned} P_l + \chi_1\{\langle \mathbf{f}_{u\mu}(\zeta_1), \zeta_l^*\rangle + \langle \mathbf{f}_{uu}(\mathbf{u}_1|\zeta_1), \zeta_l^*\rangle\} \\ + \theta_1\{\langle \mathbf{f}_{u\mu}(\zeta_2), \zeta_l^*\rangle + \langle \mathbf{f}_{uu}(\mathbf{u}_1|\zeta_2), \zeta_l^*\rangle\} = 0, \end{aligned} \quad \text{(VI.117)}$$

where

$$P_l = \tfrac{1}{3}\langle \mathbf{M}(\mathbf{u}_1), \zeta_l^*\rangle + \langle \mathbf{f}_{u\mu}(\mathbf{W}_1), \zeta_l^*\rangle + \langle \mathbf{f}_{uu}(\mathbf{u}_1|\mathbf{W}_1), \zeta_l^*\rangle,$$

is known from calculation at lower orders. We want to find χ_1 and θ_1. Inserting $\mathbf{u}_1 = \chi_0\zeta_1 + \theta_0\zeta_2$ into $\mathbf{f}_{uu}$ we find, using (VI.113), that (VI.117) has the form

$$\mathbf{P} + \mathscr{J}_0 \cdot \boldsymbol{\chi} = 0 \quad \text{where} \quad \boldsymbol{\chi} = \begin{bmatrix} \chi_1 \\ \theta_1 \end{bmatrix}, \quad \mathbf{P} = \begin{bmatrix} P_1 \\ P_2 \end{bmatrix}. \quad \text{(VI.118)}$$

These two linear equations have unique solutions if and only if

$$\det \mathscr{J}_0 \neq 0.$$

Exactly the same condition of solvability arises at higher orders.

Now we show that the stability of the bifurcating solutions $\mathbf{u}^{[k]}(\mu)$, $k = 1, 2,$ or 3, and the solution $\mathbf{u} = 0$ are determined by the eigenvalues of $\mathscr{J}_0$. Let $\mathbf{u}(\mu)$ be any one of the eventual four candidates for solutions and let $e^{\gamma t}\zeta$ be a small disturbance of $\mathbf{u}$ satisfying the spectral problem

$$\gamma\zeta = \mathbf{f}_u(\mu, \mathbf{u}(\mu)|\zeta). \quad \text{(VI.119)}$$

When $\mu = 0$, $\mathbf{u} = 0$, $\gamma(0) = 0$, and $\mathbf{f}_u(0, 0|\zeta(0)) = \mathbf{f}_u(\zeta(0)) = 0$ has two independent solutions ζ_1 and ζ_2. Hence $\zeta(0) = A\zeta_1 + B\zeta_2$, where A and B are to be determined. Differentiating (VI.119) once with respect to μ at $\mu = 0$ and replacing $\zeta(0)$ in the equation which results, we find, after applying the solvability conditions, that

$$\begin{aligned} -\gamma_\mu A + A\{\langle \mathbf{f}_{u\mu}(\zeta_1), \zeta_1^*\rangle + \langle \mathbf{f}_{uu}(\mathbf{u}_1|\zeta_1), \zeta_1^*\rangle\} \\ + B\{\langle \mathbf{f}_{u\mu}(\zeta_2), \zeta_1^*\rangle + \langle \mathbf{f}_{uu}(\mathbf{u}_1|\zeta_2), \zeta_1^*\rangle\} = 0 \end{aligned}$$

and

$$\begin{aligned} -\gamma_\mu B + A\{\langle \mathbf{f}_{u\mu}(\zeta_1), \zeta_2^*\rangle + \langle \mathbf{f}_{uu}(\mathbf{u}_1|\zeta_1), \zeta_2^*\rangle\} \\ + B\{\langle \mathbf{f}_{u\mu}(\zeta_2), \zeta_2^*\rangle + \langle \mathbf{f}_{uu}(\mathbf{u}_1|\zeta_2), \zeta_2^*\rangle\} = 0 \end{aligned}$$

Inserting $\mathbf{u}_1 = \chi_0 \zeta_1 + \theta_0 \zeta_2$ into $\mathbf{f}_{u\mu}(\mathbf{u}_1|\cdot)$ we find that

$$\gamma_\mu \mathbf{A} = \mathcal{J}_0 \cdot \mathbf{A},$$

where

$$\mathbf{A} = \begin{bmatrix} A \\ B \end{bmatrix}.$$

Hence the γ_μ are eigenvalues of $\mathcal{J}_0$, as in $\mathbb{R}^2$. A rigorous justification of this type of computation may be found in Kato (1966), cited in §VI.11.

Appendix VI.1 Examples of the Method of Projection

In this appendix we shall apply the methods of projection which were developed in this chapter to specific problems which arise as integro-differential equations or as partial differential equations. The examples could be treated as exercises for students. They are meant to teach interested readers, by example, how to place their specific problems in the frame of our general theory.

We have written the evolution problem reduced to local form as

$$\frac{d\mathbf{u}}{dt} = \mathbf{f}(\mu, \mathbf{u}) = \mathbf{f}_u(\mu|\mathbf{u}) + \frac{1}{2}\mathbf{f}_{uu}(\mu|\mathbf{u}|\mathbf{u}) + \frac{1}{3!}\mathbf{f}_{uuu}(\mu|\mathbf{u}|\mathbf{u}|\mathbf{u}) + \cdots.$$

In some of the problems studied in this appendix this series for $\mathbf{f}$ is terminating and derivatives of $\mathbf{f}$ higher than the first are independent of μ:

$$\frac{1}{2}\mathbf{f}_{uu}(\mu|\mathbf{u}|\mathbf{u}) = \mathbf{B}(\mathbf{u}, \mathbf{u})$$

$$\frac{1}{3!}\mathbf{f}_{uuu}(\mu|\mathbf{u}|\mathbf{u}|\mathbf{u}) = \mathbf{C}(\mathbf{u}, \mathbf{u}, \mathbf{u}).$$

The reader will note that some of the problems considered in the examples and exercises are framed in terms of Hilbert spaces, as in the general theory. The Hilbert space setting is convenient for purposes of rigorous demonstration and for the exposition of the principles of bifurcation in terms resembling those which arise in $\mathbb{R}^n$. It will be obvious that the *computation* of bifurcation does not require Hilbert spaces. What is really required for the formal computation of bifurcation is the Fredholm alternative. In all cases the alternative shows the way to guarantee the solvability of the equations arising in the construction of the bifurcating solution and defines the low-dimensional projections which are relevant in this construction.

EXAMPLE VI.1 (An integro-differential equation).* We consider the problem of stability and bifurcation of the solution $U = 0$ of the following integro-differential equation

$$\frac{\partial U(t, x)}{\partial t} + (1 - \mu)U(t, x) - \frac{2}{\pi}\int_0^\pi \{\sin x \sin y + b \sin 2x \sin 2y\} \times \{U(t, y) + U^3(t, y)\}\, dy = 0, \tag{VI.120}$$

where U is a real function defined for $t \geq 0$, $0 \leq x \leq \pi$, continuous and continuously differentiable once with respect to t. The parameter b is fixed and satisfies $0 < b < 1$; μ is the bifurcation parameter.

$U(t, x)$ is a continuous function of x on the interval $0 \leq x \leq \pi$ for each fixed $t \geq 0$. We distinguish the function $U(t, \cdot)$ from the value of the function $U(t, x)$ at a certain x. So we define

$$u(t) \stackrel{\text{def}}{=} [u(t)](\cdot) \stackrel{\text{def}}{=} U(t, \cdot)$$

where $U(t, \cdot) \in C[0, \pi]$, the space of continuous functions of $x \in [0, \pi]$. In this notation t plays the role of a parameter and the value of the function at x is $[u(t)](x) = U(t, x)$. With this understanding we may rewrite (VI.120) as

$$\frac{du}{dt} = f_u(\mu | u) + C(u, u, u), \quad u(t) \in C[0, \pi], \tag{VI.121}$$

where $u(t) \in C[0, \pi]$ means that $[u(t)](x)$ is a continuous function on $[0, \pi]$,

$$[f_u(\mu | u(t))](x) = -(1 - \mu)U(t, x) + \frac{2}{\pi}\int_0^\pi [\sin x \sin y + b \sin 2x \sin 2y] \times U(t, y)\, dy$$

and

$$[C(u(t), u(t), u(t))](x) = \frac{2}{\pi}\int_0^\pi [\sin x \sin y + b \sin 2x \sin 2y] U^3(t, y)\, dy.$$

We want to compute the steady solutions of (VI.121) and to test their stability. We first note that $u = 0$ is a solution of (VI.121). To study stability of $u = 0$ we consider the spectral problem

$$\sigma v = f_u(\mu | v). \tag{VI.122}$$

The values $\sigma(\mu)$ for which there are solutions $v \neq 0$ are said to be in the spectrum of $f_u(\mu | \cdot)$. More exactly, the spectrum may be defined as the set of special values for which the problem

$$\sigma v - f_u(\mu | v) = g \tag{VI.123}$$

* For more details, see Pimbley (1969), cited at the end of Chapter I.

has no unique solution (see §VI.8 on Laplace transforms for a fuller explanation). In the present case we have

$$\sigma V(x) + (1 - \mu)V(x) - \frac{2}{\pi}\int_0^\pi (\sin x \sin y + b \sin 2x \sin 2y)V(y)\, dy = F(x). \tag{VI.124}$$

Hence $F(\cdot)$ of (VI.124) is the same as g of (VI.123).

To solve (VI.124) we decompose F and V as

$$\begin{aligned} F(x) &= F_1 \sin x + F_2 \sin 2x + F_3(x) \\ V(x) &= V_1 \sin x + V_2 \sin 2x + V_3(x), \end{aligned} \tag{VI.125}$$

where

$$\int_0^\pi V_3(x) \sin x\, dx = \int_0^\pi V_3(x) \sin 2x\, dx = 0. \tag{VI.126}$$

Equation (VI.126) holds when we replace $V_3(x)$ with $F_3(x)$. Combining (VI.124) and (VI.125) we compute

$$\begin{aligned} (\sigma + 1 - \mu)V_3(x) &= F_3(x) \\ (\sigma - \mu)V_1 &= F_1 \\ (\sigma + 1 - \mu - b)V_2 &= F_2. \end{aligned} \tag{VI.127}$$

It is clear that the spectrum of $\mathbf{f}_u(\mu|\cdot)$ is entirely of eigenvalues

$$\begin{aligned} \sigma_0(\mu) &= \mu \\ \sigma_1(\mu) &= \mu + b - 1 \\ \sigma_2(\mu) &= \mu - 1. \end{aligned} \tag{VI.128}$$

We may obtain (VI.128) more directly by solving (VI.122), which in the present problem is given by (VI.124) with $F(x) = 0$.

Since $0 < b < 1$ the largest of the values (VI.128) is $\sigma_0(\mu)$. It follows that the solution $U = 0$ of (VI.120) is stable for $\mu < 0$ and is unstable for $\mu > 0$. Moreover, the Fredholm alternative applies. Consider the problem

$$f_u(0|v) = g.$$

We know that $\sigma_0(0) = 0$ is an eigenvalue of $f_u(0|\cdot)$ and the decomposition (VI.127) reduces to

$$\begin{aligned} -V_3(x) &= F_3(x) \\ 0 &= F_1 \\ (b - 1)V_2 &= F_2. \end{aligned} \tag{VI.129}$$

So if $b \neq 1$, the compatibility condition for the solvability of (VI.129) is $F_1 = 0$; that is,

$$\int_0^{\pi} F(x) \sin x \, dx = 0. \tag{VI.130$_1$}$$

If $b = 1$ then $\sigma_0(0) = \sigma_1(0) = 0$ is a double eigenvalue of $f_u(0|\cdot)$ and, in addition to (VI.130), it is necessary that

$$\int_0^{\pi} f(x) \sin 2x \, dx = 0. \tag{VI.130$_2$}$$

Though the problem (VI.120) is not set in the context of Hilbert spaces which we have used to frame the general theory of this chapter, the Fredholm alternative works well and we can compute the bifurcated steady solutions using expansions in powers of the amplitude ε:

$$U(x, \varepsilon) = \varepsilon U_1(x) + \varepsilon^2 U_2(x) + \varepsilon^2 U_3(x) + O(\varepsilon^4),$$
$$\mu = \varepsilon\mu_1 + \varepsilon^2\mu_2 + O(\varepsilon^3).$$

where

$$U_1(x) = \frac{2}{\pi}\int_0^{\pi} \{\sin x \sin y + b \sin 2x \sin 2y\} U_1(y)\, dy,$$

$$-\mu_1 U_1(x) + U_2(x) = \frac{2}{\pi}\int_0^{\pi} \{\sin x \sin y + b \sin 2x \sin 2y\} U_2(y)\, dy,$$

$$-\mu_2 U_1(x) - \mu_1 U_2(x) + U_3(x) = \frac{2}{\pi}$$
$$\times \int_0^{\pi} \{\sin x \sin y + b \sin 2x \sin 2y\}(U_3(y) + U_1^3(y))\, dy.$$

We find that

$$\begin{aligned} U(x, \varepsilon) &= \varepsilon \sin x + O(\varepsilon^3) \\ \mu &= -\tfrac{3}{4}\varepsilon^2 + O(\varepsilon^3). \end{aligned} \tag{VI.131}$$

The bifurcating solution is one-sided and subcritical; hence it is unstable.

The bifurcation problem associated with (VI.120) has special features which make possible the exact computation of all the bifurcated branches. We first note that all of the steady solutions of (VI.120) have $U(x) = 0$ at $x = 0$ and $x = \pi$. We may therefore expand $U(x)$ into a Fourier series of sines and we find that all the Fourier coefficients except the first two must vanish. Hence

$$U(x) = u_1 \sin x + u_2 \sin 2x$$

and (VI.120) reduces to

$$\begin{aligned} \mu u_1 + \tfrac{3}{4}u_1^3 + \tfrac{3}{2}u_1 u_2^2 &= 0 \\ (\mu + b - 1)u_2 + \tfrac{3}{2}b u_1^2 u_2 + \tfrac{3}{4}b u_2^3 &= 0. \end{aligned} \tag{VI.132}$$

The solutions of (VI.132) are:

$$u_1 = u_2 = 0, \quad \text{(the null solution)}$$

$$u_1 = \pm \frac{2}{\sqrt{3}}\sqrt{-\mu}, \qquad u_2 = 0, \quad \text{(the solution (VI.131))}$$

$$u_1 = 0, \qquad u_2 = \pm(2/\sqrt{3})\sqrt{-(\mu + b - 1)/b} \tag{VI.133}$$

$$u_1^2 = \frac{4}{9}\left[\frac{2}{b}(1-\mu) + \mu - 2\right], \qquad u_2^2 = \frac{4}{9}\left[\frac{\mu - 1}{b} + 1 - 2\mu\right].$$

EXERCISE

VI.1. Show that a secondary bifurcating solution branches from $(\mathrm{VI.133})_1$ provided that $b > \frac{1}{2}$ when $\mu = (b-1)/(2b-1)$.

EXAMPLE VI.2 (Bifurcation of solutions of partial differential equations). Consider the following partial differential equation:

$$\frac{\partial U}{\partial t} - \frac{\partial^2 U}{\partial x^2} + \tfrac{1}{4}(\mu - 1)U - \mu x\frac{\partial U}{\partial x} + U^2 - 4U\frac{\partial U}{\partial x} + 4U\left(\frac{\partial U}{\partial x}\right)^2 = 0 \tag{VI.134$_1$}$$

where U is a real-valued function defined for $t \geq 0$, $0 \leq x \leq \pi$ satisfying the boundary conditions

$$U(t, 0) = \frac{\partial U}{\partial x}(t, \pi) = 0. \tag{VI.134$_2$}$$

In this example we choose $H = L^2(0, \pi)$, the space of square integrable functions on $(0, \pi)$, which is a Hilbert space with the scalar product

$$\langle \mathbf{u}(t), \mathbf{v}(t)\rangle = \int_0^\pi U(t, x)\overline{V}(t, x)\,dx$$

where $\mathbf{u}(t) = U(t, \cdot)$ is a vector with different components $\mathbf{u}(t)(x)$ for each x. In what follows we suppress the variable t.

Equations (VI.134) have the following form in H:

$$\frac{d\mathbf{u}}{dt} = \mathbf{f}_u(\mu|\mathbf{u}) + \mathbf{B}(\mathbf{u}, \mathbf{u}) + \mathbf{C}(\mathbf{u}, \mathbf{u}, \mathbf{u}) \tag{VI.135}$$

where the operators $\mathbf{f}_u(\mu|\cdot)$, $\mathbf{B}$, and $\mathbf{C}$ are defined for $\mathbf{u} = U(\cdot)$ in a subspace of $\mathbf{H}$ consisting of $U(\cdot)$ satisfying the boundary conditions $U(0) = \partial U(\pi)/\partial x = 0$ and such that U, $\partial U/\partial x$, and $\partial^2 U/\partial x^2$ are square integrable on $(0, \pi)$; for example, $\langle \partial U/\partial x, \partial U/\partial x\rangle < \infty$. To further specify the operators defining (VI.135), we note that

$$\mathbf{f}_u(\mu|\cdot) = \mathbf{f}_u(0|\cdot) + \mu\mathbf{f}_{u\mu}(0|\cdot)$$

$$[\mathbf{f}_u(0|\mathbf{u})](x) = \frac{\partial^2 U(x)}{\partial x^2} + \frac{1}{4}U(x)$$

$$[\mathbf{f}_{u\mu}(0|\mathbf{u})](x) = -\frac{1}{4}U(x) + x\frac{\partial U(x)}{\partial x}$$

$$[B(\mathbf{u}, \mathbf{v})](x) = -U(x)V(x) + 2U(x)\frac{\partial V(x)}{\partial x} + 2V(x)\frac{\partial U(x)}{\partial x}$$

$$[C(\mathbf{u}, \mathbf{v}, \mathbf{w})](x) = -\frac{4}{3}\left\{U\frac{\partial V}{\partial x}\frac{\partial W}{\partial x} + W\frac{\partial U}{\partial x}\frac{\partial V}{\partial x} + V\frac{\partial W}{\partial x}\frac{\partial U}{\partial x}\right\}.$$

To study the stability of the null solution of (VI.134) we first note that the spectrum of $\mathbf{f}_u(\mu|\cdot)$ consists only of eigenvalues σ satisfying

$$\sigma U = \frac{\partial^2 U}{\partial x^2} + \frac{1}{4}U + \mu\left(-\frac{1}{4}U + x\frac{\partial U}{\partial x}\right),$$
$$U(0) = \frac{\partial U}{\partial x}(\pi) = 0 \qquad \text{(VI.136)}$$

for twice continuously differentiable functions U not identically zero. The analytic computation of eigenvalues from (VI.136) is very difficult except when $\mu = 0$; the eigenvalues of $\mathbf{A}_0$ (of (VI.136) with $\mu = 0$) are

$$\sigma(0) = \frac{1}{4} - \frac{(2k+1)^2}{4} \quad \text{for } k \in \mathbb{N}$$

and the associated eigenvectors are proportional to $\sin((2k+1)/2)x$. Zero is an eigenvalue of $\mathbf{f}_u(0|\cdot)$ for $k = 0$; the other eigenvalues are negative.

The critical eigenvalue of $\mathbf{f}_u(\mu|\cdot)$ for μ near to zero can be obtained by perturbations. We find that

$$\sigma(\mu) = \xi'(0)\mu + O(\mu^2)$$

$$\xi'(0) = \langle \mathbf{f}_{u\mu}(0|\boldsymbol{\zeta}), \boldsymbol{\zeta}^*\rangle$$

where $\boldsymbol{\zeta}(x)$ and $\boldsymbol{\zeta}^*(x)$ are proportional because $\mathbf{f}_u(0|\cdot) = \mathbf{f}_u^*(0|\cdot) \overset{\text{def}}{=} [\mathbf{f}_u(0|\cdot)]^*$. Then, with

$$\boldsymbol{\zeta}(x) = \sin\frac{x}{2}$$

and

$$\boldsymbol{\zeta}^*(x) = \frac{2}{\pi}\sin\frac{x}{2}$$

we have

$$\langle \boldsymbol{\zeta}, \boldsymbol{\zeta}^* \rangle = 1$$

and

$$\xi'(0) = \tfrac{1}{4} > 0,$$

so that the null solution is stable for $\mu < 0$ and unstable for $\mu > 0$.

Following exactly the methods used in §VI.2 to construct steady bifurcating solutions as a power series in the amplitude we write

$$\begin{bmatrix} \mathbf{u}(\varepsilon) \\ \mu(\varepsilon) \end{bmatrix} = \sum_{n \geq 1} \frac{\varepsilon}{n!} \begin{bmatrix} \mathbf{u}_n \\ \mu_n \end{bmatrix}. \tag{VI.137}$$

Combining (VI.137) and (VI.135) we find, after identifying independent powers of ε, the equations (VI.9) for the Taylor coefficients. In the present example these equations may be written as:

$$\begin{aligned} &\mathbf{f}_u(0 \mid \mathbf{u}_1) = 0 \\ &\mathbf{f}_u(0 \mid \mathbf{u}_2) + 2\mu_1 \mathbf{f}_{u\mu}(0 \mid \mathbf{u}_1) + 2\mathbf{B}(\mathbf{u}_1, \mathbf{u}_1) = 0 \\ &\mathbf{f}_u(0 \mid \mathbf{u}_3) + 3\mu_1 \mathbf{f}_{u\mu}(0 \mid \mathbf{u}_2) + 6\mathbf{B}(\mathbf{u}_1, \mathbf{u}_2) \\ &\qquad + 3\mu_2 \mathbf{f}_{u\mu}(0 \mid \mathbf{u}_1) + 6\mathbf{C}(\mathbf{u}_1, \mathbf{u}_1; \mathbf{u}_1) = 0 \\ &\mathbf{f}_u(0 \mid \mathbf{u}_n) + n\mu_{n-1} \mathbf{f}_{u\mu}(0 \mid \mathbf{u}_1) + \text{terms of lower order} = 0. \end{aligned} \tag{VI.138}$$

Equation (VI.138) implies that $\mathbf{u}_1 = \boldsymbol{\zeta}$ (see (VI.7) and (VI.11)). Hence

$$U_1(x) = \sin \frac{x}{2}.$$

We then compute the solvability condition for $(\text{VI.138})_2$:

$$\begin{aligned} \mu_1 \langle \mathbf{f}_{u\mu}(0 \mid \mathbf{u}_1), \boldsymbol{\zeta}^* \rangle &= -\langle \mathbf{B}(\mathbf{u}_1, \mathbf{u}_1), \boldsymbol{\zeta}^* \rangle \\ &= \frac{2}{\pi} \int_0^\pi \left\{ \sin^2 \frac{x}{2} - 2 \sin \frac{x}{2} \cos \frac{x}{2} \right\} \sin \frac{x}{2}\, dx = 0. \end{aligned}$$

Since $\langle \mathbf{f}_{u\mu}(0 \mid \boldsymbol{\zeta}), \boldsymbol{\zeta}^* \rangle = \xi'(0) = \frac{1}{4}$ we have $\mu_1 = 0$ and

$$\mathbf{f}_u(0 \mid \mathbf{u}_2) = -2\mathbf{B}(\boldsymbol{\zeta}, \boldsymbol{\zeta}), \ \langle \mathbf{u}_2, \boldsymbol{\zeta}^* \rangle = 0 \tag{VI.139}$$

Eq. (VI.139) is equivalent to

$$\begin{aligned} U_2'' + \tfrac{1}{4} U_2 &= 1 - \cos x - 2 \sin x \\ U_2(0) &= U_2'(\pi) \\ 0 &= \int_0^\pi U_2(x) \sin \frac{x}{2}\, dx. \end{aligned} \tag{VI.140}$$

The unique solution of (VI.140) is (exercise for the reader)

$$U_2(x) = 4 - \frac{32}{3\pi}\sin\frac{x}{2} - \frac{16}{3}\cos\frac{x}{2} + \frac{4}{3}\cos x + \frac{8}{3}\sin x. \qquad \text{(VI.141)}$$

In functional notation we write (VI.141) as

$$\mathbf{u}_2 = -2\mathbf{f}_u^{-1}(0 \mid \mathbf{B}(\zeta, \zeta))$$

with the understanding that $\mathbf{u}_2$ is in the subspace orthogonal to ζ^*.

To solve $(\text{VI.138})_3$ we must first compute

$$6\langle \mathbf{B}(\zeta, \mathbf{u}_2), \zeta^* \rangle = 20 - \frac{176}{3\pi}$$

$$6\langle \mathbf{C}(\zeta, \zeta, \zeta), \zeta^* \rangle = -\frac{3}{2}.$$

Hence, the solvability condition for $(\text{VI.138})_3$ gives

$$3\mu_2 \xi'(0) - \frac{3}{2} + 20 - \frac{176}{3\pi} = 0; \qquad \text{(VI.142)}$$

that is, $\mu_2 \cong 0.232$.

In sum, the bifurcated solution may be represented as

$$U(x) = \varepsilon \sin\frac{x}{2} + \frac{\varepsilon^2}{2} U_2(x) + O(\varepsilon^3) \qquad (\text{VI.143})_1$$

where U_2 is defined by (VI.141) and

$$\mu \cong 0.116\varepsilon^2 + O(\varepsilon^3). \qquad (\text{VI.143})_2$$

The bifurcated solution is supercritical; when $|\varepsilon|$ is small it exists only for $\mu > 0$, and has two branches: $\varepsilon > 0$ and $\varepsilon < 0$. Both branches are stable.

Example VI.3 (In which the reader may check whether she (or he) is able to go further!). Consider the following partial differential system:

$$\begin{aligned} &\frac{\partial U_1}{\partial t} - \frac{1}{\lambda}\frac{\partial^2 U_1}{\partial x^2} + \frac{\partial U_2}{\partial x} - \lambda U_1 - U_1 U_2 = 0 \\ &\frac{\partial U_2}{\partial t} - \lambda^2 \frac{\partial^2 U_2}{\partial x^2} - \lambda^4 U_2 - U_2\left(\frac{\partial U_1}{\partial x} - \frac{\lambda x}{2}\frac{\partial U_2}{\partial x}\right) = 0 \end{aligned} \qquad (\text{VI.144})_1$$

where $0 \le x \le 1$, $t \ge 0$; satisfying the boundary conditions

$$U_i(0, t) = U_i(1, t) = 0, \qquad i = 1, 2. \qquad (\text{VI.144})_2$$

(For the evolution problem some initial data would be prescribed: $U_i(x, 0)$, $i = 1, 2$.)

(1) The space H here will be $\{L^2(0, 1)\}^2 = \{(u_1, u_2)\colon u_i \in L^2(0, 1), i = 1, 2\}$ with the scalar product

$$\langle (u_1, u_2), (v_1, v_2)\rangle = \int_0^1 (u_1\bar{v}_1 + u_2\bar{v}_2)\,dx.$$

Show that the eigenvalues of the linearized operator, which arises in the study of the stability of the null solution of $(\text{VI}.144)_1$, are in the form $-\lambda^2k^2\pi^2 + \lambda^4$ and $\lambda - k^2\pi^2/\lambda$ for $k = 1, 2, \ldots$. So, if $0 < \lambda < \pi$ all the eigenvalues are real and negative, while if $\lambda > \pi$ some eigenvalues are positive. Hence the null solution is stable if $\lambda < \pi$, unstable if $\lambda > \pi$. When $\mu \overset{\text{def}}{=} \lambda - \pi = 0$, the eigenvalue zero is double, of index 1; the eigenvectors may be chosen as:

$$\zeta_1(x) = \begin{bmatrix} \sin \pi x \\ 0 \end{bmatrix}, \qquad \zeta_2(x) = \begin{bmatrix} x \sin \pi x \\ 2/\pi \sin \pi x \end{bmatrix}.$$

(2) Show that the adjoint $\mathbf{f}_u^*(0|\cdot)$ satisfies

$$\mathbf{f}_u^*(0|\mathbf{u}) = \begin{bmatrix} 1/\pi U_1'' + \pi U_1 \\ \pi^2 U_2'' + U_1' + \pi^4 U_2 \end{bmatrix}, \qquad \text{where } \mathbf{u} = \begin{bmatrix} U_1 \\ U_2 \end{bmatrix}, \qquad U_i' \equiv \frac{\partial U_i}{\partial x},$$

and U_i, $i = 1, 2$ satisfy the boundary conditions $(\text{VI}.144)_2$. Compute eigenvectors of $\mathbf{f}_u^*(0|\cdot)$, ζ_1^*, ζ_2^* such that $\langle \zeta_i, \zeta_j^*\rangle = \delta_{ij}$. Show that we may choose

$$\zeta_1^*(x) = \begin{bmatrix} 2 \sin \pi x \\ \left(-\dfrac{x}{\pi^2} + \dfrac{1}{2\pi^2} - \dfrac{\pi}{2}\right) \sin \pi x \end{bmatrix}, \qquad \zeta_2^*(x) = \begin{bmatrix} 0 \\ \pi \sin \pi x \end{bmatrix}.$$

(3) We now enter the frame of §VI.12 and look for a bifurcated steady solution of the form

$$\mathbf{u}(\mu) = \mu \mathbf{u}_1 + \tfrac{1}{2}\mu^2 \mathbf{u}_2 + O(\mu^3), \qquad \mu = \lambda - \pi, \qquad (\text{VI}.145)_1$$

where

$$\mathbf{u}_1 = \chi_0 \zeta_1 + \theta_0 \zeta_2. \qquad (\text{VI}.145)_2$$

Substituting (VI.145) into (VI.144), we obtain for χ_0, θ_0 a system of two nonlinear equations (see (VI.113)):

$$\begin{aligned} a_1\chi_0 + b_1\theta_0 + \alpha_1\chi_0^2 + 2\beta_1\chi_0\theta_0 + \gamma_1\theta_0^2 &= 0 \\ a_2\chi_0 + b_2\theta_0 + \alpha_2\chi_0^2 + 2\beta_2\chi_0\theta_0 + \gamma_2\theta_0^2 &= 0. \end{aligned} \qquad (\text{VI}.146)$$

Compute the coefficients in (VI.146) and show that $a_1 = 2$, $b_1 = 1 - \pi^2$, $a_2 = 0$, $b_2 = 2\pi^3$, $\alpha_1 = 0$, $\alpha_2 = 0$, $\beta_1 = -(16/3\pi^2) - (64/9\pi^4)$, $\beta_2 = 0$,

$\gamma_1 = (4/3\pi)(-1 + (2/\pi) + (1/\pi^3))$, $\gamma_2 = -8/3\pi$. Show that the two conics (VI.146) intersect at the origin and at one and only one other point. This other point corresponds to a bifurcated steady solution.

(4) Now you should try to find the steady, bifurcated solutions in the form

$$\mathbf{u}(\varepsilon) = \varepsilon\mathbf{u}_1 + \tfrac{1}{2}\varepsilon^2\mathbf{u}_2 + O(\varepsilon^3)$$
$$\mu = \varepsilon\mu_1 + \tfrac{1}{2}\varepsilon^2\mu_2 + O(\varepsilon^3),$$

where $\mathbf{u}_1 = \zeta_1 + \theta_0\zeta_2$. Show that all the solutions found in (3) can be represented as series in ε but that there is possibly another solution not found in (3) with $\theta_0 = 0$ and $\mu_1 = 0$. Determine whether this possibility is actually realized and send the answer to the authors. (If your work is correct you will receive a real letter of congratulation from one of us.)

EXAMPLE VI.4 (Imperfection theory). In this example we treat a problem using our analytic methods which Matkowsky and Reiss (1977; see notes to Chapter III of this book) studied using their method of matched asymptotic expansions.

The problem under consideration is given by

$$\frac{\partial U}{\partial t} = \frac{\partial^2 U}{\partial x^2} + \lambda[G(U) + \delta g(x, U)] \tag{VI.147}$$
$$U(t, x) = 0 \quad \text{at } x = 0, \pi,$$

where $0 \le x \le \pi$ and $t \ge 0$. We first study bifurcation of steady solutions with $\delta = 0$. Then we break bifurcation by perturbing it with $\delta \ne 0$.

To further specify (VI.147) we say that

$$G(U) = \sum_{n\ge 1} a_n U^n, \qquad a_1 > 0 \tag{VI.148}$$

is convergent when $|U(t, x)|$ is small enough and that

$$g(x, 0) \ne 0. \tag{VI.149}$$

The condition (VI.149) insures that $U = 0$ is not a solution of (VI.147) when $\delta \ne 0$.

In this example, as in Example VI.2, $H = L^2(0, \pi)$ with the same scalar product, and we may define the linearized "derivative" operator

$$\mathbf{f}_u(\mu|\mathbf{u}) = \frac{\partial^2 U}{\partial x^2} + \lambda a_1 U = \mathbf{f}_u(0|\mathbf{u}) + \mu\mathbf{f}_{u\mu}(0|\mathbf{u}),$$

where $\mu = \lambda - 1/a_1$, and whose domain is the space

$$\{\mathbf{u} \stackrel{\text{def}}{=} U(\cdot) \in H: U(0) = U(\pi) = 0,$$

and $U, \partial U/\partial x, \partial^2 U/\partial x^2$ are square integrable on $(0, \pi)\}$.* The eigenvalues of $\mathbf{f}_u(\mu|\cdot)$ are $\sigma_n = \lambda a_1 - n^2$, $n = 1, 2, \ldots$. Hence the null solution (VI.147) is stable for $\mu < 0$ and unstable for $\mu > 0$.

Turning next to the bifurcating solution (which exists only when $\delta = 0$) we find that

$$\mathbf{u} = \varepsilon\zeta + \frac{\varepsilon^2}{2}\mathbf{u}_2 + \frac{\varepsilon^3}{6}\mathbf{u}_3 + O(\varepsilon^4),$$

$$\mu = \varepsilon\mu_1 + \frac{\varepsilon^2}{2}\mu_2 + O(\varepsilon^3)$$

with

$$\mathbf{f}_u(0|\zeta) = 0$$

$$\mathbf{f}_u(0|\mathbf{u}_2) + 2a_1\mu_1\zeta + \frac{2a_2\zeta^2}{a_1} = 0$$

$$\mathbf{f}_u(0|\mathbf{u}_3) + 3a_1\mu_2\zeta + 3a_1\mu_1\mathbf{u}_2 + \frac{6a_2\zeta\mathbf{u}_2}{a_1} + \frac{6a_3\zeta^3}{a_1} + 6\mu_1 a_2\zeta^2 = 0$$

$$\mathbf{f}_u(0|\mathbf{u}_n) + na_1\mu_{n-1}\zeta + \text{terms of lower order} = 0.$$

We choose

$$\zeta(x) = \sin x, \qquad \zeta^*(x) = \frac{2}{\pi}\sin x,$$

so that

$$\mu_1 = -\frac{a_2}{a_1^2}\frac{8}{3\pi}.$$

If $a_2 = 0$, we obtain $\mu_1 = 0$ and $u_2 = 0$, and then

$$\mu_2 = -\frac{3a_3}{2a_1^2}.$$

Then the *bifurcated solution* of (VI.147) *with* $\delta = 0$ is two-sided if $a_2 \neq 0$:

$$U(x) = \varepsilon\sin x + O(\varepsilon^2),$$
$$\mu = \lambda - \frac{1}{a_1} = -\frac{8a_2}{3\pi a_1^2}\varepsilon + O(\varepsilon^2), \tag{VI.150}$$

or is one-sided if $a_2 = 0$ and $a_3 \neq 0$:

$$U(x) = \varepsilon\sin x + O(\varepsilon^3),$$
$$\mu = \lambda - \frac{1}{a_1} = -\frac{3a_3}{4a_1^2}\varepsilon^2 + O(\varepsilon^3). \tag{VI.151}$$

* This space is an algebra, that is, the product of two elements in this space is also in this space. Terms like $\zeta\mathbf{u}_2$ are bilinear functions in this space.

We now study the imperfect problem ($\delta \neq 0$) using the notation of §VI.10. We have

$$\begin{aligned} 0 &= \tilde{\mathscr{F}}(\mu, \mathbf{u}, \delta) \\ &\equiv \mathbf{f}_u(0|\mathbf{u}) + \mu \mathbf{f}_{u\mu}(0|\mathbf{u}) + \left(\mu + \frac{1}{a_1}\right)\left[\sum_{n \geq 2} a_n \mathbf{u}^n + \delta \mathbf{g}(\cdot, \mathbf{u})\right]. \end{aligned} \quad \text{(VI.152)}$$

Let us assume that

$$\mathbf{g}(\cdot, \mathbf{u}) = \mathbf{g}_0 + \mathbf{g}_1 \mathbf{u} + O(\mathbf{u}^2), \quad \text{(VI.153)}$$

where $\mathbf{g}_i$ are known functions on $[0, \pi]$. The condition (VI.71) gives here

$$\frac{1}{a_1} \langle \mathbf{g}_0, \boldsymbol{\zeta}^* \rangle \neq 0;$$

that is

$$\int_0^\pi \mathbf{g}_0(x) \sin x \, dx \neq 0. \quad \text{(VI.154)}$$

We seek now $\mathbf{u}(\mu, \varepsilon)$ and $\delta = \Delta(\mu, \varepsilon)$, where

$$\varepsilon \stackrel{\text{def}}{=} \langle \mathbf{u}, \boldsymbol{\zeta}^* \rangle. \quad \text{(VI.155)}$$

Formulas (VI.79), (VI.80) lead to

$$\Delta_{\varepsilon\varepsilon} = -2a_2 \frac{\langle \boldsymbol{\zeta}^2, \boldsymbol{\zeta}^* \rangle}{\langle \mathbf{g}_0, \boldsymbol{\zeta}^* \rangle}, \qquad \Delta_{\mu\varepsilon} = \frac{-a_1^2}{\langle \mathbf{g}_0, \boldsymbol{\zeta}^* \rangle},$$

so

$$\delta = -\left(\frac{8}{3\pi} a_2 \varepsilon^2 + a_1^2 \varepsilon \mu\right) \langle \mathbf{g}_0, \boldsymbol{\zeta}^* \rangle^{-1} + O[|\varepsilon|(|\varepsilon| + |\mu|)^2],$$

which describes the breaking of the bifurcation (VI.150) when $a_2 \neq 0$. In the case when $a_2 = 0$, $\Delta_{\varepsilon\varepsilon} = 0$ and $u_{\varepsilon\varepsilon} = 0$ and

$$\Delta_{\varepsilon\varepsilon\varepsilon} = -\frac{6a_3 \langle \boldsymbol{\zeta}^3, \boldsymbol{\zeta}^* \rangle}{\langle \mathbf{g}_0, \boldsymbol{\zeta}^* \rangle}, \qquad \Delta_{\mu\varepsilon\varepsilon} = \frac{2a_1^2 \langle \mathbf{g}_1 \boldsymbol{\zeta}, \boldsymbol{\zeta}^* \rangle}{\langle \mathbf{g}_0, \boldsymbol{\zeta}^* \rangle^2}$$

$$\Delta_{\mu\mu\varepsilon} = \frac{2a_1^3}{\langle \mathbf{g}_0, \boldsymbol{\zeta}^* \rangle}.$$

Hence, when $a_2 = 0$, the breaking of the bifurcation (VI.151) is given by

$$\begin{aligned} \delta = &\left[-a_1^2 \varepsilon \mu - \frac{3a_3}{4} \varepsilon^3 + a_1^2 \frac{\langle \mathbf{g}_1 \boldsymbol{\zeta}, \boldsymbol{\zeta}^* \rangle}{\langle \mathbf{g}_0, \boldsymbol{\zeta}^* \rangle} \mu \varepsilon^2 + a_1^3 \mu^2 \varepsilon\right] \\ &\times \langle \mathbf{g}_0, \boldsymbol{\zeta}^* \rangle^{-1} + O(|\varepsilon|(|\varepsilon| + |\mu|)^3). \end{aligned} \quad \text{(VI.156)}$$

We note that $\delta = 0$ gives again the bifurcated solutions (VI.150), (VI.151). If $a_2 \neq 0$, we have $\mu(\varepsilon, \delta/\varepsilon)$ in the form

$$\mu = -\frac{\langle \mathbf{g}_0, \zeta^* \rangle}{a_1^2}\frac{\delta}{\varepsilon} - \frac{8}{3\pi}\frac{a_2}{a_1^2}\varepsilon + O\left(|\varepsilon| + \left|\frac{\delta}{\varepsilon}\right|\right)^2. \tag{VI.157}$$

And if $a_2 = 0$, $a_3 \neq 0$ then

$$\mu = -\frac{\langle \mathbf{g}_0, \zeta^* \rangle}{a_1^2}\frac{\delta}{\varepsilon} - \frac{3a_3}{4a_1^2}\varepsilon^2 - \frac{\langle \mathbf{g}_1\zeta, \zeta^* \rangle}{a_1^2}\frac{\delta}{\varepsilon}\varepsilon + \frac{\langle \mathbf{g}_0, \zeta^* \rangle^2}{a_1^3}\left(\frac{\delta}{\varepsilon}\right)^2 + O\left(|\varepsilon| + \left|\frac{\delta}{\varepsilon}\right|\right)^3.$$

CHAPTER VII

Bifurcation of Periodic Solutions from Steady Ones (Hopf Bifurcation) in Two Dimensions

Up to now the only equilibrium solutions which have been introduced are steady ones. Now we shall show how a time-periodic solution may arise from bifurcation of a steady solution. In this case the symmetry of the forcing data, which is steady, is broken by the time-periodic solution. The dynamical system then has "a mind of its own" in the sense that the solution does not follow the symmetry imposed by the given data.

Following the procedure already adopted for steady solutions we start with the lowest-dimensional problems in which the characteristic bifurcation occurs and then show how this problem arises from higher-dimensional ones by the method of projection.

VII.1 The Structure of the Two-Dimensional Problem Governing Hopf Bifurcation

The problem of bifurcation of steady flow into time-periodic flow is basically two-dimensional. It is not possible for a time-periodic solution to bifurcate from a steady one in one dimension. In the two-dimensional autonomous case we again consider the evolution problem (IV.2):

$$\dot{u}_i = A_{ij}(\mu)u_j + B_{ijk}(\mu)u_j u_k + \text{higher-order terms} \tag{VII.1}$$

where

$$\dot{u}_i \overset{\text{def}}{=} \frac{du_i}{dt}$$

and $A_{ij}(\mu)$ are components of $\mathbf{A}(\mu)$.

We suppose that the discriminant $(A_{11} - A_{22})^2 + 4A_{12}A_{21}$ is negative in a neighborhood of $\mu = 0$. Then the eigenvalues $\sigma(\mu) = \xi(\mu) + i\eta(\mu)$ and eigenvectors $\boldsymbol{\zeta}(\mu)$ of $\mathbf{A}(\mu)$ are complex conjugates and

$$\sigma(\mu)\boldsymbol{\zeta} = \mathbf{A}\boldsymbol{\zeta} \qquad (\sigma\zeta_i = A_{ij}\zeta_j) \tag{VII.2$_1$}$$

and

$$\sigma(\mu)\bar{\boldsymbol{\zeta}}^* = \mathbf{A}^{\mathrm{T}}\bar{\boldsymbol{\zeta}}^*, \tag{VII.2$_2$}$$

where $\boldsymbol{\zeta}^*(\mu)$ is the adjoint eigenvector with eigenvalue $\bar{\sigma}(\mu)$ in the scalar product, $\langle \mathbf{x}, \mathbf{y} \rangle = \mathbf{x} \cdot \bar{\mathbf{y}}$. We may normalize so that

$$\begin{aligned} \langle \boldsymbol{\zeta}, \boldsymbol{\zeta}^* \rangle &= \boldsymbol{\zeta} \cdot \bar{\boldsymbol{\zeta}}^* = \zeta_k \bar{\zeta}_k^* = 1 \\ \langle \boldsymbol{\zeta}, \bar{\boldsymbol{\zeta}}^* \rangle &= \zeta_k \zeta_k^* = 0. \end{aligned} \tag{VII.3}$$

The eigenvalues $\sigma(\mu) = \xi(\mu) + i\eta(\mu)$ arise in the spectral problem for the stability of the solution $u_i = 0$ of (VII.1). We suppose that this loss of stability occurs at $\mu = 0$ so that $\xi(0) = 0$. We will get bifurcation into periodic solutions if

$$\eta(0) = \omega_0 \neq 0 \quad \text{and} \quad \frac{d\xi(0)}{d\mu} = \xi_\mu(0) \neq 0 \tag{VII.4}$$

(say $\xi_\mu(0) > 0$).

VII.2 Amplitude Equation for Hopf Bifurcation

To prove bifurcation into periodic solutions under conditions (VII.4), we note that $\boldsymbol{\zeta}$ and $\bar{\boldsymbol{\zeta}}$ are independent so that any real-valued two-dimensional vector $\mathbf{u} = (u_1, u_2)$ may be represented as

$$u_i = a(t)\zeta_i + \bar{a}(t)\bar{\zeta}_i.$$

Substitute this into (VII.1) and use (VII.2) to find

$$\begin{aligned} \dot{a}\zeta_i + \dot{\bar{a}}\bar{\zeta}_i = \sigma(\mu)\zeta_i + \bar{\sigma}(\mu)\bar{\zeta}_i &+ a^2 B_{ijk}\zeta_j\zeta_k \\ &+ 2|a|^2 B_{ijk}\zeta_i\bar{\zeta}_k + \bar{a}^2 B_{ijk}\bar{\zeta}_i\bar{\zeta}_k \\ &+ O(|a|^3). \end{aligned}$$

The orthogonality properties (VII.3) are now employed to reduce the preceding into a single, complex-valued, amplitude equation

$$\dot{a} = f(\mu, a) = \sigma(\mu)a + \alpha(\mu)a^2 + 2\beta(\mu)|a|^2 + \gamma(\mu)\bar{a}^2 + O(|a|^3) \tag{VII.5}$$

where, for example, $\alpha(\mu) = B_{ijk}(\mu)\zeta_j\zeta_k\bar{\zeta}_i^*$. (For simplicity we shall suppress cubic terms of $f(\mu, a)$ in this chapter. These terms come into the bifurcating solution at second order but do not introduce new features. In Chapter VIII we retain the terms suppressed here.) The linearized stability of the solution $a = 0$ of (VII.5) is determined by $\dot{a} = \sigma(\mu)a$, $a = \text{constant} \times e^{\sigma(\mu)t}$. At criticality ($\mu = 0$), $a = \text{constant} \times e^{i\omega_0 t}$ is 2π-periodic in $s = \omega_0 t$.

VII.3 Series Solution

We shall show that a bifurcating time-periodic solution may be constructed from the solution of the linearized problem at criticality. This bifurcating solution is in the form

$$a(t) = b(s, \varepsilon), \qquad s = \omega(\varepsilon)t, \qquad \omega(0) = \omega_0, \qquad \mu = \mu(\varepsilon), \quad \text{(VII.6)}_1$$

where ε is the amplitude of a defined by

$$\varepsilon = \frac{1}{2\pi}\int_0^{2\pi} e^{-is} b(s, \varepsilon)\, ds = [b]. \qquad \text{(VII.6)}_2$$

The solution (VII.6) of (VII.5) is unique to within an arbitrary translation of the time origin. This means that under translation $t \to t + c$ the solution $b(s + c\omega(\varepsilon), \varepsilon)$ shifts its phase. This unique solution is analytic in ε when $f(\mu, a)$ is analytic in the variables $(\mu, a, \bar{a})$ and it may be expressed as a series:

$$\begin{bmatrix} b(s, \varepsilon) \\ \omega(\varepsilon) - \omega_0 \\ \mu(\varepsilon) \end{bmatrix} = \sum_{n=1}^{\infty} \varepsilon^n \begin{bmatrix} b_n(s) \\ \omega_n \\ \mu_n \end{bmatrix}. \qquad \text{(VII.7)}$$

VII.4 Equations Governing the Taylor Coefficients

The perturbation problems which govern $b_n(s)$, ω_n, and μ_n can be obtained by identifying the coefficient of ε^n which arise when (VII.7) is substituted into the two equations: $\omega\dot{b} = f(\mu, b)$ and $\varepsilon = [b]$. We find that at order one

$$\omega_0 \dot{b}_1 - i\omega_0 b_1 = 0, \qquad [b_1] = 1, \qquad b_1(s) = e^{is}.$$

At order two we find that $[b_2] = 0$ and

$$\omega_0[\dot{b}_2 - ib_2] + \omega_1\dot{b}_1 = \mu_1\sigma_\mu b_1 + \alpha_0 b_1^2 + 2\beta_0|b_1|^2 + \gamma_0\bar{b}_1^2,$$

where $\sigma_\mu = d\sigma(0)/d\mu$ and, for example, $\alpha_0 = \alpha(0)$.

VII.5 Solvability Conditions (the Fredholm Alternative)

Equations of the form $\dot{b}(s) - ib(s) = f(s) = f(s + 2\pi)$ are solvable for $b(s) = b(s + 2\pi)$ if and only if the Fourier expansion of $f(s)$ has no term proportional to e^{is}. Hence, because $\xi_\mu \neq 0$ we obtain

$$\mu_1 = \omega_1 = 0 \qquad \text{in (VII. 7)}$$

and

$$\dot{b}_2 - ib_2 = \frac{(\alpha_0 e^{2is} + 2\beta_0 + \gamma_0 e^{-2is})}{\omega_0}.$$

We find that

$$b_2(s) = \frac{(\alpha_0 e^{2is} - 2\beta_0 - (\gamma_0 e^{-2is}/3))}{i\omega_0}.$$

The problem which governs at order three, with cubic terms in b neglected,* is

$$\dot{b}_3 - ib_3 = \frac{\{-\omega_2 \dot{b}_1 + \mu_2 \sigma_\mu b_1 + 2\alpha_0 b_1 b_2 + 2\beta_0(b_1 \bar{b}_2 + \bar{b}_1 b_2) + 2\gamma_0 \bar{b}_1 \bar{b}_2\}}{\omega_0}. \tag{VII.8}$$

To solve (VII.8) we must eliminate terms proportional to e^{is} from the right-hand side of (VII.8). This is done if $[b_3] = 0$; that is, if

$$i\omega_2 - \mu_2 \sigma_\mu = -\frac{\{4\alpha_0 \beta_0 - 4|\beta_0|^2 - 2\alpha_0 \beta_0 - (2|\gamma_0|^2/3)\}}{i\omega_0}. \tag{VII.9}$$

The real part of (VII.9) is solvable for μ_2 provided that $\xi_\mu \neq 0$. The imaginary part of (VII.9) is always solvable for ω_2.

Proceeding to higher orders, it is easy to verify that all of the perturbation problems are solvable when (VII.4) holds and, in fact, $\omega(\varepsilon) = \omega(-\varepsilon)$, $\mu(\varepsilon) = \mu(-\varepsilon)$ are even functions. It follows that periodic solutions which bifurcate from steady solutions bifurcate to one or the other side of criticality and never to both sides; periodic bifurcating solutions cannot undergo two-sided or transcritical bifurcation (see Figures II.3 and VII.2).

VII.6 Floquet Theory

Floquet theory is a linear theory of stability for solutions which depend periodically on the time. The direct object of study of Floquet theory is a linear differential equation with periodic coefficients. Such equations are generated in the study of forced T-periodic solutions leading to a non-autonomous linear equation with T-periodic coefficients, or in the study of stability of periodic solutions which bifurcate from steady (autonomous) problems. The theory of the stability of solutions which are more complicated than periodic ones, say quasi-periodic ones, is much more difficult than Floquet theory and does not admit elementary analysis.

* Such terms lead to triple products of b_1 and $\bar{b}_1$ and, in any event, can be included without difficulty.

VII.6.1 Floquet Theory in $\mathbb{R}^1$

We start by considering the stability of T-periodic solutions $U(t) = U(t + T)$ of $\dot{V} = F(V, t) = F(V, t + T)$ where $F(0, t) = F(0, t + T) \neq 0$ is a prescribed T-periodic forcing. Let $V = U(t) + u$. Then

$$\begin{aligned} \dot{u} &= F(U(t) + u, t) - F(U(t), t) \\ &= f(u, t) = f(u, t + T) \end{aligned} \tag{VII.10}$$

is in "local form," $f(0, t) = 0$, and $f(u, t)$ admits a Taylor expansion: $f(u, t) = a_1(t)u + a_2(t)u^2 + O(|u|^3)$, where $a_i(t) = a_i(t + T)$. The linearized evolution problem is

$$\dot{v} = a_1(t)v \tag{VII.11}$$

and

$$v(t) = \left(\exp \int_0^t a_1(s)\, ds\right) v_0. \tag{VII.12}$$

Let $\phi(t)$ be the solution of (VII.11) for which $v(0) = v_0 = 1$,

$$\phi(t) = \exp \int_0^t a_1(s)\, ds.$$

Since $a_1(t) = a_1(t + T)$ we have

$$\int_T^{t+T} a_1(s)\, ds = \int_0^t a_1(s)\, ds$$

and

$$\begin{aligned} \phi(t + T) &= \left(\exp \int_T^{t+T} a_1(s)\, ds\right)\left(\exp \int_0^T a_1(s)\, ds\right) \\ &= \phi(t)\phi(T) \\ \phi(2T) &= \phi(T)\phi(T) \\ \phi(nT) &= \phi^n(T). \end{aligned} \tag{VII.13}$$

The function

$$\phi(T) = e^{\sigma T} \tag{VII.14}$$

which satisfies the functional equation (VII.13) is called a *Floquet multiplier* and the numbers σ are called *Floquet exponents*. The exponents are not uniquely determined by the multiplier:

$$\sigma = \frac{1}{T} \log \phi(T) + \frac{2k\pi i}{T}, \qquad k \in \mathbb{Z}. \tag{VII.15}$$

The exponents are eigenvalues of the differential equation (VII.17), derived below. We define

$$\zeta(t) = \phi(t)e^{-\sigma t}.$$

Then

$$\zeta(t + T) = \phi(t + T)e^{-\sigma t}e^{-\sigma T} = \phi(t)e^{-\sigma t} = \zeta(t) \tag{VII.16}$$

is T-periodic and, since $\dot{\phi} = a_1(t)\phi$,

$$\sigma\zeta = -\dot{\zeta} + a_1(t)\zeta. \tag{VII.17}$$

So the general solution of (VII.11) may be written as

$$v(t) = \zeta(t)e^{\sigma t}v_0,$$

where $\zeta(t) = \zeta(t + T)$. If $\sigma < 0$ then $v(t) \to 0$ exponentially. If $\sigma > 0$ then $v(t) \to \infty$. Equivalently if $\phi(T) = \exp \int_0^T a_1(s)\, ds < 1$, then $v(t) \to 0$.

Now we show that when $\phi(T) < 1$ or, equivalently, when $\sigma < 0$, then the solution $u = 0$ of (VII.10) is conditionally stable. The proof is almost identical to the one given for the autonomous problem in §II.7. We first rewrite (VII.10) as

$$\dot{u} = a_1(t)u + b(t, u), \tag{VII.18}$$

which is equivalent to

$$\begin{aligned} u(t) &= \phi(t)u_0 + \int_0^t \phi(t)\phi^{-1}(s)b(s, u(s))\, ds \\ &= \zeta(t)e^{\sigma t}u_0 + \int_0^t e^{\sigma(t-s)}\zeta(t)\zeta^{-1}(s)b(s, u(s))\, ds \end{aligned} \tag{VII.19}$$

where $\phi^{-1} = 1/\phi$, $\zeta^{-1} = 1/\zeta$. To see this equivalence it is easiest to differentiate (VII.19). We get

$$\dot{u} = \frac{d}{dt}(\zeta(t)e^{\sigma t})u_0 + b(t, u(t)) + \int_0^t \frac{d}{dt}(\zeta(t)e^{\sigma t})e^{-\sigma s}\zeta^{-1}(s)b(s, u(s))\, ds,$$

$$\frac{d}{dt}(\zeta(t)e^{\sigma t}) = a_1(t)e^{\sigma t}\zeta(t)$$

and

$$\begin{aligned} \dot{u} &= b(t, u(t)) + a_1(t)\left\{\zeta(t)e^{\sigma t}u_0 + \int_0^t e^{\sigma(t-s)}\zeta(t)\zeta^{-1}(s)b(s, u(s))\, ds\right\} \\ &= b(t, u(t)) + a_1(t)u. \end{aligned}$$

The rest of the proof is identical to the one given in §II.7. We find that $u = 0$ is exponentially stable when u_0 is sufficiently small and $\sigma < 0$.

VII.6.2 Floquet Theory in $\mathbb{R}^2$ and $\mathbb{R}^n$

Some new features of Floquet theory must be introduced when $\mathbf{u}$ is a vector. But no new features are introduced in generalizing Floquet theory from $\mathbb{R}^2$ to $\mathbb{R}^n$ with $n > 2$. So we may work out the theory for

$$\frac{d\mathbf{v}}{dt} = \mathbf{A}(t) \cdot \mathbf{v}, \tag{VII.20}$$

where $\mathbf{v}$ is a vector with n components and

$$\mathbf{A}(t) = \mathbf{A}(t + T) \tag{VII.21}$$

is a T-periodic, $n \times n$ matrix, using $n = 2$ as an example.

The matrix $\mathbf{A}(t)$ may arise from the linearization of the equation $\dot{\mathbf{V}} = \mathbf{F}(t, \mathbf{V}) = \mathbf{F}(t + T, \mathbf{V})$, $\mathbf{F}(t, 0) \neq 0$, governing forced, T-periodic solutions $\mathbf{V} = \mathbf{U}(t) = \mathbf{U}(t + T)$. The linearization of this nonlinear problem reduced to "local form" leads to:

$$\begin{aligned} \mathbf{V} &= \mathbf{U}(t) + \mathbf{v}, \\ \dot{\mathbf{v}} &= \mathbf{F}(t, \mathbf{U}(t) + \mathbf{v}) - \mathbf{F}(t, \mathbf{U}(t)) \stackrel{\text{def}}{=} \mathbf{f}(t, \mathbf{v}) \end{aligned} \tag{VII.22}$$

where $\mathbf{f}(t, 0) = 0$ and $\mathbf{f}(t + T, \mathbf{v}) = \mathbf{f}(t, \mathbf{v})$. In this case

$$\dot{\mathbf{v}} = \mathbf{A}(t) \cdot \mathbf{v} \tag*{(VII.23)$_1$}$$

where

$$\mathbf{A}(t) = \mathbf{F}_v(t, \mathbf{U}(t) | \cdot) = \mathbf{f}_v(t | \cdot). \tag*{(VII.23)$_2$}$$

When $n > 1$, periodic solutions $\mathbf{u}(t) = \mathbf{u}(t + T)$ may bifurcate* from steady solutions $\mathbf{V} = \mathbf{V}_0$ of autonomous problems $\dot{\mathbf{V}} = \mathbf{F}(\mathbf{V})$, $\mathbf{F}(0) \neq 0$. Then the perturbation $\mathbf{v}$ in $\mathbf{V} = \mathbf{V}_0 + \mathbf{u}(t) + \mathbf{v}$ satisfies a T-periodic problem reduced to local form:

$$\begin{aligned} \dot{\mathbf{v}} &= \mathbf{F}(\mathbf{V}_0 + \mathbf{u}(t) + \mathbf{v}) - \mathbf{F}(\mathbf{V}_0 + \mathbf{u}(t)) \stackrel{\text{def}}{=} \mathbf{f}(\mathbf{u}(t), \mathbf{v}) \\ &= \mathbf{f}(\mathbf{u}(t + T), \mathbf{v}) \end{aligned} \tag{VII.24}$$

where $\mathbf{f}(\mathbf{u}(t), 0) = 0$. In this case

$$\dot{\mathbf{v}} = \mathbf{A}(t)\mathbf{v} \tag*{(VII.25)$_1$}$$

where

$$\mathbf{A}(t) = \mathbf{F}_v(\mathbf{V}_0 + \mathbf{u}(t) | \cdot) = \mathbf{f}_v(\mathbf{u}(t) | \cdot). \tag*{(VII.25)$_2$}$$

In our exposition of Floquet theory we do not usually need to maintain a distinction between periodic matrices $\mathbf{A}(t) = \mathbf{A}(t + T)$ which arise from forced T-periodic problems and those which arise from autonomous problems

* We also may have periodic solutions of autonomous problems not coming from a previous bifurcation.

having a periodic solution. But the distinction is of substance because the forced problem is invariant to $t = T$ translations of the origin of time and the autonomous problem is invariant to arbitrary translations of the origin of time. A mathematical consequence of this distinction is that $\dot{\mathbf{u}}(t)$ is a T-periodic solution of (VII.25) but $\dot{\mathbf{U}}(t)$ is not a T-periodic solution of (VII.23) (for further discussion see the penultimate paragraph of this subsection).

Let $\mathbf{v}_1$ and $\mathbf{v}_2$ be linearly independent solutions of $\dot{\mathbf{v}} = \mathbf{A}(t) \cdot \mathbf{v}$ for $\mathbf{v}(t) \in \mathbb{R}^2$. Then all solutions are linear combinations of these two solutions:

$$\mathbf{v}(t) = a\mathbf{v}_1(t) + b\mathbf{v}_2(t). \tag{VII.26}$$

Now $\mathbf{v}(t + T)$ solves $\dot{\mathbf{v}} = \mathbf{A}(t + T) \cdot \mathbf{v} = \mathbf{A}(t) \cdot \mathbf{v}$ if $\mathbf{v}(t)$ does. Hence

$$\begin{aligned} \mathbf{v}_1(t + T) &= a_1\mathbf{v}_1(t) + b_1\mathbf{v}_2(t) \\ \mathbf{v}_2(t + T) &= a_2\mathbf{v}_1(t) + b_2\mathbf{v}_2(t). \end{aligned} \tag{VII.27}$$

We have motivated the following definition of a fundamental solution matrix in $\mathbb{R}^n$:

Let $v_i^{(j)} = \mathbf{e}_j \cdot \mathbf{v}_i$ $(i, j = 1, 2, \ldots, n)$ be the jth component of the vector $\mathbf{v}_i$. A *fundamental solution matrix* is any matrix whose columns are the components of linearly independent solutions of $\dot{\mathbf{v}} = \mathbf{A}(\mathrm{t})\mathbf{v}$, $\mathbf{v} \in \mathbb{R}^n$. Suppose

$$\tilde{\mathbf{V}}(t) = [\mathbf{v}_1(t), \mathbf{v}_2(t), \ldots, \mathbf{v}_n(t)] \tag{VII.28}$$

is a fundamental solution matrix. Then

$$\dot{\tilde{\mathbf{V}}}(t + T) = \mathbf{A}(t + T) \cdot \tilde{\mathbf{V}}(t + T) = \mathbf{A}(t)\tilde{\mathbf{V}}(t + T). \tag{VII.29}$$

So $\tilde{\mathbf{V}}(t + T)$ is a fundamental solution matrix if $\tilde{\mathbf{V}}(t)$ is. It follows that we may express $\tilde{\mathbf{V}}(t + T)$ as linear combinations of the columns of $\tilde{\mathbf{V}}(t)$. Hence

$$\tilde{\mathbf{V}}(t + T) = \tilde{\mathbf{V}}(t) \cdot \mathbf{C}, \tag{VII.30}$$

where $\mathbf{C}$ is a constant $n \times n$ matrix which depends in fact on $\tilde{\mathbf{V}}(0)$ (and is of course a functional of $\mathbf{A}(t)$).

In $\mathbb{R}^2$ we have

$$\mathbf{C} = \begin{bmatrix} a_1 & a_2 \\ b_1 & b_2 \end{bmatrix}$$

and (VII.30) is the same as (VII.27).

Let $\boldsymbol{\Phi}(t)$ be a fundamental solution matrix with initial value equal to the unit matrix:

$$\boldsymbol{\Phi}(0) = \mathbf{I} \qquad (\Phi_{ij} = \delta_{ij}). \tag{VII.31}$$

Then $\boldsymbol{\Phi}(t + T) = \boldsymbol{\Phi}(t) \cdot \mathbf{C}$ so that when $t = 0$

$$\boldsymbol{\Phi}(T) = \mathbf{C}. \tag{VII.32}$$

The *monodromy matrix* is the value at $t = T$ of the fundamental solution matrix $\tilde{\mathbf{V}}(t)$ satisfying (VII.29) when $\tilde{\mathbf{V}}(0) = \mathbf{I}$. So (VII.30) can be written as

$$\begin{aligned}\mathbf{\Phi}(t+T) &= \mathbf{\Phi}(t)\mathbf{\Phi}(T)\\ \mathbf{\Phi}(2T) &= \mathbf{\Phi}^2(T)\\ \mathbf{\Phi}(3T) &= \mathbf{\Phi}(2T)\mathbf{\Phi}(T) = \mathbf{\Phi}^3(T)\end{aligned}$$

and

$$\mathbf{\Phi}(nT) = \mathbf{\Phi}^n(T). \tag{VII.33}$$

The eigenvalues of $\mathbf{\Phi}(T)$ are the *Floquet multipliers*. We find that

$$\begin{aligned}\mathbf{\Phi}(T)\cdot\boldsymbol{\psi} &= \lambda(T)\boldsymbol{\psi}\\ \dot{\mathbf{\Phi}}(nT)\cdot\boldsymbol{\psi} &= \lambda(nT)\boldsymbol{\psi}\\ \mathbf{\Phi}^n(T)\cdot\boldsymbol{\psi} &= \lambda^n(T)\boldsymbol{\psi}.\end{aligned} \tag{VII.34}$$

Since $\mathbf{\Phi}(nT) = \mathbf{\Phi}^n(T)$ we have $\lambda^n(T) = \lambda(nT)$ so that we may define a *Floquet exponent* $\sigma = \xi + i\eta$ through the relation

$$\lambda(T) = \exp \sigma T$$

and write the eigenvalue problem as

$$\mathbf{\Phi}(T)\cdot\boldsymbol{\psi} = e^{\sigma T}\boldsymbol{\psi}. \tag{VII.35}$$

If σ is a Floquet exponent belonging to $\lambda(T)$, then $\sigma + (2\pi ik/T)$, $k \in \mathbb{Z}$, is also a Floquet exponent belonging to $\lambda(T)$.

Now we derive an eigenvalue problem for the exponent. First we define

$$\mathbf{v}(t) = \mathbf{\Phi}(t)\cdot\boldsymbol{\psi}. \tag{VII.36}$$

It then follows that

$$\begin{aligned}\mathbf{v}(0) &= \boldsymbol{\psi}\\ \mathbf{v}(t+T) &= \mathbf{\Phi}(t+T)\cdot\boldsymbol{\psi} = \mathbf{\Phi}(t)\mathbf{\Phi}(T)\cdot\boldsymbol{\psi}\\ &= e^{\sigma T}\mathbf{\Phi}(t)\cdot\boldsymbol{\psi} = e^{\sigma T}\mathbf{v}(t)\end{aligned}$$

and

$$\dot{\mathbf{v}} = \mathbf{A}(t)\mathbf{v}.$$

Define

$$\boldsymbol{\zeta}(t) = e^{-\sigma t}\mathbf{v}(t)$$

so that $\boldsymbol{\zeta}$ is T-periodic:

$$\begin{aligned}\boldsymbol{\zeta}(t+T) &= e^{-\sigma(t+T)}\mathbf{v}(t+T)\\ &= e^{-\sigma t}\mathbf{v}(t) = \boldsymbol{\zeta}(t)\end{aligned} \tag{VII.37$_1$}$$

and

$$\dot{\boldsymbol{\zeta}} = -\sigma\boldsymbol{\zeta} + \dot{\mathbf{v}}e^{-\sigma t} = -\sigma\boldsymbol{\zeta} + \mathbf{A}(t)\boldsymbol{\zeta}. \tag{VII.37$_2$}$$

Equations (VII. 37) define an eigenvalue problem for the Floquet exponents.

We recall now that $\mathbf{v}(t)$ is a small disturbance of the forced T-periodic solution $\mathbf{U}(t)$ or of the periodic bifurcating solution $\mathbf{u}(t)$. From the representations (VII.36) and (VII.34) we find that

$$\mathbf{v}(t + nT) = \mathbf{\Phi}(t)\mathbf{\Phi}(nT)\boldsymbol{\psi} = \lambda^n\mathbf{\Phi}(t)\boldsymbol{\psi}$$

and conclude that $\mathbf{v}(t) \to 0$ as $t \to \infty$ (for t between nT and $(n+1)T$, $\|\mathbf{\Phi}(t)\boldsymbol{\psi}\|$ is bounded above and below) provided that $|\lambda| = e^{\xi T} < 1$ for all Floquet multipliers $\lambda_l = e^{\sigma_l T}$ in the spectrum of $\mathbf{\Phi}(T)$. The equivalent statement using exponents is that $\mathbf{v}(t) = e^{\sigma t}\boldsymbol{\zeta}(t)$, $\boldsymbol{\zeta}(t) = \boldsymbol{\zeta}(t + T)$, tends to zero at $t \to \infty$ provided that $\xi_n = \operatorname{Re} \sigma_n < 0$ for all eigenvalues σ_n of (VII.37). The stability implications of Floquet multipliers and exponents are represented graphically in Figure VII.1.

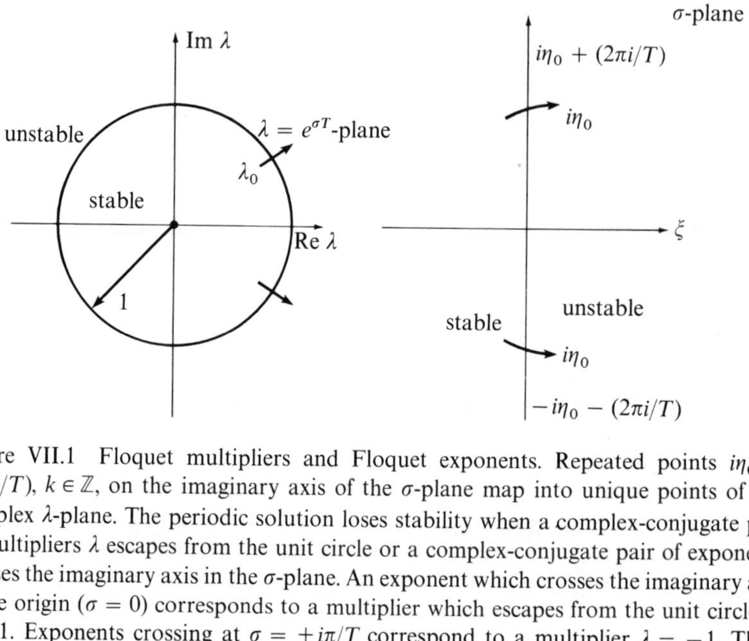

Figure VII.1 Floquet multipliers and Floquet exponents. Repeated points $i\eta_0 + (2\pi ik/T)$, $k \in \mathbb{Z}$, on the imaginary axis of the σ-plane map into unique points of the complex λ-plane. The periodic solution loses stability when a complex-conjugate pair of multipliers λ escapes from the unit circle or a complex-conjugate pair of exponents crosses the imaginary axis in the σ-plane. An exponent which crosses the imaginary axis at the origin ($\sigma = 0$) corresponds to a multiplier which escapes from the unit circle at $\lambda = 1$. Exponents crossing at $\sigma = \pm i\pi/T$ correspond to a multiplier $\lambda = -1$. There is a sense in which crossing of $\lambda = -1$ is "typical" (see Exercise XI.2).

Finally, we note again that in the autonomous case in which $\mathbf{u}(t + T) = \mathbf{u}(t)$ satisfies $\dot{\mathbf{u}} = \mathbf{f}(\mathbf{u})$, the function $\boldsymbol{\zeta} = \dot{\mathbf{u}}$ satisfying $\ddot{\mathbf{u}} = \mathbf{f}_u(\mathbf{u}|\dot{\mathbf{u}})$ is an eigenfunction of (VII.37) with eigenvalue $\sigma = 0$. The conditional stability of $\mathbf{u}(t)$ (Coddington and Levinson, 1955, p. 323; cited in §IV.2) therefore gives asymptotic stability not of a single solution but of a set of solutions $\mathbf{u}(t + \alpha)$ depending on the phase α. If small disturbances are attracted to this set, the set of periodic solutions is said to have conditional, asymptotic *orbital* stability.

Having finished this long digression on Floquet theory we are ready to return to the problem of stability of the bifurcating periodic solutions.

VII.7 Equations Governing the Stability of the Periodic Solutions

We now search for the conditions under which the bifurcating periodic solutions are stable. We consider a small disturbance $z(t)$ of $b(s, \varepsilon)$. Setting $a(t) = b(s, \varepsilon) + z(t)$ in (VII.5) we find the linearized equation $\dot{z}(t) = f_a(\mu(\varepsilon), b(s, \varepsilon))z(t)$ where $f_a = \partial f/\partial a$ and $s = \omega(\varepsilon)t$. Then, using Floquet theory, we set $z(t) = e^{\gamma t}y(s)$ where $y(s) = y(s + 2\pi)$ and find that

$$\gamma y(s) = -\omega\dot{y}(s) + f_a(\mu, b)y(s) \overset{\text{def}}{=} [J(s, \varepsilon)y](s) \tag{VII.38}$$

where $\dot{y}(s) = dy(s)/ds$.

VII.8 The Factorization Theorem

The stability result we need may be stated as a factorization theorem. To prove this theorem we use the fact that $\gamma = 0$ is always an eigenvalue of J with eigenfunction $\dot{b}(s, \varepsilon)$

$$J\dot{b} = 0 \tag{VII.39}$$

and the relation

$$\omega_\varepsilon(\varepsilon)\dot{b}(s, \varepsilon) = \mu_\varepsilon(\varepsilon)f_\mu(\mu(\varepsilon), b(s, \varepsilon)) + Jb_\varepsilon \tag{VII.40}$$

which arises from differentiating $\omega\dot{b} = f(\mu, b)$ with respect to ε at any ε.

Factorization Theorem. *The eigenfunction y of* (VII.38) *and the Floquet exponent γ are given by the following formulas:*

$$\begin{aligned} y(s, \varepsilon) &= c(\varepsilon)\left\{\frac{\tau}{\gamma}\dot{b}(s, \varepsilon) + b_\varepsilon(s, \varepsilon) + \mu_\varepsilon(\varepsilon)\varepsilon q(s, \varepsilon)\right\} \\ \tau(\varepsilon) &= \omega_\varepsilon(\varepsilon) + \mu_\varepsilon(\varepsilon)\hat{\tau}(\varepsilon) \\ \gamma(\varepsilon) &= \mu_\varepsilon(\varepsilon)\hat{\gamma}(\varepsilon), \end{aligned} \tag{VII.41}$$

where $c(\varepsilon)$ is an arbitrary constant and $q(s, \varepsilon) = q(s + 2\pi, \varepsilon)$, $\hat{\tau}(\varepsilon)$ and $\hat{\gamma}(\varepsilon)$ satisfy the equation

$$\hat{\tau}\dot{b} + \hat{\gamma}b_\varepsilon + f_\mu(\mu, b) + \varepsilon\{\gamma q - Jq\} = 0 \tag{VII.42}$$

and are smooth functions in a neighborhood of $\varepsilon = 0$. Moreover $\hat{\tau}(\varepsilon)$ and $\hat{\gamma}(\varepsilon)/\varepsilon$ are even functions and such that

$$\hat{\gamma}_\varepsilon(0) = -\xi_\mu(0), \qquad \hat{\tau}(0) = -\eta_\mu(0). \tag{VII.43}$$

Remark. If $\omega_\varepsilon(0) \neq 0$, $c(\varepsilon)$ may be chosen so that

$$y(s, \varepsilon) \to \dot{b}(s, \varepsilon) \quad \text{when } \varepsilon \to 0.$$

PROOF. Substitute the representations (VII.41) into (VII.38) utilizing (VII.39) to eliminate $J\dot{b}$ and (VII.40) to eliminate Jb_ε. This leads to (VII.42), which may be solved by series

$$\begin{bmatrix} q(s,\varepsilon) \\ \hat{\gamma}(\varepsilon)/\varepsilon \\ \hat{\tau}(\varepsilon) \end{bmatrix} = \sum_{l=0}^{\infty} \begin{bmatrix} q_l(s) \\ \hat{\gamma}_l \\ \hat{\tau}_l \end{bmatrix} \varepsilon^l \tag{VII.44}$$

where $\hat{\gamma}_0 = \hat{\gamma}_\varepsilon(0)$ and $\hat{\tau}_0 = \hat{\tau}(0)$. Using the fact that to the lowest order $b = \varepsilon e^{is}$, $\gamma = O(\varepsilon^2)$, and (from (VII.5)) $f_\mu(\mu, b) = \sigma_\mu(0)e^{is}\varepsilon$ we find that

$$e^{is}[i\hat{\tau}(0) + \hat{\gamma}_\varepsilon(0) + \sigma_\mu] - J_0 q_0 = 0, \qquad J_0 \stackrel{\text{def}}{=} J(\cdot, 0). \tag{VII.45}$$

Equation (VII.45) is solvable for $q_0(s) = q_0(s + 2\pi)$ if and only if the term in the bracket vanishes; that is if (VII.43) holds. The remaining properties asserted in the theorem may be obtained by mathematical induction using the power series (VII.44) (see D. D. Joseph, *Stability of Fluid Motions I*, (New York–Heidelberg–Berlin: Springer-Verlag, 1976), Chapter 2).

The linearized stability of the periodic solution for small values of ε may now be obtained from the spectral problem: $\mathbf{u}(s, \varepsilon) = \mathbf{u}(s + 2\pi, \varepsilon)$ is stable when $\gamma(\varepsilon) < 0$ ($\gamma(\varepsilon)$ is real) and is unstable when $\gamma(\varepsilon) > 0$ where

$$\gamma(\varepsilon) = \mu_\varepsilon(\varepsilon)\hat{\gamma}(\varepsilon) = -\mu_\varepsilon(\varepsilon)\{\xi_\mu(0)\varepsilon + O(\varepsilon^3)\}. \tag{VII.46}$$

VII.9 Interpretation of the Stability Result

We have already assumed that the solution $\mathbf{u} = 0$ of (VII.1) loses stability strictly when μ is increased past zero, $\xi_\mu(0) > 0$. So the branches for which $\mu_\varepsilon(\varepsilon)\varepsilon > 0$ are stable and the ones for which $\mu_\varepsilon(\varepsilon)\varepsilon < 0$ are unstable. There are two possibilities when ε is small: supercritical bifurcation (Figure VII.2(a)) or subcritical bifurcation (Figure VII.2(b)). It is not possible to have transcritical periodic bifurcations as in Figure II.3 because $\mu(\varepsilon) = \mu(-\varepsilon)$.

EXAMPLE VII.1 (The factorization theorem and repeated branching of periodic solutions.) Let $F(\mu, V)$ and $\omega(V)$ be analytic functions of μ and V such that $\omega(0) = 1$, $F(\mu, 0) = 0$, $F(0, V) \neq 0$ if $V \neq 0$, $F_V(\mu, 0) \lesseqgtr 0$ if $\mu \lesseqgtr 0$. Consider the following problem

$$\frac{d}{dt}\begin{bmatrix} x \\ y \end{bmatrix} = F(\mu, x^2 + y^2)\begin{bmatrix} x \\ y \end{bmatrix} + \omega(x^2 + y^2)\begin{bmatrix} 0 & -1 \\ 1 & 0 \end{bmatrix}\begin{bmatrix} x \\ y \end{bmatrix}. \tag{VII.47}$$

Every solution of (VII.47) satisfies

$$\frac{d}{dt}(x^2 + y^2) = 2(x^2 + y^2)F(\mu, x^2 + y^2). \tag{VII.48}$$

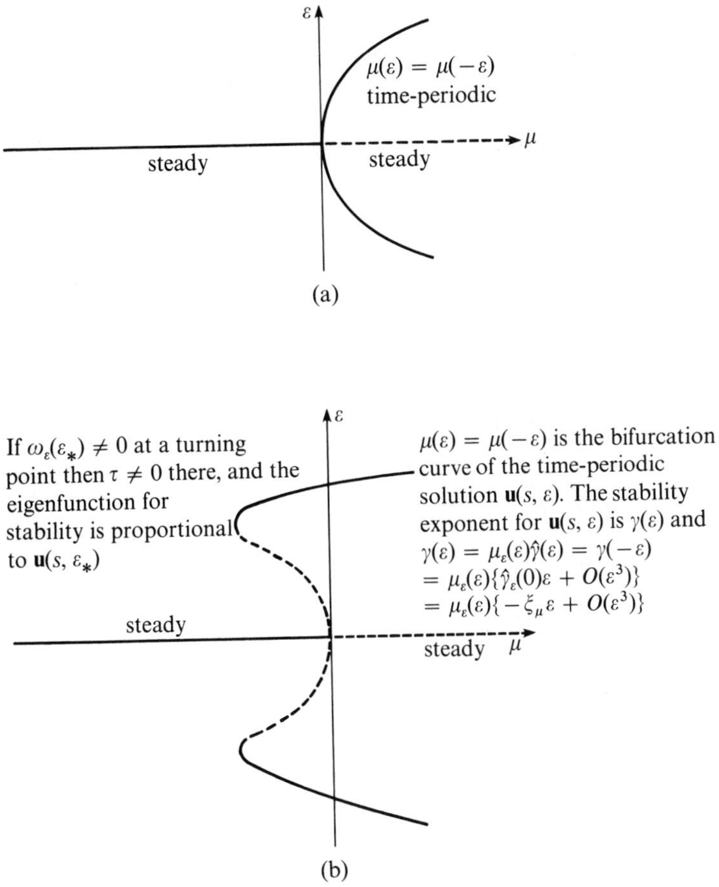
ε
μ(ε) = μ(−ε)
time-periodic
μ
steady
steady
(a)
ε
If $\omega_\varepsilon(\varepsilon_*) \neq 0$ at a turning point then $\tau \neq 0$ there, and the eigenfunction for stability is proportional to $\mathbf{u}(s, \varepsilon_*)$
$\mu(\varepsilon) = \mu(-\varepsilon)$ is the bifurcation curve of the time-periodic solution $\mathbf{u}(s, \varepsilon)$. The stability exponent for $\mathbf{u}(s, \varepsilon)$ is $\gamma(\varepsilon)$ and
$\gamma(\varepsilon) = \mu_\varepsilon(\varepsilon)\hat{\gamma}(\varepsilon) = \gamma(-\varepsilon)$
$= \mu_\varepsilon(\varepsilon)\{\hat{\gamma}_\varepsilon(0)\varepsilon + O(\varepsilon^3)\}$
$= \mu_\varepsilon(\varepsilon)\{-\xi_\mu\varepsilon + O(\varepsilon^3)\}$
steady
steady μ
(b)

Figure VII.2 (a) Supercritical (stable) Hopf bifurcation. (b) Subcritical (unstable) Hopf bifurcation with a turning point. In (b), if zero loses stability strictly as μ is increased past zero, then $\xi_\mu > 0$ and zero is unstable for $\mu > 0$ (as shown); the double eigenvalue of J_0 splits into two simple eigenvalues of $J(\cdot, \varepsilon)$: one eigenvalue is 0 and the other, $\gamma(\varepsilon)$, controls stability.

Near $x^2 + y^2 = 0$, $F(\mu, x^2 + y^2) \sim F_V(\mu, 0)(x^2 + y^2)$, so that $x^2 + y^2 = 0$ is stable when $\mu < 0$ and is unstable when $\mu > 0$. A solution $x^2 + y^2 = \varepsilon^2$ with constant radius bifurcates at the point $(\mu, \varepsilon) = (0, 0)$. This solution exists when $\mu = \mu(\varepsilon^2)$ so long as

$$F(\mu, \varepsilon^2) = 0 \tag{VII.49}$$

and is given by

$$\begin{bmatrix} x \\ y \end{bmatrix} = \begin{bmatrix} X \\ Y \end{bmatrix} = \varepsilon \begin{bmatrix} \cos s \\ \sin s \end{bmatrix} = \mathbf{X}, \qquad s = \omega(\varepsilon^2)t. \tag{VII.50}$$

Small disturbances $\psi = x^2 + y^2 - \varepsilon^2$ of (VII.50) satisfy $\dot{\psi} = 2\varepsilon^2 F_V(\mu(\varepsilon^2), \varepsilon^2)\psi$. The solution $x^2 + y^2 = \varepsilon^2$ is stable (unstable) if $F_V(\mu(\varepsilon^2), \varepsilon^2) < 0$ (> 0). It is of interest to formulate a Floquet problem for the stability of the bifurcating solution (VII.50).

We find that small disturbances $\boldsymbol{\phi}(t) = e^{\gamma t}\boldsymbol{\Gamma}(s)$, $s = \omega(\varepsilon^2)t$ of (III.50) are governed by

$$-\gamma\boldsymbol{\Gamma} + \mathscr{J}(\varepsilon)\boldsymbol{\Gamma} = 0, \qquad \boldsymbol{\Gamma}(s) = \boldsymbol{\Gamma}(s + 2\pi) \tag{VII.51}$$

where

$$\mathscr{J}(\varepsilon) = -\omega(\varepsilon^2)\frac{d}{ds} + J(\varepsilon),$$

$$J(\varepsilon) = 2\varepsilon^2 F_V(\mu, \varepsilon^2)\begin{bmatrix} \cos^2 s & \sin s \cos s \\ \sin s \cos s & \sin^2 s \end{bmatrix}$$

$$- 2\varepsilon^2\omega'(\varepsilon^2)\begin{bmatrix} \sin s \cos s & \sin^2 s \\ -\cos^2 s & -\sin s \cos s \end{bmatrix} + \omega(\varepsilon^2)\begin{bmatrix} 0 & -1 \\ 1 & 0 \end{bmatrix},$$

$$\omega'(\varepsilon^2) = \frac{d\omega(\varepsilon^2)}{d\varepsilon^2}.$$

It is easy to verify that $\dot{\mathbf{X}} \equiv d\mathbf{X}/ds$ is a solution of (VII.51) with $\gamma = 0$. This solution and $\mathbf{X}_\varepsilon \equiv d\mathbf{X}/d\varepsilon$ are independent.

The problem

$$-\gamma\overline{\boldsymbol{\Gamma}}^* + \mathscr{J}^*\overline{\boldsymbol{\Gamma}}^* = 0, \qquad \overline{\boldsymbol{\Gamma}}^*(s) = \overline{\boldsymbol{\Gamma}}^*(s + 2\pi) \tag{VII.52}$$

where

$$\mathscr{J}^*(\varepsilon) = \omega(\varepsilon^2)\frac{d}{ds} + J^*(\varepsilon)$$

$$J^*(\varepsilon) = 2\varepsilon^2 F_V(\mu, \varepsilon^2)\begin{bmatrix} \cos^2 s & \sin s \cos \\ \sin s \cos s & \sin^2 s \end{bmatrix}$$

$$-2\varepsilon^2\omega'(\varepsilon^2)\begin{bmatrix} \sin s \cos s & -\cos^2 s \\ \sin^2 s & -\sin s \cos s \end{bmatrix} + \omega(\varepsilon^2)\begin{bmatrix} 0 & 1 \\ -1 & 0 \end{bmatrix}$$

is adjoint to (VII.51).

Since (VII.47) satisfies all the conditions for Hopf bifurcation, the factorization applies. We may write the factorization as follows:

$$\begin{bmatrix} \Gamma_1 \\ \Gamma_2 \end{bmatrix} = b\left\{\frac{2\varepsilon\omega'}{\gamma}\begin{bmatrix} \dot{X} \\ \dot{Y} \end{bmatrix} + \begin{bmatrix} X_\varepsilon \\ Y_\varepsilon \end{bmatrix} + 2\varepsilon\mu_\varepsilon\begin{bmatrix} q_1 \\ q_2 \end{bmatrix}\right\} \tag{VII.53}$$

and

$$\gamma(\varepsilon^2) = 2\varepsilon^2 F_V(\mu(\varepsilon^2), \varepsilon^2) = -\varepsilon\mu_\varepsilon F_\mu(\mu, \varepsilon^2). \tag{VII.54}$$

Inserting (VII.53) and (VII.54) into (VII.51), we find that

$$-(\hat{\gamma}\mathbf{X}_\varepsilon + F_\mu(\mu, \varepsilon^2)\mathbf{X}) + 2\varepsilon\{-\gamma\mathbf{q} + \mathscr{J}(\varepsilon)\mathbf{q}\} = 0. \qquad \text{(VII.55)}$$

Equation (VII.55) may be simplified by putting $\mathbf{X} = \varepsilon\mathbf{X}_\varepsilon$.

We next note that when $\varepsilon \neq 0$ is small, $\gamma(\varepsilon^2)$ is a simple eigenvalue of (VII.51). (This follows from local analysis of Hopf bifurcation.) It follows from Fredholm theory that (VII.55) is uniquely solvable on a supplementary space of the null space of the operator $-\gamma + \mathscr{J}(\varepsilon)$ if and only if

$$(\hat{\gamma} + \varepsilon F_\mu(\mu, \varepsilon^2)) \int_0^{2\pi} (\Gamma_1^* \cos s + \Gamma_2^* \sin s)\, ds = 0. \qquad \text{(VII.56)}$$

Moreover, it is readily verified that

$$\begin{bmatrix} \Gamma_1^* \\ \Gamma_2^* \end{bmatrix} = C_1 \begin{bmatrix} \cos s \\ \sin s \end{bmatrix} \qquad \text{(VII.57)}$$

if (VII.54) holds. Hence

$$\hat{\gamma} = -\varepsilon F_\mu(\mu(\varepsilon^2), \varepsilon^2)$$

and all solutions of (VII.55) with $\gamma = 2\varepsilon^2 F_V(\mu, \varepsilon^2)$ are proportional to $\mathbf{\Gamma}$ and $q_1 = q_2 = 0$.

Returning now to (VII.53) with $\gamma = -\mu_\varepsilon \varepsilon F_\mu(\mu, \varepsilon^2) = -2\mu'(\varepsilon^2)\varepsilon^2 F_\mu(\mu, \varepsilon^2)$, we have

$$\begin{aligned} \begin{bmatrix} \Gamma_1 \\ \Gamma_2 \end{bmatrix} &= \frac{-\omega'(\varepsilon^2)}{\varepsilon\sqrt{\omega'^2 + \mu'^2 F_\mu^2}} \begin{bmatrix} \dot{X} \\ \dot{Y} \end{bmatrix} + \frac{\mu' F_\mu}{\sqrt{\omega'^2 + \mu'^2 F_\mu^2}} \begin{bmatrix} X_\varepsilon \\ Y_\varepsilon \end{bmatrix} \\ &= \frac{-\omega'(\varepsilon^2)}{\sqrt{\omega'^2 + \mu'^2 F_\mu^2}} \begin{bmatrix} -\sin s \\ \cos s \end{bmatrix} + \frac{\mu' F_\mu}{\sqrt{\omega'^2 + \mu'^2 F_\mu^2}} \begin{bmatrix} \cos s \\ \sin s \end{bmatrix} \end{aligned} \qquad \text{(VII.58)}$$

where $\mu' F_\mu = \mu'(\varepsilon^2) F_\mu(\mu(\varepsilon^2), \varepsilon^2)$.

It is of interest to consider the stability of (VII.50) from a different point of view involving the monodromy matrix and its eigenvalues, the Floquet multipliers. A small disturbance $\boldsymbol{\phi}$ of $\mathbf{X}$ satisfies

$$-\dot{\boldsymbol{\phi}} + J(\varepsilon)\boldsymbol{\phi} = 0. \qquad \text{(VII.59)}$$

There are two and only two independent solutions of (VII.59), $\boldsymbol{\phi}^{(1)}$ and $\boldsymbol{\phi}^{(2)}$. We choose $\boldsymbol{\phi}^{(1)}$ and $\boldsymbol{\phi}^{(2)}$ so that the fundamental solution matrix

$$\mathbf{\Phi}(t) = \begin{bmatrix} \phi_1^{(1)}(t) & \phi_1^{(2)}(t) \\ \phi_2^{(1)}(t) & \phi_2^{(2)}(t) \end{bmatrix}$$

satisfies $\mathbf{\Phi}(0) = \mathbf{I}$, where $\mathbf{I}$ is the unit matrix. We find that

$$\boldsymbol{\phi}^{(1)}(t) = -\frac{\omega'}{F_V}(1 - e^{\gamma t}) \begin{bmatrix} -\sin s \\ \cos s \end{bmatrix} + \begin{bmatrix} \cos s \\ \sin s \end{bmatrix} e^{\gamma t}, \qquad s = \omega t,$$

where γ satisfies (VII.54), and

$$\boldsymbol{\phi}^{(2)}(t) = \begin{bmatrix} -\sin s \\ \cos s \end{bmatrix} = \frac{1}{\varepsilon}\dot{\mathbf{X}}(s), \qquad s = \omega t.$$

The Floquet multipliers λ are the eigenvalues of the monodromy matrix $\mathbf{\Phi}(2\pi/\omega)$, that is, of the matrix

$$\begin{bmatrix} \phi_1^{(1)}\left(\dfrac{2\pi}{\omega}\right) & \phi_1^{(2)}\left(\dfrac{2\pi}{\omega}\right) \\ \phi_2^{(1)}\left(\dfrac{2\pi}{\omega}\right) & \phi_2^{(2)}\left(\dfrac{2\pi}{\omega}\right) \end{bmatrix} = \begin{bmatrix} e^{2\pi\gamma/\omega} & 0 \\ -\dfrac{\omega'}{F_V}(1 - e^{2\pi\gamma/\omega}) & 1 \end{bmatrix}.$$

It follows that $\lambda = 1$ and $\lambda = e^{2\pi\gamma/\omega}$ are algebraically simple eigenvalues of the monodromy matrix whenever $e^{2\pi\gamma/\omega} \neq 1$ and are algebraically double whenever $\gamma = 2\varepsilon^2 F_V = 0$. If $\gamma = 0$ there are still two fundamental solutions of (VII.59): $\boldsymbol{\phi}^{(2)}$ and

$$\boldsymbol{\phi}^{(1)}(t) = 2\varepsilon^2\omega'(\varepsilon^2)t\begin{bmatrix} -\sin s \\ \cos s \end{bmatrix} + \begin{bmatrix} \cos s \\ \sin s \end{bmatrix}.$$

Of these, only $\boldsymbol{\phi}^{(2)}$ is a proper 2π-periodic eigenvector. Since $\boldsymbol{\phi}^{(2)} = \dot{\mathbf{X}}/\varepsilon$, $\mathscr{J}(\dot{\mathbf{X}}) = 0$ when $\gamma = 0$ and $\mathscr{J}(\mathbf{X}_\varepsilon/\omega_\varepsilon) = \dot{\mathbf{X}}$. Hence when $\gamma = 0$, we have a two-link Jordan chain in the frame of a theorem which will be stated and proved at the end of §VIII.4.

The example exhibits the following properties.

1. It undergoes Hopf bifurcation at $\varepsilon = 0$.
2. The factorization theorem holds for all values of ε for which ω and F and its first derivatives are defined.
3. $F(\mu, \varepsilon^2)$ and $\omega(\varepsilon^2)$ are independent functions. In general, $\mu'(\varepsilon^2)$ and $\omega'(\varepsilon^2)$ do not vanish simultaneously.
4. $\gamma = 0$ is always an eigenvalue of $\mathscr{J}(\varepsilon)$. It is geometrically simple and algebraically double when $\omega'(\varepsilon^2) \neq 0$ at points at which $\gamma(\varepsilon) = 0$. If $\omega'(\varepsilon^2) = 0$ where $\gamma(\varepsilon) = 0$, then $\gamma = 0$ may be geometrically and algebraically double eigenvalues. (See §VIII.4.)
5. For suitably chosen functions $F(\mu, V)$, we get secondary and repeated bifurcation of $T(\varepsilon)$-periodic solutions in t (2π-periodic solutions in s) of constant radius ε. In fact, (VII.48) shows that the study of such bifurcation may be reduced to an equation in $\mathbb{R}^1$ whose bifurcation properties were characterized completely in Chapter II.

CHAPTER VIII

Bifurcation of Periodic Solutions in the General Case

In this chapter we shall show that the analysis of bifurcation of periodic solutions from steady ones in $\mathbb{R}^2$, which was discussed in Chapter VII, also applies in $\mathbb{R}^n$ and in infinite dimensions; say, for partial differential equations and for functional differential equations, when the steady solution loses stability at a simple, complex-valued eigenvalue. The mathematical analysis is framed in terms of the autonomous evolution equation (VI.45) reduced to local form and the analysis of the loss of stability of the solution $\mathbf{u} = 0$ given in §VII.9 is valid for the present problem.

VIII.1 Eigenprojections of the Spectral Problem

We write (VI.45) as

$$\dot{\mathbf{u}} = \mathbf{f}(\mu, \mathbf{u}) = \mathbf{f}_u(\mu|\mathbf{u}) + \mathbf{N}(\mu, \mathbf{u}), \qquad \text{(VIII.1)}$$

where $\mathbf{N}(\mu, \mathbf{u}) = O(|\mathbf{u}|^2)$. A small disturbance $\mathbf{v} = e^{\sigma t}\boldsymbol{\zeta}$ of $\mathbf{u} = 0$ satisfies

$$\sigma\boldsymbol{\zeta} = \mathbf{f}_u(\mu|\boldsymbol{\zeta}). \qquad \text{(VIII.2)}$$

The adjoint problem is (see §VI.7)

$$\sigma\bar{\boldsymbol{\zeta}}^* = \mathbf{f}_u^*(\mu|\bar{\boldsymbol{\zeta}}^*) \qquad \text{(VIII.3)}$$

and, very often in applications, there are a countably infinite number of eigenvalues $\{\sigma_n\}$ which are arranged in a sequence corresponding to the size of their real parts:

$$\xi_1 \geq \xi_2 \geq \cdots \geq \xi_n \geq \cdots,$$

clustering at $-\infty$ (see Fig. VI.1). To each eigenvalue there corresponds, at most, a finite number of eigenvectors ζ_n and adjoint eigenvectors ζ_n^*. As indicated in Appendix IV.1, in the case of a semi-simple eigenvalue σ_n we may choose the eigenvectors of $\mathbf{f}_u(\mu|\cdot)$ and $\mathbf{f}_u^*(\mu|\cdot)$ such that they form bi-orthonormal families:

$$\langle \zeta_{nk}, \zeta_{nj}^* \rangle = \delta_{kj}, \qquad k, j = 1, \ldots, m_n, \tag{VIII.4}$$

m_n being the multiplicity of the eigenvalue σ_n (assumed to be semi-simple). Taking now the scalar product of (VIII.1) with ζ_n^* we obtain

$$\begin{aligned} \frac{d}{dt}\langle \mathbf{u}, \zeta_n^* \rangle &= \langle \mathbf{f}_u(\mu|\mathbf{u}), \zeta_n^* \rangle + \langle \mathbf{N}(\mu, \mathbf{u}), \zeta_n^* \rangle \\ &= \langle \mathbf{u}, \mathbf{f}_u^*(\mu|\zeta_n^*) \rangle + \langle \mathbf{N}(\mu, \mathbf{u}), \zeta_n^* \rangle \\ &= \sigma_n \langle \mathbf{u}, \zeta_n^* \rangle + \langle \mathbf{N}(\mu, \mathbf{u}), \zeta_n^* \rangle. \end{aligned} \tag{VIII.5}$$

When $\mathbf{u}$ is small the linearized equations lead to

$$\langle \mathbf{u}(t), \zeta_n^* \rangle \simeq \langle \mathbf{u}(0), \zeta_n^* \rangle e^{\xi_n(\mu)t} e^{i\eta_n(\mu)t},$$

so that if $\xi_n(\mu) < 0$, the projection $\langle \mathbf{u}(t), \zeta_n^* \rangle$ decays to zero. In fact, for the full nonlinear problem there is a coupling between different projections, and if some of these do not decay, this last result is no longer true. Nevertheless, the important part of the evolution problem (VIII.1) is related to the part of the spectrum of $\mathbf{f}_u(\mu|\cdot)$ for which $\xi_n(\mu) \geq 0$.

In the problem of bifurcation studied in this chapter we shall assume that the real part of two complex-conjugate simple eigenvalues $\sigma(\mu)$, $\bar{\sigma}(\mu)$ changes sign when μ crosses 0 and the remainder of the spectrum stays on the left-hand side of the complex plane. Suppose ζ and $\bar{\zeta}^*$ are the eigenvectors of $\mathbf{f}_u(\mu|\cdot)$, $\mathbf{f}_u^*(\mu|\cdot)$ belonging to the eigenvalue $\sigma(\mu)$. Then, the equation governing the evolution of the projection

$$\frac{d}{dt}\langle \mathbf{u}, \zeta^* \rangle = \sigma(\mu)\langle \mathbf{u}, \zeta^* \rangle + \langle \mathbf{N}(\mu, \mathbf{u}), \zeta^* \rangle, \tag{VIII.6}$$

is complex-valued, that is, two-dimensional. So our problem is essentially two-dimensional whenever

$$\mathbf{u} - \langle \mathbf{u}, \zeta^* \rangle \zeta - \langle \mathbf{u}, \bar{\zeta}^* \rangle \bar{\zeta}$$

is an "extra little part," as in §VI.5.

VIII.2 Equations Governing the Projection and the Complementary Projection

Now we shall delineate the sense in which the essentially two-dimensional problem is strictly two-dimensional. We first decompose the bifurcating solution $\mathbf{u}$ into a real-valued sum

$$\mathbf{u}(t) = a(t)\zeta + \bar{a}(t)\bar{\zeta} + \mathbf{w}(t) \tag{VIII.7}$$

where

$$\langle \mathbf{w}, \zeta^* \rangle = \langle \bar{\zeta}, \zeta^* \rangle = \langle \zeta, \zeta^* \rangle - 1 = 0. \tag{VIII.8}$$

Substituting (VIII.7) into (VIII.1) we find, using (VIII.2), that

$$[\dot{a} - \sigma(\mu)]\zeta + [\dot{\bar{a}} - \bar{\sigma}(\mu)\bar{a}]\bar{\zeta} + \frac{d\mathbf{w}}{dt} = \mathbf{f}_u(\mu|\mathbf{w}) + \mathbf{N}(\mu, \mathbf{u}). \tag{VIII.9}$$

Projecting (VIII.9) with ζ^* leads us to an evolution problem for the "little part" $\mathbf{w}$ on a supplementary space of the space spanned by ζ and $\bar{\zeta}$:

$$\frac{d\mathbf{w}}{dt} = \mathbf{f}_u(\mu|\mathbf{w}) + (\mathbf{N}(\mu, \mathbf{u}) - \langle \mathbf{N}(\mu, \mathbf{u}), \zeta^* \rangle\zeta - \langle \mathbf{N}(\mu, \mathbf{u}), \bar{\zeta}^* \rangle\bar{\zeta}). \tag{VIII.10}$$

and to an evolution equation for the projected part

$$\dot{a} - \sigma(\mu)a = \langle \mathbf{N}(\mu, \mathbf{u}), \zeta^* \rangle. \tag{VIII.11}$$

In deriving (VIII.11) we made use of the relations

$$\left\langle \frac{d\mathbf{w}}{dt}, \zeta^* \right\rangle = \frac{d}{dt}\langle \mathbf{w}, \zeta^* \rangle = 0$$

and

$$\langle \mathbf{f}_u(\mu|\mathbf{w}), \zeta^* \rangle = \langle \mathbf{w}, \mathbf{f}_u^*(\mu|\zeta^*) \rangle = \sigma\langle \mathbf{w}, \zeta^* \rangle = 0.$$

Equation (VIII.10) now follows easily from (VIII.9) and (VIII.11).

In sum, (VIII.11) governs the evolution of the projection of the solution $\mathbf{u}$ into the eigensubspace belonging to the eigenvalue $\sigma_1(\mu) = \sigma(\mu)$, and (VIII.10) governs the evolution of the part of the solution which is orthogonal to the subspace spanned by ζ^* and $\bar{\zeta}^*$.

In bifurcation problems the complementary projection $\mathbf{w}$ plays a minor role; it arises only as a response generated by nonlinear coupling to the component of the solution spanned by ζ and $\bar{\zeta}$. To see this we note that

$$\begin{aligned}
\langle \mathbf{N}(\mu, \mathbf{u}), \zeta^* \rangle &= \tfrac{1}{2}\langle (\mathbf{f}_{uu}(\mu|\mathbf{u}|\mathbf{u}) + O(\|\mathbf{u}\|^3)), \zeta^* \rangle \\
\tfrac{1}{2}\langle \mathbf{f}_{uu}(\mu|\mathbf{u}|\mathbf{u}), \zeta^* \rangle &= \alpha(\mu)a^2 + 2\beta(\mu)|a|^2 + \gamma(\mu)\bar{a}^2 \\
&\quad + 2a\langle \mathbf{f}_{uu}(\mu|\zeta|\mathbf{w}), \zeta^* \rangle + 2\bar{a}\langle \mathbf{f}_{uu}(\mu|\bar{\zeta}|\mathbf{w}), \zeta^* \rangle \\
&\quad + \langle \mathbf{f}_{uu}(\mu|\mathbf{w}|\mathbf{w}), \zeta^* \rangle,
\end{aligned} \tag{VIII.12}$$

$$\alpha(\mu) = \tfrac{1}{2}\langle \mathbf{f}_{uu}(\mu|\zeta|\zeta), \zeta^* \rangle$$

$$\beta(\mu) = \tfrac{1}{2}\langle \mathbf{f}_{uu}(\mu|\zeta|\bar{\zeta}), \zeta^* \rangle$$

and

$$\gamma(\mu) = \tfrac{1}{2}\langle \mathbf{f}_{uu}(\mu|\bar{\zeta}|\bar{\zeta}), \zeta^* \rangle.$$

It follows that amplitude equation (VIII.11) may be written as

$$\begin{aligned}
\dot{a} - \sigma(\mu)a = \alpha(\mu)a^2 + 2\beta(\mu)|a|^2 + \gamma(\mu)\bar{a}^2 \\
+ O(|a|^3 + |a|\,\|\mathbf{w}\| + \|\mathbf{w}\|^2).
\end{aligned} \tag{VIII.13}$$

Returning now to (VIII.10) with (VIII.12) we find that after a long time $\mathbf{w} = O(|a|^2)$ and dramatize the two-dimensional structure of Hopf bifurcation in the general case by comparing (VIII.13) with the equation (VII.5) which governs the stability of the strictly two-dimensional problem.

VIII.3 The Series Solution Using the Fredholm Alternative

It is possible to construct the time-periodic solution which bifurcates from $\mathbf{u} = 0$ at criticality ($\mu = 0$) as a power series in some amplitude ε, as in (VII.44). In this construction we would compute the coefficients of the series as solutions of differential equations which arise by identification after substituting the series into (VIII.10) and (VIII.11). The strategy in this case* is to project (get (VIII.10) and (VIII.11)) and then expand. An alternative strategy, expand and then project, given below, is cleaner and easier to implement.

In the constructions, we evaluate quantities associated with the spectral problems (VIII.2) and (VIII.3) at $\mu = 0$

$$[\varepsilon, \mu, \xi(\mu), \eta(\mu), \sigma(\mu), \boldsymbol{\zeta}(\mu), \boldsymbol{\zeta}^*(\mu)] \to [0, 0, 0, \omega_0, i\omega_0, \boldsymbol{\zeta}_0, \boldsymbol{\zeta}_0^*]$$

We assume that $\pm i\omega_0$ are simple, isolated eigenvalues of $\mathbf{f}_u(0|\cdot)$, i.e., $i\omega_0 \boldsymbol{\zeta}_0 = \mathbf{f}_u(0|\boldsymbol{\zeta}_0)$, $-i\omega_0 \bar{\boldsymbol{\zeta}}_0 = \mathbf{f}_u(0|\bar{\boldsymbol{\zeta}}_0)$; and that all other eigenvalues of $\mathbf{f}_u(0|\cdot)$ have negative real parts. It is also assumed that the loss of stability of $\mathbf{u} = 0$ is strict when $\xi_\mu(0) > 0$. Noting next that the equation for the first derivative with respect to μ at $\mu = 0$ of (VIII.2),

$$\sigma_\mu(0)\boldsymbol{\zeta}_0 + i\omega_0 \boldsymbol{\zeta}_\mu = \mathbf{f}_u(0|\boldsymbol{\zeta}_\mu) + \mathbf{f}_{u\mu}(0|\boldsymbol{\zeta}_0),$$

is solvable if and only if

$$\sigma_\mu(0) = \langle \mathbf{f}_{u\mu}(0|\boldsymbol{\zeta}_0), \boldsymbol{\zeta}_0^* \rangle, \tag{VIII.14}$$

our assumption about the strict loss of stability implies that the real part of (VIII.14) is positive.

We are going to construct the periodic solution which bifurcates from $\mathbf{u} = 0$ at criticality. There are two independent periodic solutions of the linearized problem $\dot{\mathbf{v}} = \mathbf{f}_u(0|\mathbf{v})$ at criticality: $\mathbf{v}(t) = e^{i\omega_0 t}\boldsymbol{\zeta}_0$ and $\bar{\mathbf{v}}(t)$. We write $\omega_0 t = s$ and set $\mathbf{z}(s) = e^{is}\boldsymbol{\zeta}_0 = \mathbf{v}(s/\omega_0)$. Now we introduce a space of 2π-periodic functions.† We call this space of 2π-periodic functions $\mathbb{P}_{2\pi}$. Then

* This is done for partial differential equations in the paper by G. Iooss, Existence et stabilité de la solution périodique secondaire intervenant dans les problèmes d'évolution du type Navier Stokes. *Arch. Rational Mech. Anal.*, **47**, 301–329 (1972).

† Naturally, we are assuming that the functions in $\mathbb{P}_{2\pi}$ have the smoothness required in our calculations. The precise degree of smoothness is specified in the references of this chapter and will not be specified here.

$\mathbf{z}$ and $\bar{\mathbf{z}}$ are in $\mathbb{P}_{2\pi}$. We also define a scalar product in $\mathbb{P}_{2\pi}$

$$[\mathbf{a}, \mathbf{b}] \stackrel{\text{def}}{=} \frac{1}{2\pi} \int_0^{2\pi} \langle \mathbf{a}(s), \mathbf{b}(s) \rangle \, ds.$$

and an operator

$$\mathbb{J}_0 = -\omega_0 \frac{d}{ds} + \mathbf{f}_u(0|\cdot) \tag{VIII.15}$$

in whose null space are $\mathbf{z} = e^{is}\zeta_0$ and $\bar{\mathbf{z}} \in \mathbb{P}_{2\pi}$

$$\mathbb{J}_0 \mathbf{z} = \mathbb{J}_0 \bar{\mathbf{z}} = 0. \tag{VIII.16}$$

We define an adjoint operator $\mathbb{J}_0^*$ acting on arbitrary fields $\mathbf{a}(s), \mathbf{b}(s) \in \mathbb{P}_{2\pi}$,

$$[\mathbb{J}_0 \mathbf{a}, \mathbf{b}] = [\mathbf{a}, \mathbb{J}_0^* \mathbf{b}] \tag{VIII.17}$$

and find that

$$\mathbb{J}_0^* = \omega_0 \frac{d}{ds} + \mathbf{A}^*(0) \tag{VIII.18}$$

where $\mathbf{A}^*(0)$ is defined in (VI.48) and

$$\mathbb{J}_0^* \mathbf{z}^* = \mathbb{J}_0^* \bar{\mathbf{z}}^* = 0, \tag{VIII.19}$$

where

$$\mathbf{z}^* = e^{is}\zeta_0^* \in \mathbb{P}_{2\pi} \tag{VIII.20}$$

(ζ_0^* satisfies (VIII.2)). We may rewrite (VIII.14) as

$$\begin{aligned} \sigma_\mu(0) &= \langle \mathbf{f}_{u\mu}(0|\zeta_0 e^{is})e^{-is}, \zeta_0^* \rangle \\ &= \langle \mathbf{f}_{u\mu}(0|\mathbf{z}), \mathbf{z}^* \rangle \\ &= [\mathbf{f}_{u\mu}(0|\mathbf{z}), \mathbf{z}^*]. \end{aligned} \tag{VIII.21}$$

We are now ready to construct the periodic bifurcating solutions of $\dot{\mathbf{u}} = f(\mu, \mathbf{u})$ in a series of powers of the amplitude

$$\varepsilon = [\mathbf{u}, \mathbf{z}^*]. \tag{VIII.22}$$

Different definitions of the amplitude are possible. For example, we can set $\varepsilon^n = \not{f}[\mathbf{u}]$ where $\not{f}[\mathbf{u}]$ is a homogeneous functional of degree n so that if $\mathbf{u} = \varepsilon \mathbf{v}$ then $1 = \not{f}[\mathbf{v}]$. It is usually best to choose the amplitude suggested by the application. For example, in nonlinear problems involving heat transfer with boundary temperature prescribed it would be useful to define ε in terms of the integrated heat flux.

We seek 2π-periodic solutions of $s = \omega(\varepsilon)t$ in the form

$$\mathbf{u}(s, \varepsilon) = \mathbf{u}(s + 2\pi, \varepsilon), \qquad \mu(\varepsilon), \qquad \omega(\varepsilon)$$

where

$$\mathbf{u}(s, 0) = 0, \qquad \mu(0) = 0, \qquad \omega(0) = \omega_0.$$

and

$$\omega(\varepsilon)\frac{d\mathbf{u}}{ds} = \mathbf{f}(\mu(\varepsilon), \mathbf{u}). \tag{VIII.23}$$

We can find this solution by the method of power series using the Fredholm alternative

$$\begin{bmatrix} \mathbf{u}(s, \varepsilon) \\ \mu(\varepsilon) \\ \omega(\varepsilon) - \omega_0 \end{bmatrix} = \sum_{n=1}^{\infty} \frac{\varepsilon^n}{n!} \begin{bmatrix} \mathbf{u}_n(s) \\ \mu_n \\ \omega_n \end{bmatrix}. \tag{VIII.24}$$

Insert (VIII.24) into (VIII.23) and (VIII.22) to find the equations governing the coefficients on the right of (VIII.24). To keep the definition (VIII.22) of ε we must have

$$[\mathbf{u}_1, \mathbf{z}^*] - 1 = [\mathbf{u}_n, \mathbf{z}^*] = 0, \qquad n \geq 2. \tag{VIII.25}$$

Equation (VIII.23) will be satisfied if

$$\mathbb{J}_0 \mathbf{u}_1 = 0 \tag{VIII.26}$$

$$\mathbb{J}_0 \mathbf{u}_2 - 2\omega_1 \frac{d\mathbf{u}_1}{ds} + 2\mu_1 \mathbf{f}_{u\mu}(0|\mathbf{u}_1) + \mathbf{f}_{uu}(0|\mathbf{u}_1|\mathbf{u}_1) = 0 \tag{VIII.27}$$

$$\begin{aligned} \mathbb{J}_0 \mathbf{u}_3 &- 3\omega_1 \frac{d\mathbf{u}_2}{ds} + 3\mu_1 \mathbf{f}_{u\mu}(0|\mathbf{u}_2) - 3\omega_2 \frac{d\mathbf{u}_1}{ds} \\ &+ 3\mu_2 \mathbf{f}_{u\mu}(0|\mathbf{u}_1) + 3\mu_1 \mathbf{f}_{uu\mu}(0|\mathbf{u}_1|\mathbf{u}_1) \\ &+ 3\mu_1^2 \mathbf{f}_{u\mu\mu}(0|\mathbf{u}_1) + 3\mathbf{f}_{uu}(0|\mathbf{u}_1|\mathbf{u}_2) \\ &+ \mathbf{f}_{uuu}(0|\mathbf{u}_1|\mathbf{u}_1|\mathbf{u}_1) = 0 \end{aligned} \tag{VIII.28}$$

and, in general,

$$\mathbb{J}_0 \mathbf{u}_n - n\omega_{n-1} \frac{d\mathbf{u}_1}{ds} + n\mu_{n-1} \mathbf{f}_{u\mu}(0|\mathbf{u}_1) + \mathbf{R}_{n-2} = 0. \tag{VIII.29}$$

where $\mathbf{R}_{n-2}$ depends on terms of order lower than $n - 1$.

Now we must solve equations (VIII.25–29). Our assumption that $\pm i\omega_0$ are simple eigenvalues of $\mathbf{f}_u(0|\cdot)$ implies that zero is a semi-simple double eigenvalue of $\mathbb{J}_0$, with two linearly independent solutions $\mathbf{z}$ and $\bar{\mathbf{z}}$. Any other vector annihilated by $\mathbb{J}_0$, say $\mathbf{u}_1$, can be expressed as a linear combination of the independent solutions. Since $\mathbf{u}_1$ is real

$$\mathbf{u}_1 = cz + \bar{c}\bar{z} = ce^{is}\zeta_0 + \bar{c}e^{-is}\bar{\zeta}_0.$$

Now the origin of s is indeterminate, so we may just as well use another $s \sim s + \alpha$, where α can be selected so that $ce^{i\alpha} = c'$ is real-valued. Then, without losing generality,

$$\mathbf{u}_1 = c'(\mathbf{z} + \bar{\mathbf{z}}) = \mathbf{z} + \bar{\mathbf{z}}, \tag{VIII.30}$$

where $c' = 1$ is implied by $(\text{VIII.25})_1$.

Equations (VIII.27–29) are of the form

$$(\mathbb{J}_0\mathbf{u})(s) = \mathbf{g}(s) = \mathbf{g}(s + 2\pi). \tag{VIII.31}$$

We want to find $\mathbf{u}(s) = \mathbf{u}(s + 2\pi)$ solving (VIII.31). There are no solutions $\mathbf{u} \in \mathbb{P}_{2\pi}$ of (VIII.31) unless $\mathbf{g}$ satisfies certain compatability conditions which are usually framed as a Fredholm alternative:

Theorem. *Equation* (VIII.31) *is solvable for* $\mathbf{u} \in \mathbb{P}_{2\pi}$ *if and only if*

$$[\mathbf{g}, \mathbf{z}^*] = [\mathbf{g}, \bar{\mathbf{z}}^*] = 0. \tag{VIII.32}$$

If $\mathbf{g}$ *is real-valued, one complex condition* (*two real conditions*)

$$[\mathbf{g}, \mathbf{z}^*] = \overline{[\mathbf{g}, \bar{\mathbf{z}}^*]} = 0 \tag{VIII.33}$$

suffices for solvability.

The "only if" part of the Fredholm alternative is easy to prove because

$$[\mathbf{g}, \mathbf{z}^*] = [\mathbb{J}_0\mathbf{u}, \mathbf{z}^*] = [\mathbf{u}, \mathbb{J}_0^*\mathbf{z}^*] = 0.$$

For the "if" part we refer the reader to standard works which use elementary results from functional analysis not considered in this book.*

We can select ω_{n-1} and μ_{n-1} so that the equations (VIII.29) are solvable. Using (VIII.30) and (VIII.33) we find that (VIII.29) is solvable if

$$-n\omega_{n-1}\left(\left[\frac{d\mathbf{z}}{ds}, \mathbf{z}^*\right] + \left[\frac{d\bar{\mathbf{z}}}{ds}, \mathbf{z}^*\right]\right) + n\mu_{n-1}([\mathbf{f}_{u\mu}(0|\mathbf{z}), \mathbf{z}^*]$$
$$+ [\mathbf{f}_{u\mu}(0|\bar{\mathbf{z}}), \mathbf{z}^*]) + [\mathbf{R}_{n-2}, \mathbf{z}^*] = 0.$$

Now the equation $\langle\zeta, \zeta^*\rangle = 1$ implies $[\mathbf{z}, \mathbf{z}^*] = 1$. Furthermore, $[\mathbf{z}, \bar{\mathbf{z}}^*] = [\bar{\mathbf{z}}, \mathbf{z}^*] = [\mathbf{f}_{u\mu}(0|\bar{\mathbf{z}}), \mathbf{z}^*] = 0$, for example, $[\mathbf{f}_{u\mu}(0|\bar{\mathbf{z}}), \mathbf{z}^*] = [e^{-2is}\mathbf{f}_{u\mu}(0|\bar{\zeta}), \bar{\zeta}^*] = 0$ after integration on s. So using the results just listed and (VIII.21) we get one complex equation

$$n\{-i\omega_{n-1} + \mu_{n-1}\sigma_\mu\} + [\mathbf{R}_{n-2}, \mathbf{z}^*] = 0, \qquad n \geq 2$$

or two real equations

$$n\mu_{n-1}\xi_\mu + \operatorname{Re}[\mathbf{R}_{n-2}, \mathbf{z}^*] = 0 \tag{VIII.34$_1$}$$

and

$$n\{-\omega_{n-1} + \mu_{n-1}\eta_\mu\} + \operatorname{Im}[\mathbf{R}_{n-2}, \mathbf{z}^*] = 0 \tag{VIII.34$_2$}$$

* For example, see D. D. Joseph and D. H. Sattinger, Bifurcating time-periodic solutions and their stability, *Arch. Rational Mech. Anal.*, **45**, 79–108 (1972); or D. H. Sattinger, *Topics in Stability and Bifurcation Theory*, Lecture Notes in Mathematics **309** (Berlin–Heidelberg–New York: Springer-Verlag, 1972); or V. I. Iudovich, Appearance of self-oscillations in a fluid (in Russian), *Prikl. Mat. Mekh.*, **35**, 638–655 (1971).

We can solve (VIII.34)$_1$ for μ_{n-1} if $\xi_\mu \neq 0$ ($\xi_\mu > 0$ by virtue of the assumption about strict loss of stability). Then (VIII.34)$_2$ can be solved for ω_{n-1}. When $n = 2$,

$$[\mathbf{R}_0, \mathbf{z}^*] = [\mathbf{f}_{uu}(0|\mathbf{u}_1|\mathbf{u}_1), \mathbf{z}^*] = 0,$$

so that $\omega_1 = \mu_1 = 0$ and

$$\mathbb{J}_0 \mathbf{u}_2 = -\mathbf{f}_{uu}(0|\mathbf{z} + \bar{\mathbf{z}}|\mathbf{z} + \bar{\mathbf{z}}), \qquad [\mathbf{u}_2, \mathbf{z}^*] = 0.$$

When $n = 3$, we find that

$$3\{-i\omega_2 + \mu_2\sigma_\mu\} + 3[\mathbf{f}_{uu}(0|\mathbf{z} + \bar{\mathbf{z}}|\mathbf{u}_2), \mathbf{z}^*] + [\mathbf{f}_{uuu}(0|\mathbf{z} + \bar{\mathbf{z}}|\mathbf{z} + \bar{\mathbf{z}}|\mathbf{z} + \bar{\mathbf{z}}), \mathbf{z}^*] = 0 \qquad \text{(VIII.35)}$$

We can show, by mathematical induction,* that

$$\mu_{2n+1} = \omega_{2n+1} = 0, \qquad n = 0, 1, 2, \ldots,$$

so that $\mu(\varepsilon)$ and $\omega(\varepsilon)$ are even functions. The bifurcating solution which was just constructed reduces to that computed in Chapter VII when the problem is specialized to a two-dimensional one. In fact, the formulas giving $\mu(\varepsilon)$ and $\omega(\varepsilon)$ are identical for the general problem and the two-dimensional one through terms of second order in ε.

VIII.4 Stability of the Hopf Bifurcation in the General Case

We are interested in solution $\mathbf{V}(t)$ of the autonomous problem $\dot{\mathbf{V}} = \mathbf{f}(\mu, \mathbf{V})$. Let $\mathbf{V} = \mathbf{u}(s, \varepsilon) + \mathbf{v}(t)$, $s = \omega(\varepsilon)t$, $\mu = \mu(\varepsilon)$, where $\mathbf{v}(t)$ is a small disturbance of $\mathbf{u}$. Using $\omega\dot{\mathbf{u}} = \mathbf{f}(\mu(\varepsilon), \mathbf{u})$ we find, after linearizing, that

$$\dot{\mathbf{v}} = \mathbf{f}(\mu, \mathbf{u} + \mathbf{v}) - \mathbf{f}(\mu, \mathbf{u}) \sim \mathbf{f}_u(\mu(\varepsilon), \mathbf{u}(s, \varepsilon)|\mathbf{v}) \qquad \text{(VIII.36)}$$

We may study (VIII.36) using the method of Floquet (see §VII.6).

$$\mathbf{v}(t) = e^{\gamma t}\boldsymbol{\zeta}(s), \qquad s = \omega(\varepsilon)t,$$

where

$$\boldsymbol{\zeta}(s) = \boldsymbol{\zeta}(s + 2\pi), \qquad \boldsymbol{\zeta} \in \mathbb{P}_{2\pi}$$

and

$$\gamma\boldsymbol{\zeta} + \omega(\varepsilon)\frac{d\boldsymbol{\zeta}}{ds} = \mathbf{f}_u(\mu(\varepsilon), \mathbf{u}|\boldsymbol{\zeta}).$$

* Joseph and Sattinger (1972), op. cit.

It is useful to define an operator

$$\mathbb{J}(\varepsilon)(\cdot) = -\omega(\varepsilon)\frac{d(\cdot)}{ds} + \mathbf{f}_u(\mu(\varepsilon), \mathbf{u}(s, \varepsilon)|\cdot),$$

Then

$$\gamma\boldsymbol{\zeta} = \mathbb{J}\boldsymbol{\zeta}. \tag{VIII.37}$$

Differentiating $\omega(\varepsilon)\dot{\mathbf{u}}(s, \varepsilon) = \mathbf{f}(\mu(\varepsilon), \mathbf{u}(s, \varepsilon))$ first with respect to s we get

$$\mathbb{J}(\varepsilon)\dot{\mathbf{u}} = 0, \tag{VIII.38}$$

and then with respect to ε we get

$$\omega_\varepsilon \dot{\mathbf{u}} = \mu_\varepsilon \mathbf{f}_\mu(\mu(\varepsilon), \mathbf{u}(s, \varepsilon)) + \mathbb{J}\mathbf{u}_\varepsilon. \tag{VIII.39}$$

We are of course assuming that (VIII.24) is convergent on an interval $I_1(\varepsilon)$ around $\varepsilon = 0$. Now we combine (VIII.37–39) to prove the following result.

Factorization Theorem (General case). *The following representations hold:*

$$\gamma = \mu_\varepsilon(\varepsilon)\hat{\gamma}(\varepsilon)$$

$$\boldsymbol{\zeta}(s, \varepsilon) = c(\varepsilon)\left\{\frac{\tau}{\gamma}\dot{\mathbf{u}}(s, \varepsilon) + \mathbf{u}_\varepsilon(s, \varepsilon) + \varepsilon\mu_\varepsilon \mathbf{q}(s, \varepsilon)\right\} \tag{VIII.40}$$

$$\tau = \omega_\varepsilon(\varepsilon) + \mu_\varepsilon(\varepsilon)\hat{\tau}(\varepsilon),$$

where $c(\varepsilon)$ is a normalizing factor and $\hat{\tau}(\varepsilon)$ and $\hat{\gamma}$ and $\mathbf{q}(\cdot, \varepsilon) \in \mathbb{P}_{2\pi}$ satisfy the equation

$$\hat{\tau}\dot{\mathbf{u}} + \hat{\gamma}\mathbf{u}_\varepsilon + \mathbf{f}_\mu(\mu(\varepsilon), \mathbf{u}(\cdot, \varepsilon)) + \varepsilon\{\gamma\mathbf{q} - \mathbb{J}\mathbf{q}\} = 0$$

and are smooth functions in $I_2 \subseteq I_1$ containing the point $\varepsilon = 0$. Moreover, $\hat{\tau}(\varepsilon)$ and $\hat{\gamma}(\varepsilon)/\varepsilon$ are even functions and such that

$$\hat{\gamma}_\varepsilon(0) = -\xi_\mu(0), \qquad \hat{\tau}(0) = -\eta_\mu(0). \tag{VIII.41}$$

The interpretation of the stability results for the general case is exactly the one given in §VII.9 for the strictly two-dimensional problem. In particular, it shows that $\gamma(\varepsilon)$ changes sign at every point at which $\hat{\gamma}(\varepsilon) \neq 0$ and $\mu_\varepsilon(\varepsilon)$ changes sign. (In $\mathbb{R}^1$, such points are called regular turning points; see §II.8). Let us designate such points as $\varepsilon = \varepsilon_0$ and suppose $c(\varepsilon)$ to be chosen so that $c(\varepsilon) \sim \mu_\varepsilon(\varepsilon)$ as $\varepsilon \to \varepsilon_0$. Then

$$\boldsymbol{\zeta}(s, \varepsilon) \sim \frac{\omega_\varepsilon(\varepsilon)}{\hat{\gamma}(\varepsilon)}\dot{\mathbf{u}}(s, \varepsilon) + \mu_\varepsilon \mathbf{u}_\varepsilon(s, \varepsilon) + \varepsilon\mu_\varepsilon^2(\varepsilon)\mathbf{q}(s, \varepsilon). \tag{VIII.42}$$

If $\omega_\varepsilon(\varepsilon_0) \neq 0$, then $\boldsymbol{\zeta}(s, \varepsilon_0) = \omega_\varepsilon(\varepsilon_0)\dot{\mathbf{u}}(s, \varepsilon_0)/\hat{\gamma}(\varepsilon_0)$ is an eigenfunction of $\mathbb{J}(\varepsilon_0)$. The following theorem holds.*

* See D. D. Joseph and D. A. Nield, Stability of bifurcating time-periodic and steady solutions of arbitrary amplitude. *Arch. Rational Mech. Anal.*, **49**, 321 (1973). Also see D. D. Joseph, Factorization theorems, stability and repeated bifurcation. *Arch. Rational Mech. Anal.*, 99–118 (1977).

Theorem. *Assume that at* $\varepsilon_0 \neq 0$, $\gamma(\varepsilon_0) = 0$ *and* $\hat{\gamma}(\varepsilon_0) \neq 0$; *then the eigenvalue zero of* $\mathbb{J}(\varepsilon_0)$ *has at least the algebraic multiplicity* 2; $\dot{\mathbf{u}}(s, \varepsilon_0)$ *is a proper eigenvector of* $\mathbb{J}(\varepsilon_0)$,

$$\mathbb{J}(\varepsilon_0)\dot{\mathbf{u}}(\cdot, \varepsilon_0) = 0, \tag{VIII.43}$$

and when $\omega_\varepsilon(\varepsilon_0) \neq 0$ *then* $\mathbf{u}_\varepsilon(s, \varepsilon_0)$ *is a generalized eigenvector of* $\mathbb{J}(\varepsilon_0)$:

$$[\mathbb{J}(\varepsilon_0)\mathbf{u}_\varepsilon(\cdot, \varepsilon_0)](s) = \omega_\varepsilon(\varepsilon_0)\dot{\mathbf{u}}(s, \varepsilon_0). \tag{VIII.44}$$

If $\omega_\varepsilon(\varepsilon_0) = 0$, *then the geometric multiplicity of zero is at least two and* $\dot{\mathbf{u}}$ *and* $\mathbf{u}_\varepsilon$ *are both proper eigenvectors of* $\mathbb{J}(\varepsilon_0)$.

The proof of this theorem follows from the identity

$$\mathbb{J}(\varepsilon)\mathbf{u}_\varepsilon = \omega_\varepsilon \dot{\mathbf{u}} - \mu_\varepsilon \mathbf{f}_\mu(\mu(\varepsilon), \mathbf{u}),$$

where we recall that $\mu_\varepsilon(\varepsilon_0) = 0$.

Some of the results given in Example VII.1 of bifurcation and stability of periodic solutions can be viewed as applications of the foregoing theorem.

EXAMPLE VIII.1 (Periodic solutions of partial differential equations.) Let us consider the following partial differential system:

$$\begin{aligned}
\frac{\partial U_1}{\partial t} &= \frac{\partial^2 U_1}{\partial x^2} + (\pi^2 + \mu)(U_1 - U_2) \\
&\quad - U_1\left\{U_1^2 + U_2^2 + \frac{1}{\pi^2}\left[\left(\frac{\partial U_1}{\partial x}\right)^2 + \left(\frac{\partial U_2}{\partial x}\right)^2\right]\right\} \\
\frac{\partial U_2}{\partial t} &= \frac{\partial^2 U_2}{\partial x^2} + (\pi^2 + \mu)(U_1 + U_2) \\
&\quad - U_2\left\{U_1^2 + U_2^2 + \frac{1}{\pi^2}\left[\left(\frac{\partial U_1}{\partial x}\right)^2 + \left(\frac{\partial U_2}{\partial x}\right)^2\right]\right\},
\end{aligned} \tag{VIII.45$_1$}$$

where U_i, $i = 1, 2$ are real functions defined for $t \geq 0$, $0 \leq x \leq 1$ satisfying the boundary conditions

$$U_i(t, 0) = U_i(t, 1) = 0, \qquad i = 1, 2. \tag{VIII.45$_2$}$$

In this example (as in Example VI.2) we choose $H = [L^2(0, 1)]^2 = \{(u_1, u_2): u_i \in L^2(0, 1)\}$, with the scalar product

$$\langle \mathbf{u}, \mathbf{v} \rangle = \int_0^1 [U_1(x)\overline{V}_1(x) + U_2(x)\overline{V}_2(x)]\, dx, \tag{VIII.46}$$

where we use the notation $\mathbf{u} = (U_1(\cdot), U_2(\cdot))$, $\mathbf{v} = (V_1(\cdot), V_2(\cdot))$. The system (VIII.45) may be written in H as:

$$\frac{d\mathbf{u}}{dt} = \mathbf{f}(\mu, \mathbf{u}) = \mathbf{f}_u(\mu | \mathbf{u}) + \mathbf{N}(\mu, \mathbf{u}), \tag{VIII.47}$$

where the linear operator $\mathbf{f}_u(\mu|\cdot)$ and the nonlinear operator $\mathbf{N}(\mu, \cdot)$ are defined in the *subspace* of H which consists of $\mathbf{u} = (U_1(\cdot), U_2(\cdot))$ such that

$$U_i(\cdot), \quad \frac{\partial U_i(\cdot)}{\partial x}, \quad \frac{\partial^2 U_i(\cdot)}{\partial x^2} \quad \text{are in } L^2(0, 1),$$

and $U_i(0) = U_i(1) = 0$.

The evolution equation (VIII.47) in H is a Hilbert space realization of (VIII.45) provided that

$$\mathbf{f}_u(\mu|\cdot) = \mathbf{f}_u(0|\cdot) + \mu\mathbf{f}_{u\mu}(0|\cdot)$$

$$\mathbf{N}(\mu, \mathbf{u}) = \mathbf{C}(\mathbf{u}, \mathbf{u}, \mathbf{u}) = \tfrac{1}{6}\mathbf{f}_{uuu}(0|\mathbf{u}|\mathbf{u}|\mathbf{u})$$

$$[f_u(0|\mathbf{u})](x) = \left(\frac{\partial^2 U_1(x)}{\partial x^2} + \pi^2(U_1(x) - U_2(x)), \frac{\partial^2 U_2(x)}{\partial x^2} + \pi^2(U_1(x) + U_2(x))\right)$$

$$[f_{u\mu}(0|\mathbf{u})](x) = (U_1(x) - U_2(x), U_1(x) + U_2(x))$$

$$[\mathbf{C}(\mathbf{u}, \mathbf{u}, \mathbf{u})](x) = -\left\{U_1^2(x) + U_2^2(x) + \frac{1}{\pi^2}\left[\left(\frac{\partial U_1(x)}{\partial x}\right)^2 + \left(\frac{\partial U_2(x)}{\partial x}\right)^2\right]\right\} \times (U_1(x), U_2(x)).$$

The spectrum of the operator $\mathbf{f}_u(\mu|\cdot)$ may be computed exactly: it is the set of eigenvalues $\{\lambda_k^\pm = -k^2\pi^2 + (1 \pm i)(\pi^2 + \mu)$ where k is any positive integer. The eigenvector belonging to $\lambda_k^\pm$ is

$$\zeta_0^{(k)}(x) = \sin(k\pi x)(i, 1).$$

The eigenvalue with the largest real part is the one with $k = 1$. When $\mu = 0$ this largest real part vanishes; $\pm i\pi^2$ are simple eigenvalues of $\mathbf{f}_u(0|\cdot)$;

$$\zeta_0(x) = \sin(\pi x)(i, 1) \tag{VIII.48}$$

is the eigenvector for $i\pi^2$ and $\bar{\zeta}_0(x)$ is the eigenvector for $-i\pi^2$. So we have

$$\omega_0 = \pi^2, \qquad \sigma_\mu = 1 + i, \tag{VIII.49}$$

and the loss of stability is strict, $\xi_\mu = 1 > 0$.

The adjoint of $\mathbf{f}_u(\mu|\cdot)$ with respect to the scalar product in H is $\mathbf{f}_u^*(\mu|\cdot) \overset{\text{def}}{=} [\mathbf{f}_u(\mu|\cdot)]^*$. It is easily verified that

$$[\mathbf{f}_u^*(\mu|\mathbf{u})](x) = \left(\frac{\partial^2 U_1(x)}{\partial x^2} + (\pi^2 + \mu)[U_1(x) + U_2(x)], \frac{\partial^2 U_2(x)}{\partial x^2} + (\pi^2 + \mu)[U_2(x) - U_1(x)]\right). \tag{VIII.50}$$

The reader should verify that $\zeta_0 = \zeta_0^*$ where

$$\mathbf{f}_u^*(0|\zeta_0^*) = -i\pi^2\zeta_0^*, \qquad \langle\zeta_0, \zeta_0^*\rangle = 1$$

and

$$\langle\mathbf{f}_{u\mu}(0|\zeta_0), \zeta_0^*\rangle = 1 + i = \sigma_\mu.$$

To obtain the bifurcated periodic solution we follow the method of power series used in the text. In the present case the Fredholm conditions of solvability (VIII.35) becomes

$$\mu_2\sigma_\mu - i\omega_2 = -6\langle\mathbf{C}(\zeta_0, \zeta_0, \bar{\zeta}_0), \zeta_0^*\rangle = 8.$$

Hence,

$$\mu_2 = \omega_2 = 8$$

and

$$\begin{aligned} [\mathbf{u}(s, \varepsilon)](x) &= 2\varepsilon \sin \pi x\,(-\sin s, \cos s) + O(\varepsilon^3) \\ \mu(\varepsilon) &= 4\varepsilon^2 + O(\varepsilon^4) \\ \omega(\varepsilon) &= \pi^2 + 4\varepsilon^2 + O(\varepsilon^4) \end{aligned} \tag{VIII.51}$$

give the bifurcated periodic solution. In fact the higher-order terms of (VIII.51) vanish identically and the remaining explicit part is an exact solution of (VIII.45).

Example VIII.2 (Bifurcation for functional differential equations.) Let us consider the following functional differential equation

$$\frac{dU(t)}{dt} = -\left(\frac{\pi}{2} + \mu\right)U(t-1)[1 + U(t)], \tag{VIII.52}$$

called the Hutchinson–Wright equation. (A systematic study of this equation may be found in J. K. Hale, *Functional Differential Equations* (New York–Heidelberg–Berlin: Springer-Verlag, 1977.) $U(t)$ is an unknown real function depending on the values of $U(s)$ for $t - 1 \le s \le t$. So we may regard $U(s)$ as an element of the space

$$C = \{\text{continuous functions on } [-1, 0]\}$$

and the operators on this space corresponding to (VIII.52) may be defined as

$$[\mathbf{f}_u(\mu|\boldsymbol{\phi})](\theta) = \begin{cases} d\phi(\theta)/d\theta, & \text{if } -1 \le \theta < 0 \\ -(\pi/2 + \mu)\phi(-1), & \text{if } \theta = 0 \end{cases} \tag{VIII.53}$$

$$[\mathbf{N}(\mu, \boldsymbol{\phi})](\theta) = \left(\frac{\pi}{2} + \mu\right) \times \begin{cases} 0, & \text{if } -1 \le \theta < 0 \\ -\phi(-1)\phi(0), & \text{if } \theta = 0. \end{cases} \tag{VIII.54}$$

Equation (VIII.52) may be written as

$$d\mathbf{u}/dt = \mathbf{f}_u(\mu|\mathbf{u}) + \mathbf{N}(\mu, \mathbf{u}), \tag{VIII.55}$$

where

$$[\mathbf{u}(t)](\theta) \stackrel{\text{def}}{=} U(t + \theta).$$

This definition implies that (VIII.55) may be written as

$$\frac{dU(t+\theta)}{dt} = \begin{cases} dU(t+\theta)/d\theta, & \text{if } -1 \le \theta < 0 \\ -(\pi/2 + \mu)U(t-1)(1 + U(t)), & \text{if } \theta = 0. \end{cases}$$

The spectrum of the linear operator $\mathbf{f}_u(\mu|\cdot)$ is composed uniquely with eigenvalues σ of finite multiplicities, satisfying

$$\sigma + \left(\frac{\pi}{2} + \mu\right)e^{-\sigma} = 0. \tag{VIII.56}$$

For $\mu = 0$, we obtain two eigenvalues $\pm i\omega_0, \omega_0 = \pi/2$, the other eigenvalues being of negative real part. Differentiating (VIII.56) with respect to μ at $\mu = 0$, we find that

$$\sigma_\mu(0) = \left(i + \frac{\pi}{2}\right)\left(1 + \frac{\pi^2}{4}\right)^{-1}, \tag{VIII.57}$$

so the Hopf condition $\xi_\mu > 0$ is satisfied.

To study bifurcation we need to define an adjoint operator with respect to a duality product (see J. K. Hale, op. cit.)

$$\langle \boldsymbol{\phi}, \boldsymbol{\psi} \rangle = \phi(0)\bar{\psi}(0) - \frac{\pi}{2}\int_{-1}^{0} \phi(\xi)\bar{\psi}(\xi + 1)\, d\xi \tag{VIII.58}$$

between all elements $\boldsymbol{\phi} \in C$ and all elements

$$\boldsymbol{\psi} \in C^* = \{\text{continuous functions on } [0, 1]\}.$$

It then follows that the adjoint of $\mathbf{f}_u(\mu|\cdot)$ with respect to (VIII.58) is given by

$$[\mathbf{f}_u^*(\mu|\boldsymbol{\psi})](\theta) = \begin{cases} -d\psi(\theta)/d\theta, & \text{if } 0 < \theta \le 1 \\ -(\pi/2 + \mu)\psi(1), & \text{if } \theta = 0. \end{cases} \tag{VIII.59}$$

We therefore have dual eigenvalue problems

$$\mathbf{f}_u(0|\boldsymbol{\zeta}_0) = i\frac{\pi}{2}\boldsymbol{\zeta}_0$$

$$\mathbf{f}_u^*(0|\boldsymbol{\zeta}_0^*) = -i\frac{\pi}{2}\boldsymbol{\zeta}_0^*$$

$$\langle \boldsymbol{\zeta}_0, \boldsymbol{\zeta}_0^* \rangle = 1,$$

which are satisfied by

$$\zeta_0(\theta) = \exp i\frac{\pi}{2}\theta$$

and

$$\zeta_0^*(\theta) = \frac{1}{1 - i(\pi/2)}\exp i\frac{\pi}{2}\theta.$$

The reader may wish to verify that

$$\langle \mathbf{f}_{u\mu}(0|\zeta_0), \zeta_0^* \rangle = \frac{i}{(1 + i(\pi/2))} = \sigma_\mu(0).$$

We now use the Fredholm alternative to compute the series (VIII.24) giving the time-periodic solution which bifurcates from the null solution of (VIII.32). We find that $\mu_1 = \omega_1 = 0$,

$$[\mathbf{u}_1(s)](\theta) = \zeta_0(\theta)e^{is} + \bar{\zeta}_0(\theta)e^{-is}$$

$$\mathbf{f}_{uu}(0|\zeta_0|\bar{\zeta}_0) = 0$$

$$\omega_0\frac{d\mathbf{u}_2}{ds} = \mathbf{f}(0|\mathbf{u}_2) + e^{2is}\mathbf{f}_{uu}(0|\zeta_0|\zeta_0) + e^{-2is}\mathbf{f}_{uu}(0|\bar{\zeta}_0|\bar{\zeta}_0)$$

$$[\mathbf{u}_2(s)](\theta) = \zeta_2(\theta)e^{2is} + \bar{\zeta}_2(\theta)e^{-2is}$$

$$\zeta_2(\theta) = \frac{8 - 4i}{5\pi}e^{i\pi\theta}, \qquad -1 \le \theta \le 0.$$

Hence,

$$[\mathbf{u}_2(s)](\theta) = \frac{4(2 - i)}{5\pi}e^{i\pi\theta}e^{2is} + \frac{4(2 + i)}{5\pi}e^{-i\pi\theta}e^{-2is}$$

and (VIII.35) may be written as

$$\begin{aligned}\mu_2\sigma_\mu - i\omega_2 &= -4\left\langle \mathbf{f}_{uu}(0|\bar{\zeta}_0|\frac{2 - i}{5\pi}e^{i\pi\theta}, \zeta_0^*\right\rangle \\ &= \frac{2[3\pi - 2 + i(6 + \pi)]}{5\pi(1 + (\pi^2/4))} \\ \mu_2 &= \frac{4(3\pi - 2)}{5\pi^2} > 0 \\ \omega_2 &= -\frac{8}{5\pi^2}.\end{aligned}$$

Collecting our results we find that the principal part of the bifurcating solution is given by

$$\mu = \varepsilon^2\frac{\mu_2}{2} + O(\varepsilon^4),$$

$$\omega = \frac{\pi}{2} + \varepsilon^2\frac{\omega_2}{2} + O(\varepsilon^4),$$

and

$$U(t) = [\mathbf{u}(\omega t)](0) = 2\varepsilon \cos \omega t + 4\varepsilon^2 \operatorname{Re}\left\{\frac{2-i}{5\pi} e^{2i\omega t}\right\} + O(\varepsilon^3).$$

EXAMPLE VIII.3 (Bifurcation for equations which have not been reduced to local form).

$$\ddot{x} + \omega_0^2 x = f(\dot{x}, x, \mu), \tag{VIII.60}$$

where $\dot{x} \stackrel{\text{def}}{=} dx/dt$ and f is as smooth as we wish in its arguments when the arguments are small. Moreover we assume that

$$f(0, 0, 0) = f_x(0, 0, 0) = f_{\dot{x}}(0, 0, 0) = 0, \tag{VIII.61}$$

and utilize the decomposition

$$f(u_1, u_2, \mu) = \sum_{p, q_1, q_2} \mu^p u_1^{q_1} u_2^{q_2} f_{pq_1q_2} \tag{VIII.62}$$

where $f_{000} = f_{010} = f_{001} = 0$, and the decomposition is carried out to the order allowed by the smoothness of f. Equation (VIII.60) may be written in $\mathbb{R}^2$ in the following way. We define

$$\mathbf{U} = \begin{bmatrix} u_1 \\ u_2 \end{bmatrix} \stackrel{\text{def}}{=} \begin{bmatrix} x \\ \dot{x} \end{bmatrix} \in \mathbb{R}^2.$$

Then

$$\frac{d\mathbf{U}}{dt} = \mathbf{A}_0 \mathbf{U} + F(\mu, \mathbf{U}) \tag{VIII.63}$$

with

$$\mathbf{A}_0 = \begin{bmatrix} 0 & 1 \\ -\omega_0^2 & 0 \end{bmatrix} \quad \text{and} \quad \mathbf{F}(\mu, \mathbf{U}) = \begin{bmatrix} 0 \\ f(u_1, u_2, \mu) \end{bmatrix}.$$

Here $\mathbf{U} = 0$ *is not in general a steady solution* of (VIII.63), except for $\mu = 0$. But the existence of a steady solution with $\mu \neq 0$ can be guaranteed by the implicit function theorem in $\mathbb{R}^2$ and it can be computed by identification using the series representations for

$$\mathbf{U}(\mu) = \sum_{\mu \geq 1} \mu^n \mathbf{U}_n \tag{VIII.64}$$

and

$$\mathbf{F}(\mu, \mathbf{U}) = \sum_{p, q} \mu^p \mathbf{F}_{pq}(\mathbf{U}, \ldots, \mathbf{U}), \qquad \mathbf{F}_{00} = \mathbf{F}_{01} = 0, \tag{VIII.65}$$

where $\mathbf{F}_{pq}$ is q-linear in $\mathbf{U}$, and symmetric. We obtain

$$\mathbf{A}_0 \mathbf{U}_1 + \mathbf{F}_{10} = 0$$

$$\mathbf{A}_0 \mathbf{U}_2 + \mathbf{F}_{20} + \mathbf{F}_{11}(\mathbf{U}_1) + \mathbf{F}_{02}(\mathbf{U}_1, \mathbf{U}_1) = 0,$$

where

$$\mathbf{F}_{p0} = \begin{bmatrix} 0 \\ f_{p00} \end{bmatrix},$$

$$\mathbf{F}_{02}(\mathbf{U}, \mathbf{V}) = \begin{bmatrix} 0 \\ f_{020}u_1v_1 + \tfrac{1}{2}f_{011}(u_1v_2 + u_2v_1) + f_{002}u_2v_2) \end{bmatrix}$$

$$\mathbf{F}_{11}(\mathbf{U}) = \begin{bmatrix} 0 \\ f_{110}u_1 + f_{101}u_2 \end{bmatrix}.$$

Hence

$$\mathbf{U}_1 = \begin{bmatrix} f_{100}/\omega_0^2 \\ 0 \end{bmatrix}$$

and so on.

Since the eigenvalues of $\mathbf{A}_0$ are $\pm i\omega_0$, we need to consider the possibility of Hopf bifurcation into periodic solutions. In the theoretical part of this chapter we first reduced the problem to local form (see §I.3) and applied the assumption that the loss of stability of the solution $\mathbf{u} = 0$, that is, of $\mathbf{U}$, was strict. Here, $\mathbf{U} = 0$ is not a solution for all μ near zero, and we need to reformulate the condition that $\mathbf{U}(\mu)$ loses stability strictly as μ is increased past zero. First we linearize

$$\mathbf{A}_0 + \mathbf{F}_U(\mu, \mathbf{U}(\mu)|\cdot) = \mathbf{A}_0 + \mu[\mathbf{F}_{11}(\cdot) + 2\mathbf{F}_{02}(\mathbf{U}_1, \cdot)] + O(\mu^2). \quad \text{(VIII.66)}$$

The eigenvectors belonging to the eigenvalues $\pm i\omega_0$ are $\zeta_0 = (1, i\omega_0)$ and $\bar{\zeta}_0$. In the same way the adjoint eigenvectors are

$$\zeta_0^* = \left(\frac{1}{2}, \frac{i}{2\omega_0}\right) \quad \text{and} \quad \bar{\zeta}_0^*.$$

The eigenvalue $\sigma(\mu)$ which perturbs the eigenvalues $i\omega_0$ satisfies (VIII.14); hence,

$$\begin{aligned} \sigma_\mu(0) &= \langle \mathbf{F}_{11}(\zeta_0) + 2\mathbf{F}_{02}(\mathbf{U}_1, \zeta_0), \zeta_0^* \rangle \\ &= -\frac{i}{2\omega_0}\left[f_{110} + f_{101}i\omega_0 + 2f_{020}\left(\frac{f_{100}}{\omega_0^2}\right) + if_{011}\left(\frac{f_{100}}{\omega_0}\right)\right]. \end{aligned} \quad \text{(VIII.67)}$$

The Hopf condition is

$$2 \operatorname{Re} \sigma_\mu(0) = f_{101} + \frac{f_{011}f_{100}}{\omega_0^2} > 0. \quad \text{(VIII.68)}$$

We now assume that (VIII.68) is realized and

$$\begin{aligned} \mu &= \sum_{n \geq 1} \mu_n \varepsilon^n \\ \omega &= \omega_0 + \sum_{n \geq 1} \omega_n \varepsilon^n \\ \mathbf{U} &= \sum_{n \geq 1} \mathbf{V}_n \varepsilon^n, \qquad \mathbf{V}_n(s) = \mathbf{V}_n(s + 2\pi). \end{aligned} \quad \text{(VIII.69)}$$

Identifying independent powers of ε in the equation

$$\omega \frac{d\mathbf{U}}{ds} - \mathbf{A}_0 \mathbf{U} = \mathbf{F}(\mu, \mathbf{U}),$$

we find that

$$\mathbb{J}_0 \mathbf{V}_1 = \mu_1 F_{10}. \tag{VIII.70}$$

Hence

$$\mathbf{V}_1 = \zeta_0 e^{is} + \bar{\zeta}_0 e^{-is} + \mu_1 \mathbf{U}_1$$

$$\mathbb{J}_0 \mathbf{V}_2 + \omega_1 \frac{d\mathbf{V}_1}{ds} = \mu_2 \mathbf{F}_{10} + \mathbf{F}_{02}(\mathbf{V}_1, \mathbf{V}_1) + \mu_1 \mathbf{F}_{11}(\mathbf{V}_1) + \mu_1^2 \mathbf{F}_{20}, \tag{VIII.71}$$

and, using (VIII.67),

$$i\omega_1 = \mu_1 \sigma_\mu(0).$$

We find that

$$\begin{aligned} \omega_1 &= \mu_1 = 0 \\ \mathbf{V}_1 &= \zeta_0 e^{is} + \bar{\zeta}_0 e^{-is} \\ \mathbf{V}_2 &= \mathbb{J}_0^{-1} \mathbf{F}_{02}(\mathbf{V}_1, \mathbf{V}_1) + \mu_2 \mathbf{U}_1 \end{aligned} \tag{VIII.72}$$

where $\mathbb{J}_0^{-1}$ is the inverse of $\mathbb{J}_0$ on the subspace orthogonal to $\zeta_0^* e^{is}, \bar{\zeta}_0^* e^{-is}$.

We next determine ω_2 and μ_2 by applying (VIII.67) to the equation

$$\begin{aligned} \mathbb{J}_0 \mathbf{V}_3 + \omega_2 \frac{d\mathbf{V}_1}{ds} = \mu_3 \mathbf{F}_{10} &+ 2\mathbf{F}_{02}(\mathbf{V}_1, \mathbf{V}_2) + \mu_2 \mathbf{F}_{11}(\mathbf{V}_1) \\ &+ \mathbf{F}_{03}(\mathbf{V}_1, \mathbf{V}_1, \mathbf{V}_1). \end{aligned} \tag{VIII.73}$$

Hence $\mathbf{V}_2$ and $\mathbf{V}_3$ are functions of μ_2. Iterating this process, we obtain the series (VIII.69) giving the Hopf bifurcation, where as usual μ and ω are *even* in ε.

Notes

Several problems of Hopf bifurcation in special circumstances have been studied and are well understood (Figure VIII.1):

(i) Four simple eigenvalues, two conjugate pairs, cross the imaginary axis simultaneously. This problem is treated by G. Iooss, Direct bifurcation of a steady solution of the Navier–Stokes equations into an invariant torus, in *Turbulence and Navier Stokes Equations*, Lecture Notes in Mathematics, No. 565 (New York–Heidelberg–Berlin: Springer-Verlag, 1975), pp. 69–84.

(ii) Two simple conjugate eigenvalues cross at criticality, but not strictly; for example, $\xi(0) = \xi'(0) = \xi''(0) = 0$, $\xi'''(0) \neq 0$. This problem is treated by H. Kielhöfer,

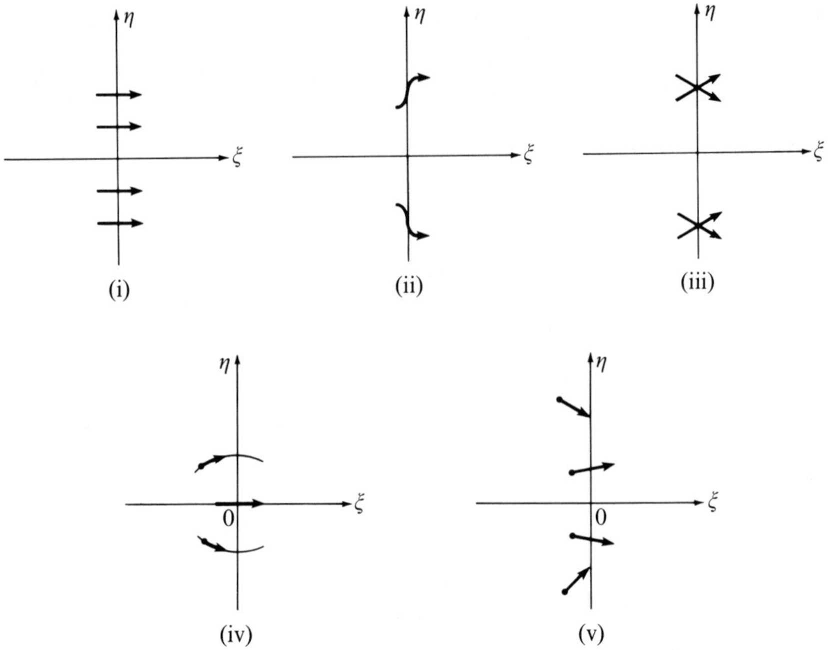

Figure VIII.1 Hopf bifurcation in special cases

Generalized Hopf bifurcation in Hilbert space, *Math. Methods in Applied Sciences*, (forthcoming).

(iii) Two multiple eigenvalues cross at criticality: H. Kielhöfer, Hopf bifurcation at multiple eigenvalues, *Arch. Rational Mech. Anal.*, **69**, 53-83 (1979).

It is more general and useful for applications to treat problems in which different eigenvalues cross the imaginary axis nearly simultaneously. In such studies it is useful to introduce two perturbation parameters as in the work of W. F. Langford, Periodic and steady-state mode interactions lead to tori, *SIAM J. Appl. Math.* **37**, 22–48 (1979). When additional symmetries are present see the results of J. Keener, Secondary bifurcation in nonlinear diffusion reaction equations. *Stud. Appl. Math.* **55**, 187–211 (1976); P. Holmes, Unfolding a degenerate nonlinear oscillator: a codimension two bifurcation, *New York Academy of Sciences Proceedings* Dec. 1979; and G. Iooss and W. Langford, Conjectures on the routes to turbulence via bifurcations, *New York Academy of Sciences Proceedings* Dec. 1979. The authors just named treat the case (iv) in which a pair of conjugate eigenvalues and a real eigenvalue, all simple, cross the imaginary axis nearly simultaneously; and the case (v) in which two conjugate pairs cross the imaginary axis nearly simultaneously which is also of interest.

Another interesting special class of problems is invariant under the action of a group and lead to wave-like solutions. In problems invariant to rotations about the axis, the bifurcating solution depends on θ and t only in the combination $\theta - \omega t$. In Problems invariant to translations through periods $2\pi/\alpha$ in x, the solution depends on x and t through $\alpha x - \omega t$, where $C = \omega/\alpha$ is the wave speed (see §XI.19).

CHAPTER IX

Subharmonic Bifurcation of Forced T-Periodic Solutions

In this chapter, and in Chapter X, we consider the bifurcation of forced T-periodic solutions. In thinking about the origin and structure of such problems it would benefit the reader to reread the explanations given in §I.2 and §I.3. Following our usual procedure we do the theory in $\mathbb{R}^n$, $n \geq 2$, and show how the analysis reduces to $\mathbb{R}^1$ or $\mathbb{R}^2$ using projections associated with the Fredholm alternative. There is a sense in which the problem in $\mathbb{R}^n$ with n finite is actually infinite-dimensional. Unlike steady problems which involve only constant vectors, we must work with vector-valued functions which depend periodically on time and hence take on infinitely many distinct values. So, in this chapter the computational simplifications which would result from considering $\mathbb{R}^2$ rather than $\mathbb{R}^n$ are not great. In $\mathbb{R}^n$ we use the same notation we would use for an evolution equation in a Banach space. So our results hold equally in $\mathbb{R}^n$ and, say, for evolution problems governed by partial differential equations, like the Navier–Stokes equations or equations governing reaction and diffusion in chemical systems, provided the writing of these partial differential equations as evolution problems in Banach space can be justified.

Notation

$$\mathbb{P}_{nT} = \{\mathbf{u}: \mathbf{u}(t) = \mathbf{u}(t + nT),\ nT\text{-periodic continuous functions}\}$$

$J(\mu)$ is a linear operator defined in §IX.2 as

$$J(\mu) = -\frac{d}{dt} + \mathbf{f}_u(t, \mu, 0|\cdot),$$

acting in $\mathbb{P}_T$, whose domain is the set of continuously differentiable T-periodic functions. As a consequence, the operator $J(\mu)$ is T-periodic. Similarly, the operator $J_0 \stackrel{\text{def}}{=} J(0)$ is T-periodic. $\mathbb{J}$ is a linear operator defined in §IX.8 which has T-periodic coefficients but which is considered acting in $\mathbb{P}_{nT}$ (a larger space).

$$\sigma(\mu) = \xi(\mu) + i\eta(\mu)$$

is the Floquet exponent for the stability of $\mathbf{u} \equiv 0$.

$$\gamma(\varepsilon) = \xi(\varepsilon) + i\eta(\varepsilon)$$

is the Floquet exponent for the stability of the bifurcating subharmonic solution $\mathbf{u}(t, \varepsilon) \not\equiv 0$. (*Note*. We use the same notation, ξ and η for different functions.)

$$\mathbf{f}_u(t|\cdot) \stackrel{\text{def}}{=} \mathbf{f}_u(t, 0, 0|\cdot)$$

is a linear operator (in $\mathbb{R}^n$) (see §I.6, 7).

$$\mathbf{f}_{uu}(t|\cdot|\cdot) \stackrel{\text{def}}{=} \mathbf{f}_{uu}(t, 0, 0|\cdot|\cdot)$$

is a bilinear symmetric operator: $\mathbf{f}_{uu}(t|\mathbf{u}_1|\mathbf{u}_2) = \mathbf{f}_{uu}(t|\mathbf{u}_2|\mathbf{u}_1)$.

$$\mathbf{f}_{uuu}(t|\cdot|\cdot|\cdot) \stackrel{\text{def}}{=} \mathbf{f}_{uuu}(t, 0, 0|\cdot|\cdot|\cdot)$$

is a trilinear symmetric operator:

$$\mathbf{f}_{uuu}(t|\mathbf{u}_1|\mathbf{u}_2|\mathbf{u}_3) = \mathbf{f}_{uuu}(t|\mathbf{u}_1|\mathbf{u}_3|\mathbf{u}_2) = \mathbf{f}_{uuu}(t|\mathbf{u}_2|\mathbf{u}_1|\mathbf{u}_3).$$

The multilinear operators arise from repeated differentiation of $\mathbf{f}(t, \mu, \mathbf{u})$ with respect to $\mathbf{u}$, at the point $\mu = 0$, $\mathbf{u} = 0$. The definitions suppress the dependence of these derivatives on the point $(\mu, \mathbf{u}) = (0, 0)$.

The work in this chapter is based on the results proved in G. Iooss and D. D. Joseph, Bifurcation and stability of nT-periodic solutions branching from T-periodic solutions at points of resonance, *Arch. Rational Mech. Anal.*, **66**, 135–172 (1977).

IX.1 Definition of the Problem of Subharmonic Bifurcation

We are interested in the nT-periodic solutions, where $n \in \mathbb{N}^*$ is a positive integer, which bifurcate from a forced T-periodic one $\mathbf{U}(t) \in \mathbb{P}_T$. When the problem is "reduced to local form" as in §I.3 we study the bifurcation of the solution $\mathbf{u} = 0$ of the evolution problem

$$\frac{d\mathbf{u}}{dt} = \mathbf{f}(t, \mu, \mathbf{u}) \tag{IX.1}$$

where $\mathbf{f}(t, \cdot, \cdot) = \mathbf{f}(t + T, \cdot, \cdot)$ has the period T of the forced solution from which it comes. Our bifurcation study applies when some measure of the amplitude of $\mathbf{u}$ is small and it is convenient to expand $\mathbf{f}$ relative to $\mathbf{u} = 0$:

$$\mathbf{f}(t, \mu, \mathbf{u}) = \mathbf{f}_u(t, \mu, 0|\mathbf{u}) + \tfrac{1}{2}\mathbf{f}_{uu}(t, \mu, 0|\mathbf{u}|\mathbf{u}) + \frac{1}{3!}\mathbf{f}_{uuu}(t, \mu, 0|\mathbf{u}|\mathbf{u}|\mathbf{u}) + O(\|\mathbf{u}\|^4). \qquad \text{(IX.2)}_1$$

We shall suppose $\mathbf{f}$ is analytic when μ and $\mathbf{u}$ are in some neighborhood of (0, 0). We may also expand (IX.2) in powers of μ, where we have suppressed higher-order terms which do not enter into the local analysis of stability and bifurcation:

$$\begin{aligned}\mathbf{f}(t, \mu, \mathbf{u}) &= \mathbf{f}_u(t|\mathbf{u}) + \mu\mathbf{f}_{u\mu}(t|\mathbf{u}) + \tfrac{1}{2}\mu^2\mathbf{f}_{u\mu\mu}(t|\mathbf{u}) \\ &\quad + \tfrac{1}{2}\{\mathbf{f}_{uu}(t|\mathbf{u}|\mathbf{u}) + \mu\mathbf{f}_{uu\mu}(t|\mathbf{u}|\mathbf{u})\} + \frac{1}{3!}\mathbf{f}_{uuu}(t|\mathbf{u}|\mathbf{u}|\mathbf{u}) \\ &\quad + O(|\mu|^3\,\|\mathbf{u}\| + \mu^2\|\mathbf{u}\|^2 + |\mu|\,\|\mathbf{u}\|^3 + \|\mathbf{u}\|^4). \qquad \text{(IX.2)}_2\end{aligned}$$

Here we omit as usual the writing of $(\mu, \mathbf{u}) = (0, 0)$ in the argument of the derivatives of $\mathbf{f}$ (see (IX.21)).

Suppose a subharmonic solution of amplitude ε

$$\begin{aligned}\mathbf{u}(t, \varepsilon) = \mathbf{u}(t + nT, \varepsilon), &\qquad \mathbf{u}(t, 0) = 0 \\ \mu = \mu(\varepsilon), &\qquad \mu(0) = 0\end{aligned} \qquad \text{(IX.3)}$$

bifurcates from $\mathbf{u} = 0$ when μ is increased past zero. To study the stability of small disturbances $\mathbf{v}$ of (IX.3) we linearize and find that

$$\begin{aligned}\frac{d\mathbf{v}}{dt} &= \mathbf{f}_u(t, \mu(\varepsilon), \mathbf{u}(t, \varepsilon)|\mathbf{v}) \\ &= \mathbf{f}_u(t, \mu(\varepsilon), 0|\mathbf{v}) + f_{uu}(t, \mu(\varepsilon), 0|\mathbf{u}(t, \varepsilon)|\mathbf{v}) \\ &\quad + \tfrac{1}{2}\mathbf{f}_{uuu}(t, \mu(\varepsilon), 0|\mathbf{u}(t, \varepsilon)|\mathbf{u}(t, \varepsilon)|\mathbf{v}) \\ &\quad + \mathbf{R}(t, \mu(\varepsilon), \mathbf{u}(t, \varepsilon)|\mathbf{v}).\end{aligned} \qquad \text{(IX.4)}$$

The linear operator $\mathbf{R}$ will not enter into local analysis because it is at least cubic in $\mathbf{u}$, and therefore, in ε.

We study (IX.4) by the spectral method of Floquet (see §VII.6.2). It is necessary to say more about the stability theory. However, for the present it will suffice to make a few preliminary remarks. To obtain the spectral equations we write

$$\mathbf{v}(t, \varepsilon) = e^{\gamma(\varepsilon)t}\mathbf{y}(t, \varepsilon), \qquad \text{(IX.5)}$$

where $\mathbf{y}(t, \varepsilon) \in \mathbb{P}_{nT}$ and

$$\gamma(\varepsilon)\mathbf{y} + \frac{d\mathbf{y}}{dt} = \mathbf{f}_u(t, \mu(\varepsilon), \mathbf{u}(t, \varepsilon)|\mathbf{y}). \qquad \text{(IX.6)}$$

In general, $\gamma(\varepsilon)$ is a complex number:

$$\gamma(\varepsilon) = \xi(\varepsilon) + i\eta(\varepsilon). \tag{IX.7}$$

In studying the stability of $\mathbf{u} = 0$ we use μ rather than ε as a parameter; we write $\mathbf{v} = e^{\sigma t}\boldsymbol{\zeta}$ and arrive at the spectral problem

$$\sigma(\mu)\boldsymbol{\zeta} + \frac{d\boldsymbol{\zeta}}{dt} = \mathbf{f}_u(t, \mu, 0|\boldsymbol{\zeta}), \qquad \boldsymbol{\zeta} \in \mathbb{P}_T, \tag{IX.8}$$

where

$$\sigma(\mu) = \xi(\mu) + i\eta(\mu). \tag{IX.9}$$

We caution the reader about the possible confusion which could result from using the same notation for the real and imaginary part of σ and γ. We shall not, in fact, need the eigenvalues $\gamma(\varepsilon)$ until §IX.13.

IX.2 Spectral Problems and the Eigenvalues $\sigma(\mu)$

Consider the linearized evolution problem for the stability of $\mathbf{u} = 0$

$$\frac{d\mathbf{v}}{dt} = \mathbf{f}_u(t, \mu, 0|\mathbf{v}) = \mathbf{f}_u(t + T, \mu, 0|\mathbf{v}) \tag{IX.10$_1$}$$

with initial values

$$\mathbf{v}(0) = \mathbf{v}_0. \tag{IX.10$_2$}$$

The solutions of (IX.10) can be expressed in terms of the special fundamental solution matrix $\boldsymbol{\Phi}(t, \mu)$ which has unit initial values $\boldsymbol{\Phi}(0, \mu) = \mathbf{I}$ as follows:

$$\mathbf{v}(t, \mu) = \boldsymbol{\Phi}(t, \mu) \cdot \mathbf{v}_0. \tag{IX.11}$$

The eigenvalues of the *monodromy matrix* $\boldsymbol{\Phi}(T, \mu)$ are the Floquet multipliers

$$\lambda(\mu) = e^{\sigma(\mu)T}, \tag{IX.12}$$

where the complex numbers $\sigma(\mu) = \xi(\mu) + i\eta(\mu)$ are Floquet exponents. The exponents are eigenvalues of (IX.8). We say $\sigma(\mu)$ is an eigenvalue of

$$J(\mu) = -\frac{d}{dt} + \mathbf{f}_u(t, \mu, 0|\cdot)$$

so that (IX.8) may be written as

$$\sigma\boldsymbol{\zeta} = J(\mu)\boldsymbol{\zeta}, \qquad \boldsymbol{\zeta} \in \mathbb{P}_T. \tag{IX.13}$$

Note that if σ is an eigenvalue of $J(\mu)$, then $\sigma + (2k\pi i/T)$ is also eigenvalue for any k in $\mathbb{Z}$ (the associated eigenvector is $\hat{\boldsymbol{\zeta}}(t) = \boldsymbol{\zeta}(t) \exp(-2\pi kit/T)$).

We next define an adjoint eigenvalue problem

$$\bar{\sigma}\boldsymbol{\zeta}^* = J^*(\mu)\boldsymbol{\zeta}^*, \qquad \boldsymbol{\zeta}^* \in \mathbb{P}_T, \tag{IX.14}$$

where

$$J^*(\mu) = \frac{d}{dt} + \mathbf{f}_u^*(t, \mu|\cdot),$$

in the following way. The linear operator $\mathbf{f}_u^*(t, \mu|\cdot)$ is adjoint to $\mathbf{f}_u(t, \mu, 0|\cdot)$; that is,

$$\langle \mathbf{a}, \mathbf{f}_u^*(t, \mu|\mathbf{b})\rangle = \langle \mathbf{f}_u(t, \mu, 0|\mathbf{a}), \mathbf{b}\rangle. \tag{IX.15}$$

In $\mathbb{R}^n$, $\mathbf{f}_u(t, \mu, 0|\cdot)$ is the matrix $\mathbf{A}(t, \mu)$ and $\mathbf{f}_u^*(t, \mu|\cdot)$ is $\mathbf{A}^T(t, \mu)$.† A second scalar product which is useful for nT-periodic functions is defined by

$$[\mathbf{a}, \mathbf{b}]_{nT} \stackrel{\text{def}}{=} \frac{1}{nT}\int_0^{nT} \langle \mathbf{a}, \mathbf{b}\rangle\, dt \tag{IX.16}$$

where $\langle \mathbf{a}, \mathbf{b}\rangle = \langle \overline{\mathbf{b}, \mathbf{a}}\rangle$. The linear operator $J^*(\mu)$ is defined by the relation

$$[\mathbf{a}, J(\mu)\mathbf{b}]_T = [J^*(\mu)\mathbf{a}, \mathbf{b}]_T. \tag{IX.17}$$

We verify that

$$[\boldsymbol{\zeta}, \bar{\sigma}\boldsymbol{\zeta}^*]_T = \sigma[\boldsymbol{\zeta}, \boldsymbol{\zeta}^*]_T = [J(\mu)\boldsymbol{\zeta}, \boldsymbol{\zeta}^*]_T = [\boldsymbol{\zeta}, J^*(\mu)\boldsymbol{\zeta}^*]_T.$$

IX.3 Biorthogonality

Eigenvectors belonging to different eigenvalues are biorthogonal:

$$[\boldsymbol{\zeta}_i, \boldsymbol{\zeta}_j^*]_T = 0$$

if $i \neq j$, as shown in Chapter VI. We equally have $[\boldsymbol{\zeta}_i, \overline{\boldsymbol{\zeta}}_j^*]_T = 0$ even when $i = j$, provided only that $\eta(\mu) \neq 0$. If $\eta = 0$, then σ, $\boldsymbol{\zeta}$, and $\boldsymbol{\zeta}^*$ are real-valued. For each and every semi-simple eigenvalue we may choose a biorthonormal set such that

$$[\boldsymbol{\zeta}_i, \boldsymbol{\zeta}_j^*]_T = \delta_{ij}. \tag{IX.18}$$

IX.4 Criticality

We assume that $\mathbf{u} = 0$ loses stability strictly as μ increases through zero from negative to positive value. So when $\mu = 0$ the real part $\xi(0)$ of $\sigma(0)$ vanishes and

$$\sigma(0) = i\eta(0) \stackrel{\text{def}}{=} i\omega_0$$

† $\mathbf{f}_u(t, \mu, 0|\cdot)$ is the derivative of $\mathbf{f}(t, \mu, \mathbf{u})$ at $\mathbf{u} = 0$, but we do not suggest that there is an $\mathbf{f}^*(t, \mu, \mathbf{u})$ with derivative $\mathbf{f}_u^*$. We *define* $\mathbf{f}_u^*$ by (IX.15).

is an eigenvalue of $J(\mu)$ and $J^*(\mu)$ at $\mu = 0$; that is,

$$\begin{aligned} i\omega_0 \zeta &= J_0 \zeta \\ -i\omega_0 \zeta^* &= J_0^* \zeta^*, \end{aligned} \tag{IX.19}$$

where $J_0 \overset{\text{def}}{=} J(0)$ and $J_0^* \overset{\text{def}}{=} J(0)^*$. The *strict* loss of stability at criticality means that

$$\xi_\mu(0) \overset{\text{def}}{=} \xi'(0) > 0. \tag{IX.20}$$

Criticality is very important. In the perturbation solution of the bifurcation problem all quantities are evaluated at criticality. To simplify our notation for the derivatives of $\mathbf{f}$ at criticality, we write

$$\begin{aligned} \mathbf{f}_u(t, 0, 0|\cdot) &\overset{\text{def}}{=} \mathbf{f}_u(t|\cdot) \\ \mathbf{f}_{uu}(t, 0, 0|\cdot|\cdot) &\overset{\text{def}}{=} \mathbf{f}_{uu}(t|\cdot|\cdot) \\ \mathbf{f}_u^*(t, 0|\cdot) &\overset{\text{def}}{=} \mathbf{f}_u^*(t|\cdot), \end{aligned} \tag{IX.21}$$

and so forth.

IX.5 The Fredholm Alternative for $J(\mu) - \sigma(\mu)$ and a Formula Expressing the Strict Crossing (IX.20)

Suppose $\sigma(\mu)$ is a semi-simple eigenvalue of $J(\mu)$ of multiplicity l and $\zeta_i^*(i = 1, 2, \ldots, l)$ are any set of independent adjoint eigenvectors. Then the equation

$$(J(\mu) - \sigma(\mu))\mathbf{a} = \mathbf{b}(t) = \mathbf{b}(t + T) \tag{IX.22}$$

can have solutions $\mathbf{a}$ in $\mathbb{P}_T$ ($\mathbf{a}(t) = \mathbf{a}(t + T)$) only if the prescribed vector $\mathbf{b}(t)$ verifies the orthogonality relations

$$[\mathbf{b}, \zeta_i^*]_T = 0. \tag{IX.23}$$

Let us assume now that $i\omega_0$ is an algebraically simple eigenvalue of J_0. To derive a formula expressing (IX.20) we differentiate (IX.13) with respect to μ at $\mu = 0$ and find that

$$\sigma_\mu(0)\zeta - \mathbf{f}_{u\mu}(t|\zeta) + (i\omega_0 - J_0)\zeta_\mu = 0, \tag{IX.24}$$

where $J_\mu(0) = \mathbf{f}_{u\mu}(t|\cdot)$ in the notation of (IX.21) and $\zeta_\mu \in \mathbb{P}_T$. Applying (IX.23) to (IX.24), we find that

$$\sigma_\mu(0) = \xi_\mu(0) + i\eta_\mu(0) = [\mathbf{f}_{u\mu}(t|\zeta), \zeta^*]_T. \tag{IX.25}$$

IX.6 Spectral Assumptions

It is perhaps necessary to emphasize that $\sigma(\mu)$ is here and henceforth taken to be eigenvalue of $J(\mu)$ with the largest real part. At criticality, then, there are no eigenvalues of J_0 with positive real parts. If $\eta(\mu) \neq 0$ there are always at least the two families of eigenvalues $\sigma(\mu) + (2k\pi i/T)$ and $\bar{\sigma}(\mu) - (2k\pi i/T)$, for any k in $\mathbb{Z}$, of largest real part and if $i\omega_0$ is an eigenvalue of J_0 so is $-i\omega_0$.

Now we state the basic spectral assumptions for the study of T-periodic solutions. They are essentially the same assumptions under which we have discussed double-point bifurcation and Hopf bifurcation. The assumptions are framed in terms of the eigenvalues at criticality. Since we are thinking only of problems in which the quantities vary continuously with μ and the eigenvalues are assumed to be isolated, statements about the eigenvalues at $\mu = 0$ hold also in a small (possibly large) neighborhood $\mu = 0$. First we state spectral assumptions for J_0:

(I) $i\omega_0$ is an isolated, algebraically simple eigenvalue of J_0.

(II) $\pm i(\omega_0 + (2\pi k/T)), k \in \mathbb{Z}$ are the only eigenvalues of J_0 on the imaginary axis of the σ-plane. (If ζ is a T-periodic eigenvector of J_0, $i\omega_0\zeta = J_0\zeta$, then $e^{-2\pi ikt/T}\zeta(t) = \hat{\zeta}(t) = \hat{\zeta}(t + T)$ is also a T-periodic eigenvector of J_0, $J_0\hat{\zeta} = i(\omega_0 + (2\pi k/T))\hat{\zeta}$.) All the other eigenvalues of J_0 are on the left-hand side of the σ-plane.

(III) Equation (IX.20) holds.

The three assumptions just laid down may also be stated relative to the eigenvalues λ_0 of the monodromy operator $\mathbf{\Phi}(T, 0)$ (the multipliers).

(I) λ_0 is an isolated, algebraically simple eigenvalue of $\mathbf{\Phi}(T, 0)$.

(II) λ_0 and $\bar{\lambda}_0$ are the only eigenvalues of $\mathbf{\Phi}(T, 0)$ on the unit circle in the λ-plane. All the other eigenvalues of $\mathbf{\Phi}(T, 0)$ are inside the unit circle.

(III) Equation (IX.20) holds so that $|\lambda|_\mu > 0$.

To make $\lambda_0 = e^{i\omega_0 T}$ a single-valued function for ω_0 we may, without losing generality, require that

$$0 \leq \omega_0 < \frac{2\pi}{T}. \tag{IX.26}$$

IX.7 Rational and Irrational Points of the Frequency Ratio at Criticality

Almost periodic functions are a generalization of the periodic functions which leave intact the property of completeness of Fourier series. An almost periodic function on the line $-\infty < x < \infty$ can wiggle more or less arbitrarily but in such a way that any value of the function is very nearly repeated at least once in every sufficiently large but finite interval.

A (complex) continuous function $f(t)$ $(-\infty < x < \infty)$ is *almost periodic* if for each $\varepsilon > 0$ there exists $l = l(\varepsilon) > 0$ such that each real interval of length $l(\varepsilon)$ contains at least one number τ for which $|f(t+\tau) - f(t)| \le \varepsilon$. Each such τ is called a translation number. If $\varepsilon = 0$, then $f(t)$ is a periodic function and τ is a period.

A *quasi-periodic* function of n variables is a function $g(\omega_1 t, \omega_2 t, \ldots, \omega_n t) = f(t)$ containing a finite number n of rationally independent frequencies $\omega_1, \omega_2, \ldots, \omega_n$ which is periodic with period 2π in each of its variables. All quasi-periodic functions are almost periodic; in the most general case n is ∞. For example, the function $g(\omega_1 t, \omega_2 t) = \cos t \cos \pi t = f(t)$ is a quasi-periodic function with two frequencies $\omega_1 = 2\pi$ and $\omega_2 = 2$. The value $f(t) = 1$ occurs when $t = 0$ but not again; though $g(t) < 1$ when $t \ne 0$, there is always $t(\varepsilon) > 0$ such that $|f(t) - f(0)| < \varepsilon$ for preassigned $\varepsilon > 0$.

The solution $\mathbf{v}(t) = \mathbf{Z}(t)$

$$\mathbf{Z}(t) = e^{i\omega_0 t}\boldsymbol{\zeta}(t) = \mathbf{g}\left(\omega_0 t, \frac{2\pi}{T} t\right) \tag{IX.27}$$

of the linearized equation (IX.10) at criticality ($d\mathbf{Z}/dt = \mathbf{f}_u(t|\mathbf{Z})$) is quasi-periodic if $\omega_0 T/2\pi$ is irrational. If $\omega_0 T/2\pi$ is rational, ω_0 and $2\pi/T$ are rationally dependent and the solution is periodic with period a multiple of T. The set of points ω_0 on the interval (IX.26) for which ω_0 and $2\pi/T$ are rationally dependent are called *rational points at criticality*. These points are a dense set on the interval (IX.26). The set of points for which ω_0 and $2\pi/T$ are rationally independent are called *irrational points at criticality*.

Rational points of the frequency ratio at criticality are necessarily in the form

$$0 \le \frac{\omega_0 T}{2\pi} = \frac{m}{n} < 1, \qquad n \ne 0, \tag{IX.28}$$

where m and n are integers and m/n is a fraction. A *subharmonic* solution is an nT-periodic solution with $n \ge 1$. The solution (IX.27) is said to be subharmonic if ω_0 is a rational point. Every nT-periodic solution (IX.27) satisfies the relation

$$e^{i\omega_0(t+nT)}\boldsymbol{\zeta}(t+nT) = e^{i\omega_0 t}\boldsymbol{\zeta}(t).$$

Hence $e^{i\omega_0 nT} = 1$ and $\omega_0 nT = 2\pi m$, where ω_0 satisfies (IX.26). Since

$$\lambda_0^n = (e^{i\omega_0 T})^n = (e^{2\pi i m/n})^n = 1 = \bar{\lambda}_0^n,$$

λ_0 and $\bar{\lambda}_0$ are roots of unity when ω_0 is a rational point. There are two rational points for which $\lambda_0 = \bar{\lambda}_0$ is real: $m/n = 0/1$, $\lambda_0 = 1$; and $m/n = 1/2$, $\lambda_0 = -1$.

When $m/n = 0/1$, the subharmonic solution (IX.27) is T-periodic. When $m/n = \frac{1}{2}$ the subharmonic solution (IX.27) is $2T$-periodic.

IX.8 The Operator $\mathbb{J}$ and Its Eigenvectors

We define the operator

$$\mathbb{J} = -\frac{d}{dt} + \mathbf{f}_u(t\,|\cdot) \tag{IX.29}$$

whose domain consists of differentiable nT-periodic vectors.

Of course, the set of T-periodic vectors is smaller than the set of nT-periodic vectors, dom $J_0 \subset$ dom $\mathbb{J}$. The spectrum of $\mathbb{J}$ corresponds to the spectrum of $\mathbf{\Phi}(nT, 0)$, and since $\mathbf{\Phi}(nT, 0) = \{\mathbf{\Phi}(T, 0)\}^n$, 1 is the only eigenvalue of $\mathbf{\Phi}(nT, 0)$ on the unit circle. When $n = 1$ ($m/n = 0$, where m and n are as in (IX.28) and when $n = 2$ ($m/n = \frac{1}{2}$), this eigenvalue is simple. When $n \geq 3$, this eigenvalue is double. Unit eigenvalues of $\mathbf{\Phi}(nT, 0)$ correspond to zero eigenvalues of $\mathbb{J}$. The nT-periodic vector

$$\mathbf{Z}(t) = \exp\left(\frac{2\pi imt}{nT}\right)\zeta(t), \qquad \zeta \in \mathbb{P}_T \tag{IX.30}$$

and its conjugate $\bar{\mathbf{Z}}$ are in the null space of $\mathbb{J}$:

$$\mathbb{J}\mathbf{Z} = \mathbb{J}\bar{\mathbf{Z}} = 0. \tag{IX.31}$$

When $n = 2$ ($m/n = \frac{1}{2}$) there is only one eigenvector and we can *choose* it to be real.

It is useful to state a lemma about the eigenvalue zero of the operator $\mathbb{J}$. We recall that $\mathbb{J}$ is defined at criticality.

Lemma. *Assume that hypotheses (I) and (II) about the eigenvalues of J_0 (or, equivalently, the eigenvalue λ_0 of $\mathbf{\Phi}(T, 0)$) hold. Then when*

$$n = 1 \tag{IX.32}$$

zero is a simple eigenvalue of $\mathbb{J}$ with one real eigenvector $\mathbf{Z} = \zeta = \bar{\mathbf{Z}} = \bar{\zeta}$; when

$$n = 2 \tag{IX.33}$$

zero is a simple eigenvalue of $\mathbb{J}$ with one real eigenvector $\mathbf{Z} = e^{i\pi t/2T}\zeta(t) = \bar{\mathbf{Z}}$; when

$$n > 2, \tag{IX.34}$$

zero is a double eigenvalue of $\mathbb{J}$ and any solution $\mathbf{v}$ of $\mathbb{J}\,\mathbf{v} = 0$ can be composed as a linear combination of the two independent vectors $(\mathbf{Z}, \bar{\mathbf{Z}})$ on the null space of $\mathbb{J}$.

IX.9 The Adjoint Operator $\mathbb{J}^*$, Biorthogonality, Strict Crossing, and the Fredholm Alternative for $\mathbb{J}$

We may define an adjoint operator in $\mathbb{P}_{nT}$ as follows. Let **a** and **b** be any two smooth vectors lying in $\mathbb{P}_{nT}$. Then, $\mathbb{J}^*$ is the unique linear operator satisfying the equation

$$[\mathbb{J}\mathbf{a}, \mathbf{b}]_{nT} = [\mathbf{a}, \mathbb{J}^*\mathbf{b}]_{nT},$$

where the scalar product $[\cdot, \cdot]_{nT}$ is defined by (IX.16) and

$$\mathbb{J}^* = \frac{d}{dt} + \mathbf{f}_u^*(t|\cdot). \tag{IX.35}$$

Of course, if zero is an eigenvalue of $\mathbb{J}$ it is also an eigenvalue of $\mathbb{J}^*$ such that

$$\mathbb{J}^*\mathbf{Z}^* = 0, \mathbf{Z}^*(t) = \exp\left(\frac{2\pi imt}{nT}\right)\zeta^*(t). \tag{IX.36}$$

When $n = 1$ or $n = 2$, there is just one adjoint eigenvector $\mathbf{Z}^*$; when $n > 2$, $\mathbf{Z}^*$ and $\bar{\mathbf{Z}}^*$ are independent eigenvectors of $\mathbb{J}^*$.

The biorthogonality relations

$$[\mathbf{Z}, \mathbf{Z}^*]_{nT} = 1, \qquad [\mathbf{Z}, \bar{\mathbf{Z}}^*]_{nT} = 0 \tag{IX.37}$$

follow from a direct computation using (IX.18). Moreover, the relation

$$[\mathbf{f}_{u\mu}(t|\zeta), \zeta^*]_T = [\mathbf{f}_{u\mu}(t|\mathbf{Z}), \mathbf{Z}^*]_{nT}$$

is an identity. Hence, using (IX.30) we may write (IX.25) as

$$\sigma_\mu(0) = [\mathbf{f}_{u\mu}(t|\mathbf{Z}), \mathbf{Z}^*]_{nT} \tag{IX.38}$$

and

$$[\mathbf{f}_{u\mu}(t|\bar{\mathbf{Z}}), \mathbf{Z}^*]_{nT} = [e^{-4i\pi mt/nT}\mathbf{f}_{u\mu}(t|\bar{\zeta}), \zeta^*]_{nT} = 0$$

when $n \geq 3$.

Theorem (Fredholm alternative for $\mathbb{J}$). *Suppose that hypotheses (I) and (II) of §IX.6 about the simplicity of the eigenvalue $i\omega_0$ of J_0 hold, and consider the equation*

$$\mathbb{J}\mathbf{u} = \mathbf{g} \in \mathbb{P}_{nT}. \tag{IX.39}$$

Then there exists $\mathbf{u} \in \mathbb{P}_{nT}$, unique to within linear combinations of eigenvectors $\mathbf{Z}$ and $\bar{\mathbf{Z}}$ of $\mathbb{J}$, if and only if

$$[\mathbf{g}, \mathbf{Z}^*]_{nT} = [\mathbf{g}, \bar{\mathbf{Z}}^*]_{nT} = 0. \tag{IX.40}$$

In general, there are as many orthogonality relations (IX.40) as there are adjoint eigenvectors; in the present case, as a consequence of hypotheses (I) and (II) we have just the two relations (IX.40) when $n > 2$ and just one ($\mathbf{Z}^* = \bar{\mathbf{Z}}^*$) when $n = 1$ or $n = 2$.

If **g** is real-valued, then one orthogonality relation

$$\overline{[g, \bar{\mathbf{Z}}^*]_{nT}} = [g, \mathbf{Z}^*]_{nT} = 0 \tag{IX.41}$$

will suffice for solvability.

It is useful to note that we get a *unique* **u** if we require that **u** be orthogonal to vectors on the null space of $\mathbb{J}^*$; that is, $[\mathbf{u}, \mathbf{Z}^*]_{nT} = [\mathbf{u}, \bar{\mathbf{Z}}^*]_{nT} = 0$.

IX.10 The Amplitude ε and the Biorthogonal Decomposition of Bifurcating Subharmonic Solutions

Now we look for real subharmonic solutions (IX.3), of amplitude ε, bifurcating from $\mathbf{u} = 0$ at points of resonance. The amplitude ε may be defined in various equivalent ways consistent with the requirement that $\mathbf{u}(t, \varepsilon)/\varepsilon$ is bounded when $\varepsilon \to 0$. Moreover, we may always extract from such solutions the part of the solution which lies on the null space of $\mathbb{J}$ and the other part:

$$\mathbf{u}(t, \varepsilon) = a(\varepsilon)\mathbf{Z}(t) + \bar{a}(\varepsilon)\bar{\mathbf{Z}}(t) + \mathbf{W}(t, \varepsilon), \tag{IX.42}$$

where

$$0 = [\mathbf{W}, \mathbf{Z}^*]_{nT}. \tag{IX.43}$$

and

$$a(\varepsilon) = [\mathbf{u}, \mathbf{Z}^*]_{nT}.$$

It is convenient to consider two cases separately, (i) $n = 1$ and $n = 2$ when there is just one eigenvector $\mathbf{Z} = \bar{\mathbf{Z}}$ of $\mathbb{J}$; and (ii) $n > 2$ when $\mathbf{Z}$ and $\bar{\mathbf{Z}}$ are independent eigenvectors.

For case (i) we define

$$\varepsilon = a(\varepsilon) = [\mathbf{u}(t, \varepsilon), \mathbf{Z}^*]_{nT} \tag{IX.44}$$

and, as we shall see,

$$\mathbf{W}(t, \varepsilon) = \varepsilon^2\mathbf{w}(t, \varepsilon), \tag{IX.45}$$

where $\mathbf{w}(t, 0)$ is bounded. Therefore in case (i) we may write the decomposition (IX.42) in the form

$$\mathbf{u}(t, \varepsilon) = \varepsilon\mathbf{Z} + \varepsilon^2\mathbf{w}(t, \varepsilon). \tag{IX.46}$$

For case (ii) we find it convenient to define ε by requiring that

$$a(\varepsilon) = \varepsilon e^{i\phi(\varepsilon)} = [\mathbf{u}, \mathbf{Z}^*]_{nT}. \tag{IX.47}$$

This definition is consistent with the fact that the principal part of any bifurcated solution lies in the eigenspace of the linearized operator relative to the eigenvalue zero. For case (ii) we may write the decomposition (IX.42) as

$$\mathbf{u}(t, \varepsilon) = \varepsilon(A(\varepsilon)\mathbf{Z} + \bar{A}(\varepsilon)\bar{\mathbf{Z}}) + \varepsilon^2\mathbf{w}(t, \varepsilon). \tag{IX.48}$$

IX.11 The Equations Governing the Derivatives of Bifurcating Subharmonic Solutions with Respect to ε at $\varepsilon = 0$

We are going to compute derivatives of $\mathbf{u}(t, \varepsilon) \in \mathbb{P}_{nT}$ and $\mu(\varepsilon)$ with respect to ε at $\varepsilon = 0$. When everything is analytic these derivatives are the coefficients for the Taylor series representation

$$\begin{bmatrix} \mathbf{u}(t, \varepsilon) \\ \mu(\varepsilon) \end{bmatrix} = \sum_{p=1}^{\infty} \frac{\varepsilon^p}{p!} \begin{bmatrix} \mathbf{u}_p(t) \\ \mu_p \end{bmatrix} \tag{IX.49}$$

of the solution. The coefficients satisfy equations which arise after differentiating (IX.1) and (IX.2), using the simplified notation given by (IX.21) or by identification using (IX.49) and (IX.2):

$$0 = \mathbb{J}\mathbf{u}_1$$

$$0 = \mathbb{J}\mathbf{u}_2 + 2\mu_1\mathbf{f}_{u\mu}(t|\mathbf{u}_1) + \mathbf{f}_{uu}(t|\mathbf{u}_1|\mathbf{u}_1) \tag{IX.50}$$

$$\begin{aligned} 0 = \mathbb{J}\mathbf{u}_3 &+ 3\mu_1\mathbf{f}_{uu\mu}(t|\mathbf{u}_1|\mathbf{u}_1) + 3\mu_1^2\mathbf{f}_{u\mu\mu}(t|\mathbf{u}_1) \\ &+ 3\mu_1\mathbf{f}_{u\mu}(t|\mathbf{u}_2) + 3\mathbf{f}_{uu}(t|\mathbf{u}_1|\mathbf{u}_2) \\ &+ 3\mu_2\mathbf{f}_{u\mu}(t|\mathbf{u}_1) + \mathbf{f}_{uuu}(t|\mathbf{u}_1|\mathbf{u}_1|\mathbf{u}_1), \end{aligned} \tag{IX.51}$$

and, for $p > 3$,

$$\begin{aligned} 0 = \mathbb{J}\mathbf{u}_p &+ p\mu_{p-1}\mathbf{f}_{u\mu}(t|\mathbf{u}_1) + p\{\mu_1\mathbf{f}_{u\mu}(t|\mathbf{u}_{p-1}) + \mathbf{f}_{uu}(t|\mathbf{u}_1|\mathbf{u}_{p-1})\} \\ &+ \frac{p(p-1)}{2}\Big\{\mu_2\mathbf{f}_{u\mu}(t|\mathbf{u}_{p-2}) + \mathbf{f}_{uuu}(t|\mathbf{u}_1|\mathbf{u}_1|\mathbf{u}_{p-2}) \\ &+ \mu_1^2\mathbf{f}_{u\mu\mu}(t|\mathbf{u}_{p-2}) + 2\mu_1\mathbf{f}_{uu\mu}(t|\mathbf{u}_1|\mathbf{u}_{p-2}) \\ &+ \mathbf{f}_{uu}(t|\mathbf{u}_2|\mathbf{u}_{p-2}) + \mu_{p-2}\mathbf{f}_{u\mu}(t|\mathbf{u}_2) + \mu_{p-2}\mathbf{f}_{uu\mu}(t|\mathbf{u}_1|\mathbf{u}_1) \\ &+ 2\mu_1\mu_{p-2}\mathbf{f}_{u\mu\mu}(t|\mathbf{u}_1)\Big\} + \mathbf{g}_p, \end{aligned} \tag{IX.52}$$

where $\mathbf{g}_p$ depends on lower-order terms, that is, $\mathbf{g}_p(\mu_m, \mathbf{u}_l)$, $l < p-2, m < p-2$. Of course, $\mathbf{u}_p(t)$ is nT-periodic.

It is also useful to note that the expansion of (IX.48) may be written as

$$\mathbf{u}_p = p[A_{p-1}\mathbf{Z} + \bar{A}_{p-1}\bar{\mathbf{Z}}] + p(p-1)\mathbf{w}_{p-2} \tag{IX.53}$$

where

$$A(\varepsilon) \stackrel{\text{def}}{=} e^{i\phi(\varepsilon)} = \sum_{p=0}^{\infty} \frac{A_p}{p!} \varepsilon^p$$

and

$$A_p = \frac{d^p}{d\varepsilon^p} e^{i\phi(\varepsilon)}\bigg|_{\varepsilon=0} = i\phi_p e^{i\phi_0} + b_p e^{i\phi_0}$$

where b_p depends on $\phi_l = d^l\phi/d\varepsilon^l|_{\varepsilon=0}$ of lower orders $l < p$.

For local studies of the stability of subharmonic bifurcating solutions near $\varepsilon = 0$ it is useful to expand the spectral problem (IX.6) in powers of ε. We find the expansion of the right-hand side of (IX.4) which is induced by (IX.49) and deduce that

$$\begin{aligned}\gamma \mathbf{y} + \frac{d\mathbf{y}}{dt} = {} & \mathbf{f}_u(t|\mathbf{y}) + \varepsilon\{\mu_1 \mathbf{f}_{u\mu}(t|\mathbf{y}) + \mathbf{f}_{uu}(t|\mathbf{u}_1|\mathbf{y})\} \\ & + \tfrac{1}{2}\varepsilon^2\{\mathbf{f}_{uu}(t|\mathbf{u}_2|\mathbf{y}) + 2\mu_1 \mathbf{f}_{uu\mu}(t|\mathbf{u}_1|\mathbf{y}) + \mu_2 \mathbf{f}_{u\mu}(t|\mathbf{y}) \\ & + \mathbf{f}_{uuu}(t|\mathbf{u}_1|\mathbf{u}_1|\mathbf{y}) + \mu_1^2 \mathbf{f}_{u\mu\mu}(t|\mathbf{y})\} + O(\varepsilon^3),\end{aligned} \tag{IX.54}$$

where $\mathbf{y} \in \mathbb{P}_{nT}$.

IX.12 Bifurcation and Stability of T-Periodic and $2T$-Periodic Solutions

This is case (i) specified in §IX.10 as $n = 1$ and $n = 2$. The normalizing condition (IX.44) requires that

$$[\mathbf{u}_1, \mathbf{Z}^*]_{nT} = 1$$

and

$$[\mathbf{u}_p, \mathbf{Z}^*]_{nT} = 0, \qquad p \geq 2.$$

Since $\mathbb{J}\mathbf{u}_1 = 0$ and $\mathbf{Z}$ satisfying $\mathbb{J}\mathbf{Z} = 0$, $[\mathbf{Z}, \mathbf{Z}^*] = 1$ is unique we get

$$\mathbf{u}_1 = \mathbf{Z}. \tag{IX.55}$$

The Fredholm alternative of $\mathbb{J}$ in case (i) states that we may solve $\mathbb{J}\mathbf{u} = \mathbf{g} \in \mathbb{P}_{nT}$ ($n = 1, 2$) if and only if $[\mathbf{g}, \mathbf{Z}] = 0$. Hence, (IX.50) is solvable if

$$2\mu_1 \sigma_\mu(0) + [\mathbf{f}_{uu}(t|\mathbf{Z}|\mathbf{Z}), \mathbf{Z}^*]_{nT} = 0, \tag{IX.56}$$

where $\sigma_\mu(0)$ is given by (IX.38) and is real since $\mathbf{Z}$ and $\mathbf{Z}^*$ are real. When (IX.56) holds, (IX.50) is solvable for $\mathbf{u}_2$, and is uniquely solvable for $\mathbf{u}_2$ such that $[\mathbf{u}_2, \mathbf{Z}^*]_{nT} = 0$. Similarly, all the problems in the form (IX.52) are solvable when μ_{p-1} is selected so that

$$p\mu_{p-1}\sigma_\mu(0) + [\mathbf{f}_p, \mathbf{Z}^*] = 0,$$

where $\mathbf{f}_p$ is independent of $\mathbf{u}_p$ and μ_{p-1}.

When $n = 1$

$$\mu_1 = -\frac{[\mathbf{f}_{uu}(t|\zeta|\zeta), \zeta^*]_T}{2\sigma_\mu(0)} \tag{IX.57}$$

is, in general, not zero. It follows from (IX.49) that near $\varepsilon = 0$ the bifurcation of T-periodic solutions from T-periodic solutions is two-sided (transcritical), as shown in Figure IX.1. The bifurcation of T-periodic into T-periodic solutions is very important in nature. It is the analogue for problems undergoing periodic forcing, of the bifurcation under steady forcing of steady solutions into other steady solutions. In physical examples this type of bifurcation is often associated with the breakup of spatial symmetry.

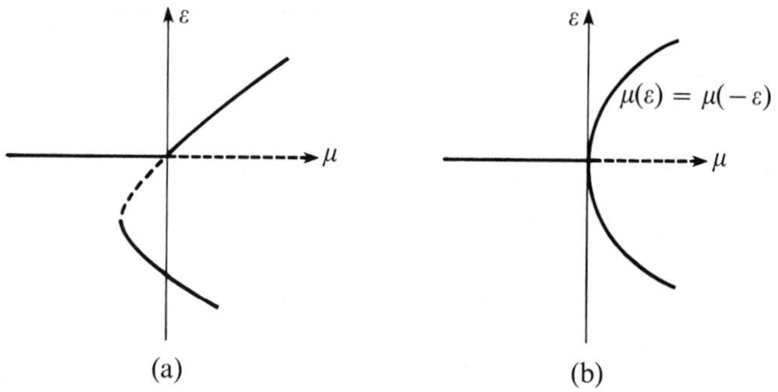

Figure IX.1 (a) the bifurcation of T-periodic solutions into T-periodic solutions is two-sided. (b) the bifurcation into $2T$-periodic solutions is one-sided (supercritical in the sketch)

When $n = 2$, a short computation using (IX.30) and (IX.36) shows that

$$\mu_1 = [e^{\pi i t/T}\mathbf{f}_{uu}(t|\zeta|\zeta), \zeta^*]_{2T}/2\sigma_\mu = 0.$$

Then, (IX.50) is solvable for $\mathbf{u}_2$ (with $\mu_1 = 0$) and (IX.51) is solvable for $\mathbf{u}_3$ if and only if

$$3\mu_2\sigma_\mu(0) + 3[\mathbf{f}_{uu}(t|\mathbf{Z}|\mathbf{u}_2), \mathbf{Z}^*]_{2T} + [\mathbf{f}_{uuu}(t|\mathbf{Z}|\mathbf{Z}|\mathbf{Z}), \mathbf{Z}^*]_{2T} = 0.$$

In general, $\mu_2 \neq 0$.

It can be shown by mathematical induction that all odd-order derivatives of $\mu(\varepsilon)$ vanish when $n = 2$. So for $2T$-periodic subharmonic bifurcation we get

$$\mu(\varepsilon) = \mu(-\varepsilon). \tag{IX.58}$$

It follows that, unlike T-periodic bifurcation with $n = 1$, two-sided or transcritical bifurcation is impossible and the bifurcation is one-sided, supercritical if it bifurcates to the right and subcritical if it bifurcates to the left.

Now we demonstrate that advancing the time origin by T in the $2T$-periodic solution is exactly the same as changing the sign of amplitude ε. This means that the direction of $\mathbf{u}(t, \varepsilon)$ changes each period T. If $\mathbf{u}$ is interpreted as a motion it goes one way for one-half of the period $2T$ and the other way for the other half. To prove this we write

$$\mathbf{u}(t, \varepsilon) = \sum_{p=1}^{\infty} \frac{\varepsilon^p}{p!} \mathbf{u}_p(t).$$

Now $\mathbf{u}_n$ is a polynomial whose terms are the composition of vectors in $\mathbb{P}_T$ with exponentials of the form

$$\exp \frac{i\pi t r_p}{T} = k_p \tag{IX.59}$$

where r_p is an odd integer when p is, and an even integer when p is. So

$$\exp \frac{i\pi(t + T)}{T} r_p = \begin{cases} -k_p & \text{if } p \text{ is odd} \\ k_p & \text{if } p \text{ is even.} \end{cases}$$

Hence,

$$\mathbf{u}(t + T, \varepsilon) = \sum_1 \frac{\varepsilon^p}{p!} \mathbf{u}_p(t + T) = \sum_{p=1}^{\infty} \frac{(-\varepsilon)^p}{p!} \mathbf{u}_p(t) = \mathbf{u}(t, -\varepsilon)$$

We summarize our results so far as follows.

Theorem *When* $\mathbf{f}$ *is analytic and hypotheses (I), (II) and (III) of* §IX.6 *hold with* $n = 1, 2$, *then there is a unique nontrivial bifurcating solution of* (IX.1). *When* $n = 1$ *the bifurcation is, in general, two-sided: when* $n = 2$ *it is one-sided. To leading order*

$$\begin{aligned} \mathbf{u}(t, \varepsilon) &= \varepsilon\zeta(t)e^{i\theta t} + O(\varepsilon^2) \\ \mu(\varepsilon) &= \begin{cases} \varepsilon\mu_1 + O(\varepsilon^2) & \text{when } n = 1 \\ \varepsilon^2\mu_2 + O(\varepsilon^4) & \text{when } n = 2, \end{cases} \end{aligned} \tag{IX.60}$$

where $\theta = 0$ *if* $n = 1$, $\theta = \pi/T$ *if* $n = 2$. *Moreover, in the case* $n = 2$, μ *is an analytic* function of* ε^2 *and* $\mathbf{u}(t + T, \varepsilon) = \mathbf{u}(t, -\varepsilon)$.

* If $\mathbf{f}$ is analytic in $(\mu, \mathbf{u})$.

We conclude this section with another factorization theorem.

Theorem (Stability of the subharmonic bifurcating solutions when $n = 1$ and $n = 2$). *Referring to* (IX.6), *we claim that*

$$\mathbf{y}(t, \varepsilon) = b(\varepsilon)\left[\frac{\partial \mathbf{u}(t, \varepsilon)}{\partial \varepsilon} + \frac{d\mu(\varepsilon)}{d\varepsilon}\mathbf{g}(t, \varepsilon)\right], \tag{IX.61}$$

$$\mathbf{g}(\cdot, \varepsilon) \in \mathbb{P}_{nT}$$

and

$$\gamma(\varepsilon) = \frac{d\mu(\varepsilon)}{d\varepsilon}\hat{\gamma}, \tag{IX.62}$$

where $b(\varepsilon)$ is a normalizing factor and $\hat{\gamma}(\varepsilon)$ and $\mathbf{g}(t, \varepsilon)$ satisfy

$$\hat{\gamma}\frac{\partial \mathbf{u}(t, \varepsilon)}{\partial \varepsilon} + \mathbf{f}_\mu(t, \mu, \mathbf{u}(t, \varepsilon)) = -\left(\gamma\mathbf{g} + \frac{d\mathbf{g}}{dt}\right)$$
$$+ \mathbf{f}_u(t, \mu(\varepsilon), \mathbf{u}(t, \varepsilon)|\mathbf{g}). \tag{IX.63}$$

When ε is small

$$\hat{\gamma}(\varepsilon) = -\sigma_\mu(0)\varepsilon + O(\varepsilon^p),$$

where $p = 2$ when $n = 1$ and $p = 3$ when $n = 2$.

We leave the proof of this factorization theorem as an exercise for the reader. The proof follows exactly along the lines laid out in §VII.8. The factorization theorem shows that subcritical solutions are unstable and supercritical solutions are stable when ε is small, and it implies the change of stability at regular turning points if no other eigenvalue than $\gamma(\varepsilon)$ (possibly complex) crosses the imaginary axis when ε increases from the bifurcation point to the turning point.

IX.13 Bifurcation and Stability of nT-Periodic Solutions with $n > 2$

nT-periodic solutions with $n > 2$ fall under case (ii) specified in §IX.10. The normalizing condition (IX.47) requires that

$$e^{i\phi_0} = [\mathbf{u}_1, \mathbf{Z}^*]_{nT}.$$

So we may take $\mathbf{u}_1$ satisfying $\mathbb{J}\mathbf{u}_1 = 0$ as

$$\begin{aligned}\mathbf{u}_1 &= e^{i\phi_0}\mathbf{Z} + e^{-i\phi_0}\overline{\mathbf{Z}}\\ &= \exp i\left(\phi_0 + \left(\frac{2\pi mt}{nT}\right)\right)\zeta(t) + \exp\left(-i\left(\phi_0 + \left(\frac{2\pi mt}{nT}\right)\right)\right)\overline{\zeta}(t).\end{aligned} \tag{IX.64}$$

Application of the Fredholm alternative to (IX.50), using (IX.41), shows that (IX.50) is solvable provided that

$$2\mu_1[\mathbf{f}_{u\mu}(t|\mathbf{u}_1), \mathbf{Z}^*]_{nT} + [\mathbf{f}_{uu}(t|\mathbf{u}_1|\mathbf{u}_1), \mathbf{Z}^*]_{nT} = 0. \qquad \text{(IX.65)}$$

To facilitate the computation of integrals like those in (IX.65) we recall that

$$\begin{aligned}[\mathbf{a}(t), \mathbf{Z}^*]_{nT} &= \frac{1}{nT}\int_0^{nT} \langle \mathbf{a}(t), \mathbf{Z}^*(t)\rangle\, dt \\ &= \frac{1}{nT}\int_0^{nT} e^{-2\pi imt/nT}\langle \mathbf{a}(t), \boldsymbol{\zeta}^*(t)\rangle\, dt \\ &= [e^{-2\pi imt/nT}\mathbf{a}(t), \boldsymbol{\zeta}^*(t)]_{nT}\end{aligned}$$

because $\overline{\langle \mathbf{a}, \mathbf{b}\rangle} = \langle \mathbf{b}, \mathbf{a}\rangle$. It follows then that

$$[\mathbf{f}_{u\mu}(t|\bar{\mathbf{Z}}), \mathbf{Z}^*]_{nT} = [e^{-4\pi imt/nT}\mathbf{f}_{u\mu}(t|\bar{\boldsymbol{\zeta}}), \boldsymbol{\zeta}^*]_{nt} = 0$$

when $m/n \neq \frac{0}{1}, \frac{1}{2}$, and $m/n < 1$. We therefore have (IX.65), using (IX.64), in the form

$$\begin{aligned}&2\mu_1\sigma_\mu(0)e^{i\phi_0} + e^{2i\phi_0}[e^{2\pi imt/nT}\mathbf{f}_{uu}(t|\boldsymbol{\zeta}|\boldsymbol{\zeta}), \boldsymbol{\zeta}^*]_{nT} \\ &\quad + 2[e^{-2\pi imt/nT}\mathbf{f}_{uu}(t|\bar{\boldsymbol{\zeta}}|\boldsymbol{\zeta}), \boldsymbol{\zeta}^*]_{nT} \\ &\quad + e^{-2i\phi_0}[e^{-6\pi imt/nT}\mathbf{f}_{uu}(t|\bar{\boldsymbol{\zeta}}|\bar{\boldsymbol{\zeta}}), \boldsymbol{\zeta}^*]_{nT} = 0. \qquad \text{(IX.66)}\end{aligned}$$

Since $\boldsymbol{\zeta}, \boldsymbol{\zeta}^* \in \mathbb{P}_T$ and $0 < m/n < 1, n > 2$, the three scalar products appearing in (IX.66) vanish except when $n = 3$. So if $n > 3$, $\mu_1 = 0$.

When $\mu_1 = 0$, ϕ_0 is not determined by (IX.66). In these cases we may determine μ_2 and ϕ_0 from the solvability condition for (IX.51):

$$3\mu_2\sigma_\mu(0)e^{i\phi_0} + 3[\mathbf{f}_{uu}(t|\mathbf{u}_1|\mathbf{u}_2), \mathbf{Z}^*]_{nT} + [\mathbf{f}_{uuu}(t|\mathbf{u}_1|\mathbf{u}_1|\mathbf{u}_1), \mathbf{Z}^*]_{nT} = 0 \qquad \text{(IX.67)}$$

where $\mathbf{u}_2$ is given by (IX.50).

IX.14 Bifurcation and Stability of $3T$-Periodic Solutions

When $n = 3$, m can be 1 or 2 (so $m/3 < 1$) and (IX.66) reduces to

$$2\mu_1\sigma_\mu(0)e^{i\phi_0} + \Lambda_1 e^{-2i\phi_0} = 0, \qquad \mu_1 = \frac{1}{2}\left|\frac{\Lambda_1}{\sigma_\mu}\right| > 0, \qquad \text{(IX.68)}$$

where

$$\Lambda_1 = [e^{-2\pi imt/T}\mathbf{f}_{uu}(t|\bar{\boldsymbol{\zeta}}|\bar{\boldsymbol{\zeta}}), \boldsymbol{\zeta}^*]_{3T}.$$

When (IX.68) holds we may solve (IX.50) for $\mathbf{w}_0 \in \mathbb{P}_{3T}$ where, using the decomposition (IX.48),

$$\mathbf{u}_2(t) = 2i\phi_1 e^{i\phi_0}\mathbf{Z} - 2i\phi_1 e^{-i\phi_0}\bar{\mathbf{Z}} + 2\mathbf{w}_0(t). \qquad \text{(IX.69)}$$

The Fredholm alternative, without normalization, determines the part of $\mathbf{u}_2$ which is orthogonal to $\mathbf{Z}^*$, $\bar{\mathbf{Z}}^*$ (that is, $2\mathbf{w}_0$, $\mathbb{J}\mathbf{u}_2 = 2\mathbb{J}\mathbf{w}_0$) and leaves the second term ϕ_1 in the expansion $\phi(\varepsilon) = \phi_0 + \varepsilon\phi_1 + O(\varepsilon^2)$, undetermined. To determine ϕ_1 we apply the Fredholm alternative to (IX.51) and find that

$$\begin{aligned} 3\mu_1[\mathbf{f}_{u\mu}(t|\mathbf{u}_2), \mathbf{Z}^*]_{3T} &+ 3[\mathbf{f}_{uu}(t, \mathbf{u}_1|\mathbf{u}_2), \mathbf{Z}^*]_{3T} \\ &+ 3\mu_2[\mathbf{f}_{u\mu}(t|\mathbf{u}_1), \mathbf{Z}^*]_{3T} + \text{terms independent of } \mu_2 \text{ and } \mathbf{u}_2 \\ &= 3\mu_1(2i\phi_1 e^{i\phi_0})\sigma_\mu(0) - 3(2i\phi_1 e^{-2i\phi_0})\Lambda_1 \\ &+ 3\mu_2 e^{i\phi_0}\sigma_\mu(0) + \text{terms independent of } \mu_2 \text{ and } \phi_1 = 0. \end{aligned}$$

When this last relation is combined with (IX.68) we get

$$e^{i\phi_0}\sigma_\mu(0)\{\mu_2 + 6i\phi_1\} = h(\mu_1, \mathbf{u}_1, \mathbf{w}_0), \tag{IX.70}$$

where $h(\mu_1, \mathbf{u}_1, \mathbf{w}_0)$ is known. Since $e^{i\phi_0}\sigma_\mu(0)$ is never zero we may always solve this complex-valued equation for μ_2 and ϕ_1. Exactly the same type of equation (IX.69) appears at higher orders and determines, sequentially, the values of μ_n and ϕ_{n-1}.

If we tried to solve this problem using the implicit function theorem we would come up with an equation determining $\mu(\varepsilon)$ and $\phi(\varepsilon)$ of the form (IX.70). In other words we get the same information from the Fredholm solvability condition at higher order and from the implicit function theorem for a system of two equations in two unknown functions μ and ϕ of one variable ε (see Appendix V.1). So the series solution we construct is unique. There are no other small solutions which bifurcate.

We now summarize our results so far and state a few new implications of the equations.

Theorem. *Suppose that the hypotheses (I), (II) and (III) of* §IX.6 *hold and* $\Lambda_1 \neq 0$. *Then there is a unique nontrivial* $3T$*-periodic solution of* (IX.1) *bifurcating when* μ *is close to zero. The bifurcation is two-sided and is given to lowest order by*

$$\begin{aligned} \mathbf{u}(t, \varepsilon) &= \varepsilon \exp i\left(\phi(\varepsilon) + \left(\frac{2\pi imt}{3T}\right)\right)\zeta(t) \\ &\quad + \varepsilon \exp\left(-i\left(\phi(\varepsilon) + \left(\frac{2\pi imt}{3T}\right)\right)\right)\bar{\zeta}(t) + O(\varepsilon^2), \qquad m = 1, 2, \\ \mu(\varepsilon) &= \varepsilon\mu_1 + O(\varepsilon^2), \qquad \mu_1 = \left|\frac{\Lambda_1}{\sigma_\mu(0)}\right| \\ \phi(\varepsilon) &= \frac{1}{3}\arg\left(-\frac{\Lambda_1}{\sigma_\mu(0)}\right) + \frac{2k\pi}{3} + O(\varepsilon), \qquad k = 0, 1, 2, \end{aligned} \tag{IX.71}$$

where $\mathbf{u}$, ϕ, μ *are analytic* in* ε *in a neighborhood of zero, and* $k = 0, 1, 2$ *corresponds to translations of the origin in* t: $0, T, 2T$ *if* $m = 1$ *and* $0, 2T, T$ *if* $m = 2$, *and where* Λ_1 *is defined by* (IX.68).

* If $\mathbf{f}$ is analytic in $(\mu, \mathbf{u})$.

Equation (IX.71)$_3$ solves (IX.68). From our construction and the decomposition (IX.48) it is suggested that $\mathbf{u}(t, \varepsilon)$ depends on t through two times, $\tau(t)$ and t:

$$\mathbf{u}(t, \varepsilon) = \mathscr{U}(\tau(t), t) = \mathscr{U}(\tau(t), t + T).$$

$\mathscr{U}$ is T-periodic in its second argument. This T-periodicity has its origin in the T-periodicity of $\boldsymbol{\zeta}(t)$. In the first argument of $\mathscr{U}(\cdot, \cdot)$ we pose

$$\tau(t) = \phi(\varepsilon) + \left(\frac{2\pi m t}{3T}\right)$$

and

$$\tau(t + T) = \phi(\varepsilon) + \frac{2\pi m}{3} + \frac{2\pi m t}{3T}.$$

From this property the last statement in the theorem may be proven.*

We turn next to the study of stability of the 3T-periodic solutions and expand $\gamma(\varepsilon)$ and $\mathbf{y}(\cdot, \varepsilon) \in \mathbb{P}_{3T}$:

$$\gamma(\varepsilon) = \gamma_1 \varepsilon + o(\varepsilon)$$

and

$$\mathbf{y}(t, \varepsilon) = \mathbf{y}_0(t) + \varepsilon \mathbf{y}_1(t) + o(\varepsilon).$$

Inserting these expansions into (IX.54) we find that

$$\mathbb{J}\mathbf{y}_0 = 0, \qquad \mathbf{y}_0 \in \mathbb{P}_{3T}$$

and

$$\gamma_1 \mathbf{y}_0 = \mathbb{J}\mathbf{y}_1 + \mu_1 \mathbf{f}_{u\mu}(t | \mathbf{y}_0) + \mathbf{f}_{uu}(t | \mathbf{u}_1 | \mathbf{y}_0) \tag{IX.72}$$

where $\mathbf{y}_1 \in \mathbb{P}_{3T}$. Hypotheses (I) and (II) about the eigenvalues zero of $\mathbb{J}$ imply that

$$\mathbf{y}_0 = A\mathbf{Z} + B\bar{\mathbf{Z}},$$

where A and B are to be determined. Recalling that $\mathbf{u}_1 = a_0 \mathbf{Z} + \bar{a}_0 \bar{\mathbf{Z}}$ where $a_0 = e^{i\phi_0}$ we may write (IX.72) as

$$\begin{aligned}\gamma_1(A\mathbf{Z} + B\bar{\mathbf{Z}}) = \mathbb{J}\mathbf{y}_1 &+ \mu_1\{A\mathbf{f}_{u\mu}(t | \mathbf{Z}) + B\mathbf{f}_{u\mu}(t, \bar{\mathbf{Z}})\} \\ &+ a_0 A\mathbf{f}_{uu}(t | \mathbf{Z} | \mathbf{Z}) + a_0 B\mathbf{f}_{uu}(t | \mathbf{Z} | \bar{\mathbf{Z}}) \\ &+ \bar{a}_0 A\mathbf{f}_{uu}(t | \bar{\mathbf{Z}} | \mathbf{Z}) + \bar{a}_0 B\mathbf{f}_{uu}(t | \bar{\mathbf{Z}} | \bar{\mathbf{Z}}).\end{aligned} \tag{IX.73}$$

To solve (IX.73) we must make the inhomogeneous terms orthogonal to $\mathbf{Z}^*$ and $\bar{\mathbf{Z}}^*$. Both projections are required here because the inhomogeneous terms are complex-valued. Writing (IX.73) as $\mathbb{J}\mathbf{y}_1 = \mathbf{g}$, we find, using (IX.68) that the two equations $[\mathbf{g}, \mathbf{Z}^*]_{3T} = [\mathbf{g}, \bar{\mathbf{Z}}^*]_{3T} = 0$ may be written as

$$\gamma_1 A = \mu_1 \sigma_\mu(0) A + \bar{a}_0 B \Lambda_1$$

and

$$\gamma_1 B = a_0 A \bar{\Lambda}_1 + \mu_1 \bar{\sigma}_\mu(0) B.$$

* For a rigorous proof see G. Iooss and D. D. Joseph, *Arch. Rational Mech. Anal.*, **66**, 135–172 (1977).

It follows that $\gamma_1^{(1)}$ and $\gamma_1^{(2)}$ are eigenvalues of the matrix

$$\mathscr{M} = \begin{bmatrix} \mu_1\sigma_\mu(0) & e^{-i\phi_0}\Lambda_1 \\ e^{i\phi_0}\bar{\Lambda}_1 & \mu_1\bar{\sigma}_\mu(0) \end{bmatrix}$$

where

$$\gamma_1^{(1)} + \gamma_1^{(2)} = \text{tr}\,\mathscr{M} = 2\mu_1 \,\text{Re}\,\sigma_\mu(0) = 2\mu_1\xi_\mu(0) > 0 \tag{IX.74}$$

because $\mu_1 > 0$ and $\xi_\mu(0) > 0$, and

$$\gamma_1^{(1)}\gamma_1^{(2)} = \det \mathscr{M} = \mu_1^2|\sigma_\mu(0)|^2 - |\Lambda_1|^2. \tag{IX.75}$$

But (IX.67) shows that $|\Lambda_1|^2 = 4\mu_1^2|\sigma_\mu|^2$ so that

$$\gamma_1^{(1)}\gamma_1^{(2)} = -3\mu_1^2|\sigma_\mu(0)|^2 < 0 \tag{IX.76}$$

and one of the two eigenvalues is positive and the other is negative. Since $\gamma_1^{(1)} \neq \gamma_1^{(2)}$, the two eigenvalues $\gamma^{(1)}(\varepsilon)$ and $\gamma^{(2)}(\varepsilon)$ are regular functions of ε. It follows that one of the two eigenvalues

$$\begin{bmatrix} \gamma^{(1)}(\varepsilon) \\ \gamma^{(2)}(\varepsilon) \end{bmatrix} = \varepsilon \begin{bmatrix} \gamma_1^{(1)} \\ \gamma_1^{(2)} \end{bmatrix} + O(\varepsilon^2)$$

is positive on both sides of criticality, that is, for both positive and negative ε, as in Figure IX.2.

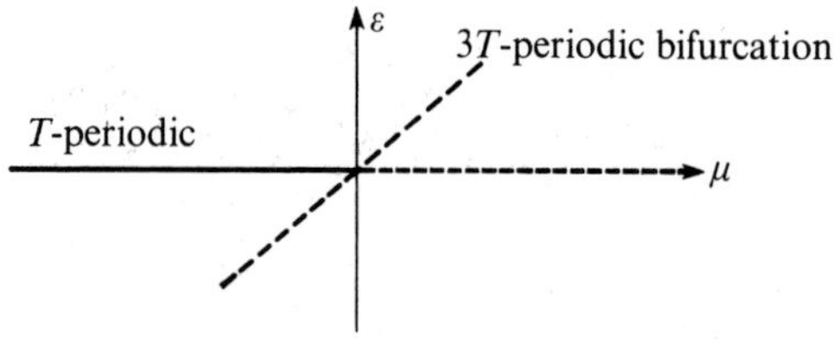

Figure IX.2 The $3T$-periodic bifurcating solution is unstable on both sides of criticality

IX.15 Bifurcation of $4T$-Periodic Solutions

$4T$-periodic solutions fall under case (ii) with $n = 4$ and $m = 1, 3$ specified in §IX.10. The normalizing condition (IX.47) requires that

$$e^{i\phi_0} = [\mathbf{u}_1, \mathbf{Z}^*]_{4T}.$$

So we may take $\mathbf{u}_1$ satisfying $\mathbb{J}\mathbf{u}_1 = 0$ in the form given by (IX.64). We already decided, in §IX.13, that $\mu_1 = 0$ when $n = 4$. With $\mu_1 = 0$, we can solve (IX.50), not for $\mathbf{u}_2$, but for $\mathbf{w}_0$ in the decomposition (IX.69) of $\mathbf{u}_2$. The terms proportional to ϕ_1 in $\mathbf{u}_2 = 2i\phi_1 e^{i\phi_0}\mathbf{Z} - 2i\phi_1 e^{-i\phi_0}\bar{\mathbf{Z}} + 2\mathbf{w}_0(t)$ vanish after integration and

$$[\mathbf{f}_{uu}(t|\mathbf{u}_1|\mathbf{u}_2), \mathbf{Z}^*]_{4T} = 2[\mathbf{f}_{uu}(t|\mathbf{u}_1|\mathbf{w}_0), \mathbf{Z}^*]_{4T}. \tag{IX.77}$$

(In fact, (IX.77) holds when $\mathbf{u}_2$ is replaced by $\mathbf{u}_n$ and $2\mathbf{w}_0$ is replaced by $n(n-1)\mathbf{w}_{n-2}$ (see (IX.53)).)

To get μ_2 and ϕ_0 we need to work out the integrals in (IX.67) and (IX.77) and to solve the reduced equations for μ_2 and ϕ_0. As a preliminary for reduction we note that with $\mu_1 = 0$, (IX.50) may be written as

$$\begin{aligned}\mathbb{J}2\mathbf{w}_0 &= -\mathbf{f}_{uu}(t|\mathbf{u}_1|\mathbf{u}_1)\\ &= -\{e^{2i\phi_0}\mathbf{f}_{uu}(t|\mathbf{Z}|\mathbf{Z})\\ &\quad + 2\mathbf{f}_{uu}(t|\bar{\mathbf{Z}}|\mathbf{Z}) + e^{-2i\phi_0}\mathbf{f}_{uu}(t|\bar{\mathbf{Z}}|\bar{\mathbf{Z}})\},\end{aligned} \tag{IX.78}$$

where $\mathbf{w}_0$ contains no terms proportional to eigenvectors of $\mathbb{J}$; that is, $[\mathbf{w}_0, \mathbf{Z}^*]_{4T} = 0$. In fact, the solutions of (IX.78) which are orthogonal to $\mathbf{Z}^*$ (and $\bar{\mathbf{Z}}^*$) are unique and in the form

$$\begin{aligned}2\mathbf{w}_0(t) &= \mathbf{w}_{02}(t) + \exp i\left(2\phi_0 + \left(\frac{m\pi t}{T}\right)\right)\mathbf{w}_{01}(t)\\ &\quad + \exp\left(-i\left(2\phi_0 + \left(\frac{m\pi t}{T}\right)\right)\right)\bar{\mathbf{w}}_{01}(t), \qquad \mathbf{w}_{0j} \in \mathbb{P}_T.\end{aligned} \tag{IX.79}$$

Returning now to (IX.67) with (IX.64), (IX.77), and (IX.79) we find that many terms integrate to zero and that

$$\mu_2\sigma_\mu(0)e^{i\phi_0} + \Lambda_2 e^{i\phi_0} + \Lambda_3 e^{-3i\phi_0} = 0 \tag{IX.80}$$

where

$$\begin{aligned}\Lambda_2 &= [\mathbf{f}_{uu}(t|\bar{\zeta}|\mathbf{w}_{01}), \zeta^*]_T + [\mathbf{f}_{uu}(t|\zeta|\mathbf{w}_{02}), \zeta^*]_T\\ &\quad + [\mathbf{f}_{uuu}(t|\zeta|\zeta|\bar{\zeta}), \zeta^*]_T,\\ \Lambda_3 &= [e^{-2\pi imt/T}\mathbf{f}_{uu}(t|\bar{\zeta}|\mathbf{w}_{01}), \zeta^*]_T\\ &\quad + \tfrac{1}{3}[e^{-2\pi imt/T}\mathbf{f}_{uuu}(t|\bar{\zeta}|\bar{\zeta}|\bar{\zeta}), \zeta^*]_T\end{aligned}$$

and $m = 1$ or 3.

We may write (IX.80) as

$$e^{-4i\phi_0} = -\frac{\mu_2 + (\Lambda_2/\sigma_\mu)}{\Lambda_3/\sigma_\mu}. \tag{IX.81}$$

Since the modulus of $e^{-4i\phi_0}$ is unity we have

$$\left|\frac{\Lambda_3}{\sigma_\mu}\right| = \left|\mu_2 + \frac{\Lambda_2}{\sigma_\mu}\right|. \tag{IX.82}$$

Real values of μ_2 solving (IX.82) can, of course, exist only if

$$\left|\frac{\Lambda_3}{\sigma_\mu}\right| \geq \left|\operatorname{Im}\frac{\Lambda_2}{\sigma_\mu}\right|. \tag{IX.83}$$

Suppose (IX.83) holds; then, squaring each side of (IX.82) we find and solve a quadratic equation for μ_2

$$\begin{bmatrix}\mu_2^{(1)}\\ \mu_2^{(2)}\end{bmatrix} = -\operatorname{Re}\left(\frac{\Lambda_2}{\sigma_\mu}\right)\begin{bmatrix}1\\1\end{bmatrix} + \left\{\left|\frac{\Lambda_3}{\sigma_\mu}\right|^2 - \left[\operatorname{Im}\frac{\Lambda_2}{\sigma_\mu}\right]^2\right\}^{1/2}\begin{bmatrix}1\\-1\end{bmatrix}. \tag{IX.84}$$

For each of the values $\mu_2^{(1)}$ and $\mu_2^{(2)}$ we get four values of ϕ_0 solving (IX.80)

$$\phi_0^{(l)} = \frac{1}{4}\arg\left\{-\frac{\Lambda_3}{(\sigma_\mu \mu_2^{(l)} + \Lambda_2)}\right\} + \left(\frac{k\pi}{2}\right) \tag{IX.85}$$
$$l = 1, 2, \qquad k = 0, 1, 2, 3.$$

To determine ϕ_p and μ_{p+2} it is necessary to consider the solvability condition $[(\text{IX.52}), \mathbf{Z}^*]_{nT}$ for (IX.52):

$$\begin{aligned} p\mu_{p-1}\sigma_\mu e^{i\phi_0} &+ p(p-1)\{\mu_2[\mathbf{f}_{u\mu}(t|\mathbf{u}_{p-2}), \mathbf{Z}^*]_{4T} \\ &+ [\mathbf{f}_{uu}(t|\mathbf{u}_2|\mathbf{u}_{p-2}), \mathbf{Z}^*]_{4T} \\ &+ [\tfrac{1}{2}\mathbf{f}_{uuu}(t|\mathbf{u}_1|\mathbf{u}_1|\mathbf{u}_{p-2}), \mathbf{Z}^*]_{4T}\} \\ &+ p(p-1)(p-2)[\mathbf{f}_{uu}(t|\mathbf{u}_1|\mathbf{w}_{p-3}), \mathbf{Z}^*]_{4T} + [\mathbf{g}_p, \mathbf{Z}^*]_{4T} = 0, \end{aligned} \tag{IX.86}$$

where ϕ_{p-3} enters (IX.86) through $\mathbf{u}_{p-2}$ defined by (IX.53). It is possible, by mathematical induction, to establish that

$$\mu_{2p-1} = \phi_{2p-1} = 0 \quad \text{for } p \geq 1. \tag{IX.87}$$

The solvability of the equations at higher order is equivalent to a solution of the bifurcation problems by the implicit function theorem (see remark at the conclusion of Appendix V.1). The regularity of the solution as a function of ε depends on the regularity of $\mathbf{f}(\cdot, \cdot)$ and is analytic if $\mathbf{f}$ is.

We summarize the results about $4T$-periodic subharmonic bifurcating solutions which have been proved so far.

Theorem. *Suppose that the hypotheses (I), (II) and (III) of* §IX.6 *hold with* $n = 4$, *the coefficients,* $\sigma_\mu(0)$, Λ_2, *and* Λ_3 *being defined by* (IX.38) *and under* (IX.80). *Then if* $|\text{Im } \Lambda_2/\sigma_\mu| > |\Lambda_3/\sigma_\mu|$, *there is no small-amplitude,* $4T$*-periodic bifurcated solution of* (IX.1), *for* μ *near zero. If* $|\Lambda_3/\sigma_\mu| > |\text{Im}\,(\lambda_2/\sigma_\mu)|$, *two nontrivial* $4T$*-periodic solutions of* (IX.1) *bifurcate, each on one side of criticality. If* $|\Lambda_2| < |\Lambda_3|$, *one solution exists only for* $\mu \geq 0$; *the other exists only for* $\mu \leq 0$. *If* $|\Lambda_2| > |\Lambda_3|$ *the two solutions bifurcate on the same side of* $\mu = 0$: $\mu \geq 0$ *if* $\text{Re}\,(\Lambda_2/\sigma_\mu) < 0$, $\mu \leq 0$ *if* $\text{Re}\,(\Lambda_2/\sigma_\mu) > 0$. *The principal parts of the bifurcating solutions are given by*

$$\begin{aligned} \mathbf{u}^{(j)}(t, \varepsilon) &= \varepsilon \exp i\left[\phi^{(j)}(\varepsilon^2) - \left(\frac{m\pi t}{2T}\right)\right]\boldsymbol{\zeta}(t) \\ &\quad + \varepsilon \exp\left(-i\left[\phi^{(j)}(\varepsilon^2) - \left(\frac{m\pi t}{2T}\right)\right]\right)\bar{\boldsymbol{\zeta}}(t) + O(\varepsilon^2) \\ \mu^{(j)}(\varepsilon^2) &= \varepsilon^2\mu_2^{(j)} + O(\varepsilon^4), \qquad \mu_2^{(1)} = 0 \quad \text{if } |\Lambda_2| = |\Lambda_3|, \\ \phi^{(j)}(\varepsilon^2) &= \frac{1}{4}\arg\left[-\frac{\Lambda_3}{(\sigma_\mu\mu_2^{(j)} + \Lambda_2)}\right] + \left(\frac{k\pi}{2}\right) + O(\varepsilon^2) \\ &\qquad m = 1, 3, \qquad j = 1, 2. \end{aligned} \tag{IX.88}$$

The values $k = 0, 1, 2, 3$ correspond to translations of t through period T: $0, T, 2T, 3T$ if $m = 1$; $0, 3T, 2T, T$ if $m = 3$. The functions $\mu^{(j)}$ are analytic in ε^2, and $\mathbf{u}^{(j)}$ is analytic in ε.*

The recursive construction of our solution in series shows that the invariance properties of $\mathbf{u}(t, \varepsilon)$ with respect to period T translates of t can be deduced from the transformation properties of the coefficient

$$e^{i(\phi_0 + (\pi m t/T))} \stackrel{\text{def}}{=} e^{i\theta_0 t}$$

in the expression for $\varepsilon\mathbf{u}_1(t) = \varepsilon e^{i\theta_0 t}\zeta(t) + \varepsilon e^{-i\theta_0 t}\bar{\zeta}(t)$, $\zeta(\cdot) \in \mathbb{P}_T$. This expression; and the coefficients $\mathbf{u}_n(t)$ which depend on $\mathbf{u}_1(t)$ recursively, are unchanged under the first group of translates $\phi_0 \mapsto \phi_0 + (\pi/2)$, $t \mapsto t - T (m = 1)$ and $t \mapsto t + T$ $(m = 3)$. On the other hand, the group of translates $\phi_0 \mapsto \phi_0 + \pi$, $t \mapsto t - 2T$ $(m = 1$ or $3)$ induces the transformation $\varepsilon\mathbf{u}_1(t) \mapsto \varepsilon\mathbf{u}_1(t - 2T) = (-\varepsilon)\mathbf{u}_1(t)$. This transformation is equivalent to $\mathbf{u}(t, \varepsilon) \mapsto \mathbf{u}(t - 2T, \varepsilon) = \mathbf{u}(t, -\varepsilon)$ because the t translate changes the sign of the odd-order coefficients $\mathbf{u}_{2n-1}(t)$ which is the same as changing the sign of ε in the expansion of $\mathbf{u}(t, \varepsilon)$.

IX.16 Stability of 4*T*-Periodic Solutions

To determine the stability of the $4T$-periodic solutions near $\varepsilon = 0$ we consider the spectral problem (IX.54) and determine the coefficients in the expansion† of

$$\gamma(\varepsilon) = \gamma_1\varepsilon + \tfrac{1}{2}\gamma_2\varepsilon^2 + o(\varepsilon^2) \qquad \text{(IX.89)}_1$$

and

$$\mathbf{y}(t, \varepsilon) = \mathbf{y}_0(t) + \mathbf{y}_1(t)\varepsilon + \tfrac{1}{2}\mathbf{y}_2(t)\varepsilon^2 + o(\varepsilon^2) \in \mathbb{P}_{4T} \qquad \text{(IX.89)}_2$$

for each of the two independent bifurcating solutions. We find $\gamma_1 = 0$, so that stability is determined by the sign of γ_2.

We proceed in the usual way by combining (IX.89) and (IX.54), identifying independent powers of ε, and find that

$$\mathbb{J}\mathbf{y}_0 = 0, \qquad \mathbf{y}_0 \in \mathbb{P}_{4T} \qquad \text{(IX.90)}$$

$$\gamma_1\mathbf{y}_0 = \mathbb{J}\mathbf{y}_1 + \mathbf{f}_{uu}(t|\mathbf{u}_1|\mathbf{y}_0), \qquad \mathbf{y}_1 \in \mathbb{P}_{4T} \qquad \text{(IX.91)}$$

and

$$\begin{aligned} 2\gamma_1\mathbf{y}_1 + \gamma_2\mathbf{y}_0 = {} & \mathbb{J}\mathbf{y}_2 + 2\mathbf{f}_{uu}(t|\mathbf{u}_1|\mathbf{y}_1) \\ & + \{\mu_2\mathbf{f}_{u\mu}(t|\mathbf{y}_0) + \mathbf{f}_{uu}(t|\mathbf{u}_2|\mathbf{y}_0) \\ & + \mathbf{f}_{uuu}(t|\mathbf{u}_1|\mathbf{u}_1|\mathbf{y}_0)\}, \qquad \mathbf{y}_2(\cdot) \in \mathbb{P}_{4T}. \end{aligned} \qquad \text{(IX.92)}$$

* If $\mathbf{f}$ is analytic in $(\mu, \mathbf{u})$.

† We find that $\gamma(0) = 0$ is *semi-simple*, $\gamma_1 = 0$ and $\gamma(\varepsilon)$ is well separated at order ε^2.

On the other hand, we may decompose

$$\mathbf{y}(t, \varepsilon) = A(\varepsilon)\mathbf{Z} + B(\varepsilon)\bar{\mathbf{Z}} + \varepsilon\boldsymbol{\psi}(t, \varepsilon) \tag{IX.93}$$

where A, B and $\boldsymbol{\psi}$ are complex-valued and

$$[\boldsymbol{\psi}, \mathbf{Z}^*]_{4T} = [\boldsymbol{\psi}, \bar{\mathbf{Z}}^*]_{4T} = 0.$$

It follows from (IX.90) and (IX.93) that

$$\mathbf{y}_0 = A_0\mathbf{Z} + B_0\bar{\mathbf{Z}}, \qquad |A_0|^2 + |B_0|^2 \neq 0 \tag{IX.94}$$

and

$$\mathbf{y}_1 = A_1\mathbf{Z} + B_1\bar{\mathbf{Z}} + \boldsymbol{\psi}_0(t). \tag{IX.95}$$

Now, the following identity holds:

$$[\mathbf{f}_{uu}(t|\mathbf{u}_1|\mathbf{y}_0), \mathbf{Z}^*]_{4T} = 0 \tag{IX.96}$$

because $[\mathbf{f}_{uu}(t|\mathbf{Z}_l|\mathbf{Z}_j), \mathbf{Z}_k^*]_{4T} = 0$ where l, j, $k = 1$ or 2 and $\mathbf{Z}_2 = \bar{\mathbf{Z}}_1 = \mathbf{Z}$, $\mathbf{Z}_2^* = \bar{\mathbf{Z}}_1^* = \mathbf{Z}^*$. The Fredholm alternative applied to (IX.91) leads to:

$$\gamma_1 A_0 = \gamma_1 B_0 = 0 \quad \text{and} \quad \gamma_1 = 0.$$

Returning now to (IX.91) with $\gamma_1 = 0$ and (IX.95), we find that

$$\begin{aligned}\mathbb{J}\boldsymbol{\psi}_0 &+ A_0 e^{i\phi_0}\mathbf{f}_{uu}(t|\mathbf{Z}|\mathbf{Z}) + B_0 e^{-i\phi_0}\mathbf{f}_{uu}(t|\mathbf{Z}|\bar{\mathbf{Z}})\\ &+ (B_0 e^{i\phi}{}_0 + \Lambda_0 e^{-i\phi_0})\mathbf{f}_{uu}(t|\mathbf{Z}|\bar{\mathbf{Z}}) = 0.\end{aligned} \tag{IX.97}$$

Comparing (IX.97) with (IX.78), we find that

$$\begin{aligned}\boldsymbol{\psi}_0(t) &= A_0 e^{i\phi_0}e^{im\pi t/T}\mathbf{w}_{01}(t) + B_0 e^{-i\phi_0}e^{-im\pi t/T}\bar{\mathbf{w}}_{01}(t)\\ &+ \tfrac{1}{2}(A_0 e^{-i\phi_0} + B_0 e^{i\phi_0})\mathbf{w}_{02}(t),\end{aligned} \tag{IX.98}$$

where $m = 1, 3$ and $\mathbf{w}_{0l} \in \mathbb{P}_T$ are the functions defined in (IX.79).

Turning next to the conditions $[\mathbb{J}\mathbf{y}_2, \mathbf{Z}_l^*]_{4T} = 0$ where $l = 1, 2$ and $\mathbf{Z}_2^* = \bar{\mathbf{Z}}_1^* = \mathbf{Z}^*$ for the solvability of (IX.92) we find, using (IX.95), (IX.96), and (IX.98), that

$$\begin{aligned}&[\mathbf{f}_{uu}(t|\mathbf{u}_1|\mathbf{y}_1), \mathbf{Z}^*]_{4T}\\ &\quad= [\mathbf{f}_{uu}(t|\mathbf{u}_1|\boldsymbol{\psi}_0), \mathbf{Z}^*]_{4T}\\ &\quad= \tfrac{1}{2}(A_0 + B_0 e^{2i\phi_0})[\mathbf{f}_{uu}(t|\boldsymbol{\zeta}|\mathbf{w}_{02}), \boldsymbol{\zeta}^*]_T + A_0[\mathbf{f}_{uu}(t|\bar{\boldsymbol{\zeta}}|\mathbf{w}_{01}), \boldsymbol{\zeta}^*]_T\\ &\quad\quad+ B_0 e^{-2i\phi_0}[e^{-2\pi imt/T}\mathbf{f}_{uu}(t|\bar{\boldsymbol{\zeta}}|\bar{\mathbf{w}}_{01}), \boldsymbol{\zeta}^*]_T.\end{aligned} \tag{IX.99}$$

This same expression, (IX.99), holds when $(\mathbf{Z}^*, A_0, B_0)$ are replaced by $(\bar{\mathbf{Z}}^*, B_0, A_0)$ and all the other quantities are replaced by their conjugates:

$$\begin{aligned}[\mathbf{f}_{uu}(t|\mathbf{u}_1|\mathbf{y}_1), \bar{\mathbf{Z}}^*]_{4T} &= \tfrac{1}{2}(B_0 + A_0 e^{-2i\phi_0})[\mathbf{f}_{uu}(t|\bar{\boldsymbol{\zeta}}|\mathbf{w}_{02}), \bar{\boldsymbol{\zeta}}^*]_T\\ &\quad+ B_0[\mathbf{f}_{uu}(t|\boldsymbol{\zeta}|\bar{\mathbf{w}}_{01}), \bar{\boldsymbol{\zeta}}^*]_T\\ &\quad+ A_0 e^{2i\phi_0}[e^{2\pi imt/T}\mathbf{f}_{uu}(t|\boldsymbol{\zeta}|\mathbf{w}_{01}), \bar{\boldsymbol{\zeta}}^*]_T.\end{aligned}$$

Similarly, using (IX.69), (IX.79), and (IX.94), we find that

$$\begin{aligned}[\mathbf{f}_{uu}(t|\mathbf{u}_2|\mathbf{y}_0), \mathbf{Z}^*]_{4T} &= [\mathbf{f}_{uu}(t|2\mathbf{w}_0|\mathbf{y}_0), \mathbf{Z}^*]_{4T} \\ &= A_0[\mathbf{f}_{uu}(t|\mathbf{w}_{02}|\zeta), \zeta^*]_T \\ &\quad + B_0 e^{2i\phi_0}[\mathbf{f}_{uu}(t|\mathbf{w}_{01}|\bar{\zeta}), \zeta^*]_T \\ &\quad + B_0 e^{-2i\phi_0}[e^{-2\pi imt/T}\mathbf{f}_{uu}(t|\bar{\mathbf{w}}_{01}|\bar{\zeta}), \zeta^*]_T\end{aligned}$$

and

$$\begin{aligned}[\mathbf{f}_{uu}(t|\mathbf{u}_2|\mathbf{y}_0), \bar{\mathbf{Z}}^*]_{4T} &= B_0[\mathbf{f}_{uu}(t|\bar{\mathbf{w}}_{02}|\bar{\zeta}), \bar{\zeta}^*]_T \\ &\quad + A_0 e^{-2i\phi_0}[\mathbf{f}_{uu}(t|\bar{\mathbf{w}}_{01}|\zeta), \bar{\zeta}^*]_T \\ &\quad + A_0 e^{2i\phi_0}[e^{2\pi imt/T}\mathbf{f}_{uu}(t|\mathbf{w}_{01}|\zeta), \bar{\zeta}^*]_T.\end{aligned}$$

In the same way, we find that

$$\begin{aligned}[\mathbf{f}_{uuu}(t|\mathbf{u}_1|\mathbf{u}_1|\mathbf{y}_0), \mathbf{Z}^*]_{4T} &= 2A_0[\mathbf{f}_{uuu}(t|\zeta|\bar{\zeta}|\zeta), \zeta^*]_T \\ &\quad + B_0 e^{2i\phi_0}[\mathbf{f}_{uuu}(t|\zeta|\zeta|\bar{\zeta}), \zeta^*]_T \\ &\quad + B_0 e^{-2i\phi_0}[e^{-2\pi imt/T}\mathbf{f}_{uuu}(t|\bar{\zeta}|\bar{\zeta}|\bar{\zeta}), \zeta^*]_T\end{aligned}$$

and

$$\begin{aligned}[\mathbf{f}_{uuu}(t|\mathbf{u}_1|\mathbf{u}_1|\mathbf{y}_0), \bar{\mathbf{Z}}^*]_{4T} &= 2B_0[\mathbf{f}_{uuu}(t|\bar{\zeta}|\zeta|\bar{\zeta}), \bar{\zeta}^*]_T \\ &\quad + A_0 e^{-2i\phi_0}[\mathbf{f}_{uuu}(t|\bar{\zeta}|\bar{\zeta}|\zeta), \bar{\zeta}^*]_T \\ &\quad + A_0 e^{2i\phi_0}[e^{2\pi imt/T}\mathbf{f}_{uuu}(t|\zeta|\zeta|\zeta), \bar{\zeta}^*]_T.\end{aligned}$$

Finally, we use (IX.38) to compute

$$[\mathbf{f}_{u\mu}(t|\mathbf{y}_0), \mathbf{Z}^*]_{4T} = \sigma_\mu A_0$$

and

$$[\mathbf{f}_{u\mu}(t|\mathbf{y}_0), \bar{\mathbf{Z}}^*]_{4T} = \bar{\sigma}_\mu B_0.$$

Putting all these results together, we find that the two solvability conditions for (IX.92) are

$$\gamma_2 A_0 = (\sigma_\mu \mu_2 + 2\Lambda_2)A_0 + B_0(\Lambda_2 e^{2i\phi_0} + 3\Lambda_3 e^{-2i\phi_0})$$

and

$$\gamma_2 B_0 = (\bar{\sigma}_\mu \mu_2 + 2\bar{\Lambda}_2)B_0 + A_0(\bar{\Lambda}_2 e^{-2i\phi_0} + 3\bar{\Lambda}_3 e^{2i\phi_0}).$$

So γ_2 are the eigenvalues of the matrix

$$\mathbf{S} \stackrel{\text{def}}{=} \begin{bmatrix} \mu_2\sigma_\mu + 2\Lambda_2 & \Lambda_2 e^{2i\phi_0} + 3\Lambda_3 e^{-2i\phi_0} \\ \bar{\Lambda}_2 e^{-2i\phi_0} + 3\bar{\Lambda}_3 e^{2i\phi_0} & \mu_2\bar{\sigma}_\mu + 2\bar{\Lambda}_2 \end{bmatrix}, \tag{IX.100}$$

where $\mu_2\sigma_\mu + \Lambda_2 + \Lambda_3 e^{-4i\phi_0} = 0$. The eigenvalues $\gamma_2^{(1)}$ and $\gamma_2^{(2)}$ of (IX.100) satisfy

$$(\gamma_2^{(1)} + \gamma_2^{(2)}) = \text{tr}\,\mathbf{S} = 2(\mu_2\xi_\mu + 2\,\text{Re}\,\Lambda_2)$$

and

$$\gamma_1^{(1)}\gamma_2^{(1)} = \det \mathbf{S} = |\mu_2\sigma_\mu + 2\Lambda_2|^2 - |\Lambda_2 + 3\Lambda_3 e^{-4i\phi_0}|^2$$
$$= |\sigma_\mu|^2\left\{\left|\mu_2 + \frac{2\Lambda_2}{\sigma_\mu}\right|^2 - \left|3\mu_2 + \frac{2\Lambda_2}{\sigma_\mu}\right|^2\right\}$$
$$= -8\mu_2|\sigma_\mu|^2\left\{\mu_2 + \operatorname{Re}\frac{\Lambda_2}{\sigma_\mu}\right\}.$$

If $|\Lambda_2| < |\Lambda_3|$ we know from the theorem in §IX.15 that $\mu_1^{(1)}\mu_2^{(2)} < 0$ and we note that for $\mu_2^{(1)} < 0 < \mu_2^{(2)}$

$$\mu_2^{(1)} + \operatorname{Re}\frac{\Lambda_2}{\sigma_\mu} = -\mu_2^{(2)} - \operatorname{Re}\frac{\Lambda_2}{\sigma_\mu} < 0.$$

Hence $\gamma_2^{(1)}\gamma_2^{(2)} < 0$ for each of the two bifurcating solutions. This means that the two $4T$-periodic bifurcating solutions are unstable. On the other hand, if $|\Lambda_2| > |\Lambda_3|$ and $|\operatorname{Im}(\Lambda_2/\sigma_\mu)| < |\Lambda_3/\sigma_\mu|$, then $\mu_2^{(1)}\mu_2^{(2)} > 0$, and $\gamma_2^{(1)}\gamma_2^{(2)}$ is

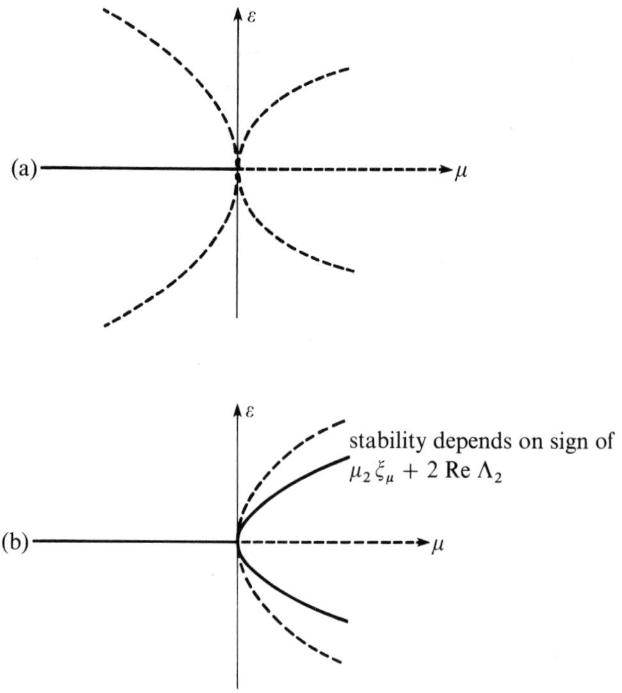

Figure IX.3 $4T$-periodic bifurcating solutions at small amplitude. $4T$-periodic solutions bifurcate when $|\operatorname{Im}(\Lambda_2/\sigma_\mu)| < |\Lambda_3/\sigma_\mu|$. (a) $|\Lambda_2| < |\Lambda_3|$. Two $4T$-periodic solutions bifurcate and both are unstable. (b) $|\Lambda_2| > |\Lambda_3|$, Re $(\Lambda_2/\sigma_\mu) < 0$. Two solutions bifurcate, one is unstable and the stability of the other depends on the problem. If Re $(\Lambda_2/\sigma_\mu) > 0$, the two solutions bifurcate to $\mu < 0$ and one of them is unstable negative for one of the two bifurcating solutions. For the other solution $\gamma_2^{(1)}\gamma_2^{(2)} > 0$ and stability is determined by the sign of $\mu_2 \xi_\mu + 2 \operatorname{Re} \Lambda_2$ (stable if <0, unstable if >0). (See Figure IX.3.)

IX.17 Nonexistence of Higher-Order Subharmonic Solutions and Weak Resonance

We now suppose that $n \geq 5$. Analysis of (IX.66) shows that $\mu_1 = 0$ so that μ_2 and ϕ_0 are to be determined by (IX.67), with $\mathbf{u}_1$ given by (IX.64) with $n \geq 5$, $m < n$, and $\mathbf{u}_2$ given by (IX.69). A short calculation of a now familiar type shows that

$$\mu_2 \sigma_\mu + \Lambda_2 = 0, \tag{IX.101}$$

where Λ_2 is given under (IX.80). In general (IX.101) is not solvable because μ_2 is real-valued and $\operatorname{Im}(\Lambda_2/\sigma_\mu) \neq 0$.

Subharmonic solutions with $n \geq 5$ can bifurcate in the special case in which $\mu_2 = -\Lambda_2/\sigma_\mu$, ϕ_0 may be determined by higher-order solvability conditions, and in fact two nT-periodic solutions with $n \geq 5$ may bifurcate. If $n = 5$, the condition $\operatorname{Im}(\Lambda_2/\sigma_\mu) = 0$ is in general a sufficient condition for the existence of these two $5T$-periodic solutions. It may be shown that this bifurcation is one-sided; both solutions are unstable when they are subcritical, one solution is stable when they bifurcate supercritically (see Figure IX.4).

Subharmonic solutions with $n \geq 5$ require special conditions (the vanishing of certain scalar products) beyond the ones required for the strongly resonant cases $n = 1, 2, 3, 4$. These exceptional solutions are called *weakly resonant* (terminology due to V. I. Arnold). Detailed analysis of weak resonance is given in G. Iooss, *Bifurcation of Maps and Applications* (Amsterdam: North-Holland, 1979), Chapter III.

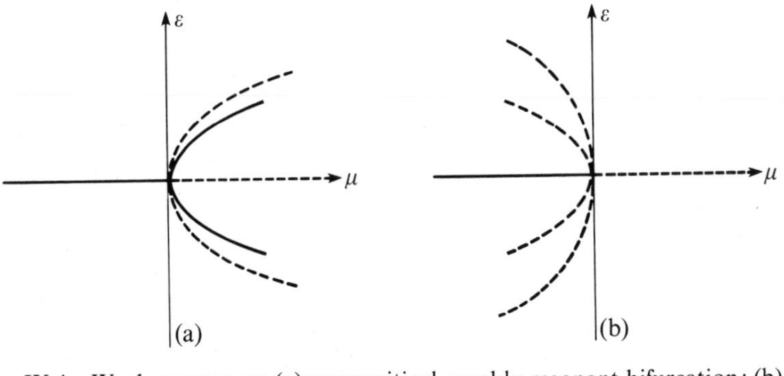

Figure IX.4 Weak resonance: (a) supercritical, weakly resonant bifurcation; (b) subcritical, weakly resonant bifurcation

IX.18 Summary of Results about Subharmonic Bifurcation

Suppose the hypotheses (I), (II), and (III) of §IX.6 hold.

(i) When $n = 1$ a single, one-parameter (ε) family of T-periodic solutions of (IX.1) bifurcates on both sides of criticality. When $n = 2$ a single, one-parameter (ε) family of $2T$-periodic solutions of (IX.1) bifurcates on one side of criticality. Supercritical ($\mu(\varepsilon) > 0$) bifurcating solutions are stable; subcritical ($\mu(\varepsilon) < 0$) bifurcating solutions are unstable.

(ii) When $n = 3$ a single, one-parameter family of $3T$-periodic solutions of (IX.1) bifurcates and is unstable on both sides of criticality.

(iii) When $n = 4$ and $|\Lambda_3/\sigma_\mu| > |\text{Im}\,(\Lambda_2/\sigma_\mu)|$, Λ_2 and Λ_3 being defined under (IX.80), two one-parameter (ε) families of $4T$-periodic solutions of (IX.1) bifurcate. If $|\Lambda_2| < |\Lambda_3|$, one of the two bifurcating solutions bifurcates on the subcritical side ($\mu(\varepsilon^2) < 0$) and the other on the supercritical side ($\mu(\varepsilon^2) > 0$), and both solutions are unstable. If $|\Lambda_2| > |\Lambda_3|$ the two solutions bifurcate on the same side of criticality and at least one of the two is unstable; the stability of the other solution depends on the details of the problem.

(iv) When $n \geq 5$ and $\text{Im}\,(\Lambda_2/\sigma_\mu) \neq 0$, Λ_2 being defined under (IX.80), there is in general no small-amplitude nT-periodic solution of (IX.1) near criticality.

IX.19 Imperfection Theory with a Periodic Imperfection

We are going to perturb steady bifurcating solutions with a T-periodic imperfection. We frame the mathematical problem for this study as follows:

$$\frac{d\mathbf{u}}{dt} = \mathscr{F}(\mu, \mathbf{u}, \delta, t). \tag{IX.102}$$

$$\begin{aligned} &\mathscr{F}(\mu, \mathbf{u}, 0, t) \overset{\text{def}}{=} \mathbf{f}(\mu, \mathbf{u}) \quad \text{is independent of } t. \\ &\mathscr{F}(\mu, \mathbf{u}, \delta, t) = \mathscr{F}(\mu, \mathbf{u}, \delta, t + T) \quad \text{when } \delta \neq 0. \\ &\mathscr{F}(\mu, 0, 0, t) \equiv 0. \end{aligned} \tag{IX.103}$$

We also assume that zero is a simple eigenvalue of $\mathbf{f}_u(0|\cdot)$; the other ones have negative real parts. We recall that $\mathbf{f}_u(0|\zeta_0) = 0$, $\mathbf{f}_u^*(0|\zeta_0^*) = 0$, $\langle\zeta_0, \zeta_0^*\rangle = 1$, and note that the assumption that $\mathbf{u} \equiv 0$ loses stability is a condition on $\mathbf{f}_{u\mu}$:

$$\sigma_\mu(0) = \langle \mathbf{f}_{u\mu}(0|\zeta_0), \zeta_0^* \rangle > 0. \tag{IX.104}$$

These assumptions are enough to guarantee the existence of steady bifurcating solutions ($\mu(\varepsilon)$, $\mathbf{u}(\varepsilon)$) which may be computed by the methods of Chapter VI.

Now we regard the steady bifurcating solution as a T-periodic one (for any T) and we look for a T-periodic solution of (IX.102) close to zero. We again define

$$\mathbb{J}(\mu) = -\frac{d}{dt} + \mathbf{f}_u(\mu|\cdot),$$

where $\mathbb{J}$ is defined only when it operates on T-periodic vectors $\mathbf{u}(t) = \mathbf{u}(t + T)$. By virtue of our assumptions about $\mathbf{f}_u(0|\cdot)$ the imaginary eigenvalues of $\mathbb{J}_0$ are all simple and, except for the eigenvalues (Floquet exponents) $\sigma(0) = \pm 2\pi k i/T$, $k \in \mathbb{Z}$, all have negative real parts. Condition (IX.104) holds for the eigenvalue $\sigma(\mu)$ of $\mathbb{J}(\mu)$ satisfying $\sigma(0) = 0$, and ζ_0 and ζ_0^* are steady and such that $\mathbb{J}_0\zeta_0 = \mathbb{J}_0^*\zeta_0^* = 0$, $[\zeta_0, \zeta_0^*]_T = 1$.

We can use the methods for studying imperfections given in §VI.10 if the analogue

$$[\mathscr{F}_\delta(0, 0, 0, t), \zeta_0^*]_T \overset{\text{def}}{=} \frac{1}{T}\int_0^T \langle \mathscr{F}_\delta(0, 0, 0, t), \zeta_0^* \rangle \, dt \neq 0 \qquad \text{(IX.105)}$$

of (VI.71) holds. Then we may compute the series

$$\delta = \varepsilon \sum_{p+q \geq 0} \varepsilon^p \mu^q \Delta_{p+1,q}$$

$$\mathbf{u}(t) = \varepsilon\zeta_0 + \varepsilon \sum_{p+q \geq 1} \mathbf{u}_{p+1,q}(t)\varepsilon^p \mu^q, \qquad \text{(IX.106)}$$

where the $\mathbf{u}_{p,q}(\cdot)$ are T-periodic.

When $\delta \neq 0$ the bifurcation picture for steady solutions is broken when (IX.105) holds and is replaced by two nonintersecting branches of T-periodic solutions, close to the steady bifurcating one, as in Figure III.5.

Exercise

$$\frac{du}{dt} = \mu u - u^2 + \delta(a + \cos t), \qquad u \in \mathbb{R}^1.$$

Prove that the bifurcation curves for $\delta = 0$ are split into nonintersecting branches of 2π-periodic solutions assuming that $a \neq 0$. Find a series for $\mu(\varepsilon, \delta/\varepsilon)$ where ε and δ/ε are both small which is valid when $a \neq 0$ and when $a = 0$. Repeat the exercise when u^2 is replaced by u^3.

CHAPTER X

Bifurcation of Forced T-Periodic Solutions into Asymptotically Quasi-Periodic Solutions

In Chapter IX we determined the conditions under which subharmonic solutions, nT-periodic solutions with integers $n \geq 1$, could bifurcate from forced T-periodic solutions. That is to say, we looked for the conditions under which nonautonomous, T-periodic differential equations give rise to subharmonic solutions when the Floquet exponents at criticality lie in the set of rational points ($\omega_0 = 2\pi m/nT, 0 \leq m/n < 1$) or, equivalently, when the Floquet multipliers at criticality are the nth roots of unity, $\lambda_0^n = (e^{i\omega_0 T})^n = 1$. We found that unless certain very special (weak resonance) conditions were satisfied such subharmonic solutions could bifurcate only when $n = 1, 2, 3, 4$. (The case $n = 4$ is special in that there are in general two possibilities depending on the parameters; see §IX.15.) So we now confront the problem of finding out what happens for all the values of ω_0, $0 \leq \omega_0 < 2\pi/T$ such that

$$\frac{\omega_0 T}{2\pi} \neq 0, \frac{1}{2}, \frac{1}{3}, \frac{2}{3}, \frac{1}{4}, \frac{3}{4}.$$

We shall show that, unless highly exceptional conditions are satisfied, the solutions which bifurcate lie on a torus and are asymptotic to quasi-periodic solutions near criticality. The subharmonic solutions which bifurcate when the exceptional conditions hold are also on the stable (supercritical) torus. The exceptional subharmonic solutions bifurcate in pairs; one solution is stable and the other one is unstable.

X.1 The Biorthogonal Decomposition of the Solution and the Biorthogonal Decomposition of the Equations

We start with an evolution equation with T-periodic coefficients reduced to local form, as in (I.21). In fact, this is exactly the problem studied in Chapter IX, but now we want to know what happens when nT-periodic solutions with $n = 1, 2, 3$ or 4 cannot bifurcate. We analyze this problem in the spirit of Chapter IX, using the method of power series and the Fredholm alternative together with a method using two times in Appendix X.1 and more directly in Appendices X.2 and X.3. But we prefer to begin with an analysis using an entirely different method, which involves an extension of the method of averaging and allows us to reduce the equations with T-periodic coefficients to autonomous ones. To start, we write (I.21) in a slightly different way

$$\frac{d\mathbf{u}}{dt} = \mathbf{f}_u(t, \mu|\mathbf{u}) + \mathbf{N}(t, \mu, \mathbf{u}) \tag{X.1}$$

where

$$\mathbf{N}(t, \mu, \mathbf{u}) = \mathbf{f}(t, \mu, \mathbf{u}) - \mathbf{f}_u(t, \mu|\mathbf{u})$$

are the nonlinear terms, and, of course, $\mathbf{u} = 0$ is a solution. The spectral problem for the stability of $\mathbf{u} = 0$ is given by (IX.8), and since we are excluding the points of strong resonance corresponding to $n = 1, 2, 3, 4$, all the Floquet multipliers $e^{\sigma(\mu)T}$ and all of the exponents $\sigma(\mu)$ corresponding to the critical one are complex.

Without loss of generality we may decompose

$$\mathbf{u} = Z\boldsymbol{\zeta} + \overline{Z}\overline{\boldsymbol{\zeta}} + \mathbf{W}, \tag{X.2}$$

where $\boldsymbol{\zeta} = \boldsymbol{\zeta}(\mu, t) = \boldsymbol{\zeta}(\mu, T + t)$ is an eigenfunction of the spectral problem (IX.8). To define Z we project using the adjoint eigenfunction $\boldsymbol{\zeta}^*$ satisfying (IX.14) and the orthogonality properties of the time-dependent scalar product $\langle \cdot, \cdot \rangle$, which are established in Exercise X.1.

Exercise

X.1 Assume that σ is a simple eigenvalue of the linear operator

$$-\frac{d}{dt} + \mathbf{f}_u(t, \mu|\cdot)$$

in the space of T-periodic vector functions. Let $\boldsymbol{\zeta}(\cdot)$ be the eigenfunction belonging to σ, and $\boldsymbol{\zeta}^*(\cdot)$ the eigenfunction belonging to $\bar{\sigma}$ for the adjoint operator. Show that

$\langle \boldsymbol{\zeta}(t), \boldsymbol{\zeta}^*(t) \rangle =$ constant independent of t

$\langle \boldsymbol{\zeta}(t), \overline{\boldsymbol{\zeta}}^*(t) \rangle = Ce^{(\bar{\sigma} - \sigma)t}$, C is constant.

Deduce that we can choose $\zeta(t)$ and $\zeta^*(t)$ such that

$$\langle \zeta(t), \zeta^*(t) \rangle = 1$$

and that if $\bar{\sigma} - \sigma \neq 2k\pi i/T$, $k \in \mathbb{Z}$, then

$$\langle \zeta(t), \bar{\zeta}^*(t) \rangle = 0.$$

Show that this condition is realized in our frame for μ close to 0 (verify it at $\mu = 0$ and perturb).

$$Z = Z(\mu, t) = \langle \mathbf{u}(t), \zeta^*(t) \rangle \tag{X.3}$$

and

$$\langle \mathbf{W}(t), \zeta^*(t) \rangle = 0. \tag{X.4}$$

It is always possible to decompose the vector **u** as in (X.2). When, in addition, **u** solves (X.1), we call (X.2) the biorthogonal decomposition of the solution. The vectors **u** and **W** are real-valued.

To get the biorthogonal decomposition of the equations, we combine (X.1) and (X.2) and find that

$$\begin{aligned}\dot{Z}\zeta + Z\dot{\zeta} + \dot{\bar{Z}}\bar{\zeta} + \bar{Z}\dot{\bar{\zeta}} + \dot{\mathbf{W}} \\ = Z\mathbf{f}_u(t, \mu | \zeta) + \bar{Z}\mathbf{f}_u(t, \mu | \bar{\zeta}) + \mathbf{f}_u(t, \mu | \mathbf{W}) + \mathbf{N}(t, \mu, \mathbf{u}).\end{aligned}$$

After making use of the equation (IX.8) satisfied by ζ, we find that

$$\dot{Z}\zeta + \dot{\bar{Z}}\bar{\zeta} + \dot{\mathbf{W}} = \sigma Z\zeta + \bar{\sigma}\bar{Z}\bar{\zeta} + \mathbf{f}_u(t, \mu | \mathbf{W}) + \mathbf{N}(t, \mu, \mathbf{u}). \tag{X.5}$$

We next form the scalar product of (X.5) with ζ^* and use (X.4) to get

$$\langle \dot{\mathbf{W}}, \zeta^* \rangle = \frac{d}{dt}\langle \mathbf{W}, \zeta^* \rangle - \langle \mathbf{W}, \dot{\zeta}^* \rangle = -\langle \mathbf{W}, \dot{\zeta}^* \rangle$$

and, using (IX.14),

$$\begin{aligned}\langle [-\dot{\mathbf{W}} + \mathbf{f}_u(t, \mu | \mathbf{W})], \zeta^* \rangle &= \langle \mathbf{W}, \dot{\zeta}^* + \mathbf{f}_u^*(t, \mu | \zeta^*) \rangle \\ &= \langle \mathbf{W}, \bar{\sigma}\zeta^* \rangle = 0.\end{aligned}$$

Since $\langle \zeta, \zeta^* \rangle = 1$ and $\langle \zeta, \bar{\zeta}^* \rangle = 0$, we find that

$$\dot{Z} = \sigma(\mu)Z + \langle \mathbf{N}(t, \mu, \mathbf{u}), \zeta^* \rangle. \tag{X.6}$$

Returning to (X.5) with (X.6) we find that

$$\dot{\mathbf{W}} = \mathbf{f}_u(t, \mu | \mathbf{W}) + \mathbf{N}(t, \mu, \mathbf{u}) - \langle \mathbf{N}(t, \mu, \mathbf{u}), \zeta^* \rangle\zeta - \langle \mathbf{N}(t, \mu, \mathbf{u}), \bar{\zeta}^* \rangle\bar{\zeta}. \tag{X.7}$$

Equations (X.2), (X.6), and (X.7) give the biorthogonal decomposition of the equations. To further prepare these equations for analysis we note that when $\mathbf{f}(t, \mu, \cdot)$ possesses derivatives through order $k + 1$ at $\mathbf{u} = 0$, we may write

$$\mathbf{N}(t, \mu, \mathbf{u}) = \frac{1}{2}\mathbf{f}_{uu}(t, \mu, 0|\mathbf{u}|\mathbf{u}) + \frac{1}{3!}\mathbf{f}_{uuu}(t, \mu, 0|\mathbf{u}|\mathbf{u}|\mathbf{u}) + \cdots + \frac{1}{k!}\underbrace{\mathbf{f}_{u, \cdots, u}}_{k \text{ times}}(t, \mu, 0\underbrace{|\mathbf{u}|\mathbf{u}|\cdots|\mathbf{u})}_{k \text{ times}} + O(\|\mathbf{u}\|^{k+1}).$$

Since $\mathbf{u} = Z\boldsymbol{\zeta} + \bar{Z}\bar{\boldsymbol{\zeta}} + \mathbf{W}$, $\mathbf{N}$ is in the form

$$\mathbf{N}(t, \mu, Z\boldsymbol{\zeta} + \bar{Z}\bar{\boldsymbol{\zeta}} + \mathbf{W}) = \mathbf{n}_0(t, \mu, Z, \bar{Z}) + \mathbf{n}_1(t, \mu, Z, \bar{Z}, \mathbf{W}),$$

where

$$\mathbf{n}_0(t, \mu, Z, \bar{Z}) = \mathbf{N}(t, \mu, Z\bar{\boldsymbol{\zeta}} + \bar{Z}\boldsymbol{\zeta}) = O(|Z|^2)$$

and

$$\mathbf{n}_1(t, \mu, Z, \bar{Z}, \mathbf{W}) = O(|Z|\,\|\mathbf{W}\| + \|\mathbf{W}\|^2).$$

It follows that (X.6) and (X.7) may be written as

$$\dot{Z} = \sigma Z + b(t, \mu, Z, \bar{Z}, \mathbf{W}) \tag{X.8}$$

and

$$\dot{\mathbf{W}} = \mathbf{f}_u(t, \mu|\mathbf{W}) + \mathbf{B}(t, \mu, Z, \bar{Z}, \mathbf{W}), \tag{X.9}$$

where

$$b(t, \mu, Z, \bar{Z}, \mathbf{W}) = \langle \mathbf{n}_0 + \mathbf{n}_1, \boldsymbol{\zeta}^* \rangle$$

and

$$\mathbf{B}(t, \mu, Z, \bar{Z}, \mathbf{W}) = \mathbf{n}_0 + \mathbf{n}_1 - \langle \mathbf{n}_0 + \mathbf{n}_1, \boldsymbol{\zeta}^* \rangle \boldsymbol{\zeta} - \langle \mathbf{n}_0 + \mathbf{n}_1, \bar{\boldsymbol{\zeta}}^* \rangle \bar{\boldsymbol{\zeta}}.$$

The T-periodicity of b and $\mathbf{B}$ follows from the T-periodicity of $\mathbf{N}(t, \cdot, \cdot) = \mathbf{N}(t + T, \cdot, \cdot)$, and $\mathbf{B}$ is orthogonal to $\boldsymbol{\zeta}^*$.

We have the following decomposition:

$$\begin{aligned} b &= b_0 + b_1, \\ \mathbf{B} &= \mathbf{B}_0 + \mathbf{B}_1 \end{aligned} \tag{X.10}$$

where

$$b_0 \stackrel{\text{def}}{=} \langle \mathbf{n}_0, \boldsymbol{\zeta}^* \rangle, \qquad b_1 \stackrel{\text{def}}{=} \langle \mathbf{n}_1, \boldsymbol{\zeta}^* \rangle$$

$$\mathbf{B}_0 = \mathbf{n}_0 - b_0\boldsymbol{\zeta} - b_0\bar{\boldsymbol{\zeta}}, \qquad \mathbf{B}_1 = \mathbf{n}_1 - b_1\boldsymbol{\zeta} - \bar{b}_1\bar{\boldsymbol{\zeta}}.$$

Moreover

$$\begin{aligned} b_0 &= b(t, \mu, Z, \bar{Z}, 0) = \langle \mathbf{n}_0, \boldsymbol{\zeta}^* \rangle \\ &= \tfrac{1}{2}\{Z^2\langle \mathbf{f}_{uu}(t, \mu, 0|\boldsymbol{\zeta}|\boldsymbol{\zeta}), \boldsymbol{\zeta}^* \rangle + 2|Z|^2\langle \mathbf{f}_{uu}(t, \mu, 0|\boldsymbol{\zeta}|\bar{\boldsymbol{\zeta}}), \bar{\boldsymbol{\zeta}}^* \rangle \\ &\quad + \bar{Z}^2\langle \mathbf{f}_{uu}(t, \mu, 0|\bar{\boldsymbol{\zeta}}|\boldsymbol{\zeta}), \boldsymbol{\zeta}^* \rangle\} + O(|Z|^3) \\ &= \sum_{p+q\geq 2} Z^p\bar{Z}^q\hat{b}_{pq}(t, \mu), \end{aligned} \tag{X.11}$$

where $\hat{b}_{pq}(t, \mu) = \hat{b}_{pq}(t + T, \mu)$. Similarly,

$$\mathbf{B}_0 = \mathbf{B}(t, \mu, Z, \bar{Z}, 0) = \mathbf{n}_0 - b_0 \boldsymbol{\zeta} - \bar{b}_0 \bar{\boldsymbol{\zeta}} = \sum_{p+q\geq 2} Z^p \bar{Z}^q \hat{\mathbf{B}}_{pq}(t, \mu), \quad \text{(X.12)}$$

where $\hat{\mathbf{B}}_{pq}(t, \mu) = \hat{\mathbf{B}}_{pq}(t + T, \mu)$. It is easy to ascertain that

$$\begin{aligned} b_1 &= b_1(t, \mu, Z, \bar{Z}, \mathbf{W}) = O(|Z| \|\mathbf{W}\| + \|\mathbf{W}\|^2) \\ \mathbf{B}_1 &= \mathbf{B}_1(t, \mu, Z, \bar{Z}, \mathbf{W}) = O(|Z| \|\mathbf{W}\| + \|\mathbf{W}\|^2). \end{aligned} \quad \text{(X.13)}$$

X.2 Change of Variables

Mathematicians usually like to change variables; it's like buying new clothes; you may not actually look better in the new ones, but you think you do. The variables which we think look best are

$$y = Z + \gamma(t, \mu, Z, \bar{Z}) \quad \text{(X.14)}$$

and

$$\mathbf{Y} = \mathbf{W} + \boldsymbol{\Gamma}(t, \mu, Z, \bar{Z}) \quad \text{(X.15)}$$

where $\langle \mathbf{Y}, \boldsymbol{\zeta}^* \rangle = \langle \boldsymbol{\Gamma}, \boldsymbol{\zeta}^* \rangle = 0$. The complex-valued function γ and real-vector-valued function $\boldsymbol{\Gamma}$ are to be determined, are T-periodic, and they vanish when Z does:

$$\gamma \text{ and } \boldsymbol{\Gamma} \text{ are } O(|Z|^2).$$

Our goal is to show that the solutions which bifurcate from T-periodic ones when $n \neq 1, 2, 3, 4$ are typically not subharmonic and lie on a torus in phase space, and that the trajectories on the torus are asymptotically quasi-periodic. We determine γ and $\boldsymbol{\Gamma}$ so that we can apply a method of averaging to reduce (X.8) and (X.9) to essential terms.

To start the analysis we differentiate (X.14) and (X.15) with respect to time and replace the derivatives of $Z, \bar{Z}$ and $\mathbf{W}$ with the expressions (X.8) and (X.9). After elimination we get

$$\dot{y} = \sigma Z + b + \frac{\partial \gamma}{\partial t} + \frac{\partial \gamma}{\partial Z}(\sigma Z + b) + \frac{\partial \gamma}{\partial \bar{Z}}(\bar{\sigma}\bar{Z} + \bar{b}) \quad \text{(X.16)}_1$$

$$\dot{\mathbf{Y}} = \mathbf{f}_u(t, \mu | \mathbf{W}) + \mathbf{B} + \frac{\partial \boldsymbol{\Gamma}}{\partial t} + \frac{\partial \boldsymbol{\Gamma}}{\partial Z}(\sigma Z + b) + \frac{\partial \boldsymbol{\Gamma}}{\partial \bar{Z}}(\bar{\sigma}\bar{Z} + \bar{b}), \quad \text{(X.16)}_2$$

where b and $\mathbf{B}$ are given by (X.10–13) and Z and $\mathbf{W}$ are to-be-determined functions of y and $\mathbf{Y}$.

We shall construct γ and $\boldsymbol{\Gamma}$ as polynomials of degree N in Z and $\bar{Z}$ with T-periodic coefficients

$$\begin{bmatrix} \gamma(t, \mu, Z, \bar{Z}) \\ \boldsymbol{\Gamma}(t, \mu, Z, \bar{Z}) \end{bmatrix} = \sum_{p+q\geq 2}^{N} Z^p \bar{Z}^q \begin{bmatrix} \gamma_{pq}(t, \mu) \\ \boldsymbol{\Gamma}_{pq}(t, \mu) \end{bmatrix} \quad \text{(X.17)}$$

The truncation number N in (X.17) is *unrestricted* (though there may be a "best" N for the asymptotic approximation). Since $\mathbf{\Gamma}$ is real-valued, $\mathbf{\Gamma}_{pq} = \bar{\mathbf{\Gamma}}_{qp}$ and

$$\langle \mathbf{\Gamma}, \zeta^* \rangle = \langle \mathbf{\Gamma}, \bar{\zeta}^* \rangle = \langle \mathbf{\Gamma}_{pq}, \zeta^* \rangle = \langle \mathbf{\Gamma}_{pq}, \bar{\zeta}^* \rangle = 0.$$

We next truncate the series (X.11) and (X.12) after N terms, as in (X.17), and substitute the resulting polynomials into (X.16), keeping only terms through degree N.

$$\dot{y} = \sigma Z + \sum_{p+q\geq 2}^{N} Z^p \bar{Z}^q \left[\frac{\partial \gamma_{pq}}{\partial t} + (\sigma p + \bar{\sigma} q)\gamma_{pq} + \hat{b}_{pq} \right] + b_1 + \left(b \frac{\partial \gamma}{\partial Z} + \bar{b} \frac{\partial \gamma}{\partial \bar{Z}} \right) + O(|Z|^{N+1}) \qquad \text{(X.18)}_1$$

$$\dot{\mathbf{Y}} = \mathbf{f}_u(t, \mu | \mathbf{W}) + \sum_{p+q\geq 2}^{N} Z^p \bar{Z}^q \left[\frac{\partial \mathbf{\Gamma}_{pq}}{\partial t} + (\sigma p + \bar{\sigma} q)\mathbf{\Gamma}_{pq} + \hat{\mathbf{B}}_{pq} \right] + \mathbf{B}_1 + \left(b \frac{\partial \mathbf{\Gamma}}{\partial Z} + \bar{b} \frac{\partial \mathbf{\Gamma}}{\partial \bar{Z}} \right) + O(|Z|^{N+1}) \qquad \text{(X.18)}_2$$

where b_1 and $\mathbf{B}_1$ are defined under (X.10); $b = b_0 + b_1$; and b_0, γ, and $\mathbf{\Gamma}$ stand for the polynomial representations of these quantities through terms of order N.

We can eliminate Z and $\mathbf{W}$ from (X.18) by inverting the transformations (X.14) and (X.15):

$$Z = y + \sum_{p+q\geq 2}^{N} \gamma'_{pq}(t, \mu) y^p \bar{y}^q + O(|y|^{N+1}). \qquad \text{(X.19)}$$

$$\mathbf{W} = \mathbf{Y} + \sum_{p+q\geq 2}^{N} \mathbf{\Gamma}'_{pq}(t, \mu) y^p \bar{y}^q + O(|y|^{N+1}). \qquad \text{(X.20)}$$

This transformation is always possible when y and Z are small, and γ'_{pq} and Γ'_{pq} may be determined in terms of γ_{ln} and Γ_{ln}, $l + n \leq p + q$, by identifying independent powers of $Z^p \bar{Z}^q$. For example,

$$\begin{aligned}
\gamma'_{pq} &= -\gamma_{pq} \quad \text{for } p + q = 2 \\
\gamma'_{30} &= -\gamma_{30} + 2\gamma_{20}^2 + \gamma_{11}\bar{\gamma}_{02} \\
\gamma'_{21} &= -\gamma_{21} + 3\gamma_{11}\gamma_{20} + |\gamma_{11}|^2 + 2|\gamma_{02}|^2 \\
\gamma'_{12} &= -\gamma_{12} + 2\gamma_{02}\gamma_{20} + \gamma_{11}\bar{\gamma}_{20} + \gamma_{11}^2 + 2\bar{\gamma}_{11}\gamma_{02} \\
\gamma'_{03} &= -\gamma_{03} + \gamma_{02}\gamma_{11} + 2\gamma_{02}\bar{\gamma}_{20}.
\end{aligned} \qquad \text{(X.21)}$$

More generally, we have

$$\gamma'_{pq} = -\gamma_{pq} + \text{functions of } \gamma_{ln} \text{ with } l + n \leq p + q - 1.$$

Using (X.15), (X.17), and (X.20) we find that

$$\begin{aligned}
\mathbf{\Gamma}'_{pq} &= -\mathbf{\Gamma}_{pq} \\
\mathbf{\Gamma}'_{30} &= -\mathbf{\Gamma}_{30} + 2\gamma_{20}\mathbf{\Gamma}_{20} + \bar{\gamma}_{02}\mathbf{\Gamma}_{11} \\
\mathbf{\Gamma}'_{21} &= -\mathbf{\Gamma}_{21} + 2\gamma_{11}\mathbf{\Gamma}_{20} + (\gamma_{20} + \bar{\gamma}_{11})\mathbf{\Gamma}_{11} + 2\bar{\gamma}_{02}\mathbf{\Gamma}_{02} \\
\mathbf{\Gamma}'_{12} &= -\mathbf{\Gamma}_{12} + 2\gamma_{02}\mathbf{\Gamma}_{20} + (\gamma_{11} + \bar{\gamma}_{20})\mathbf{\Gamma}_{11} + 2\bar{\gamma}_{11}\mathbf{\Gamma}_{02} \\
\mathbf{\Gamma}'_{03} &= -\mathbf{\Gamma}_{03} + \gamma_{20}\mathbf{\Gamma}_{11} + 2\bar{\gamma}_{20}\mathbf{\Gamma}_{02}.
\end{aligned} \tag{X.22}$$

More generally, we have

$$\mathbf{\Gamma}'_{pq} = -\mathbf{\Gamma}_{pq} + \text{functions of } (\mathbf{\Gamma}_{ln}, \gamma_{ln}) \text{ with } l + n \le p + q - 1.$$

Finally, we substitute (X.19) and (X.20) into (X.18) and eliminate γ'_{pq} and $\mathbf{\Gamma}'_{pq}$ with γ_{pq} and $\mathbf{\Gamma}_{pq}$ using formulas of the type just listed. We get

$$\dot{y} = \sigma y + \sum_{p+q\ge 2}^{N} y^p \bar{y}^q \{\dot{\gamma}_{pq} + [(p-1)\sigma + q\bar{\sigma}]\gamma_{pq} + b_{pq}\} + \tilde{b}_1(t, \mu, y, \bar{y}, \mathbf{Y}) \tag{X.23}$$

and

$$\begin{aligned}
\dot{\mathbf{Y}} = \mathbf{f}_u(t, \mu | \mathbf{Y}) + \sum_{p+q\ge 2}^{N} y^p \bar{y}^q \{&\dot{\mathbf{\Gamma}}_{pq} - \mathbf{f}_u(t, \mu | \mathbf{\Gamma}_{pq}) \\
&+ (p\sigma + q\bar{\sigma})\mathbf{\Gamma}_{pq} + \mathbf{B}_{pq}\} + \tilde{\mathbf{B}}_1(t, \mu, y, \bar{y}, \mathbf{Y}),
\end{aligned} \tag{X.24}$$

where

$$\begin{aligned}
|\tilde{b}_1(t, \mu, y, \bar{y}, \mathbf{Y})| &= O(|y|^{N+1} + |y|\,\|\mathbf{Y}\| + \|\mathbf{Y}\|^2) \\
\|\tilde{\mathbf{B}}_1(t, \mu, y, \bar{y}, \mathbf{Y})\| &= O(|y|^{N+1} + |y|\,\|\mathbf{Y}\| + \|\mathbf{Y}\|^2),
\end{aligned}$$

and b_{pq} and $\mathbf{B}_{pq}$ are functions of γ_{ln} and $\mathbf{\Gamma}_{ln}$ with $l + n \le p + q - 1$, T-periodic coefficients, and such that terms in (X.24) orthogonal to $\boldsymbol{\zeta}^*$ and $\bar{\boldsymbol{\zeta}}^*$ are the factors of $y^p\bar{y}^q$ and $\tilde{\mathbf{B}}_1$.

X.3 Normal Form of the Equations

We want to choose γ_{pq} and $\mathbf{\Gamma}_{pq}$ to make the coefficients of $y^p\bar{y}^q$ in (X.23) and (X.24) vanish whenever possible. In Lemma 1 below we assert that it is always possible to simplify the $\mathbf{Y}$ equation by choosing $\mathbf{\Gamma}_{pq}$ to make the coefficients in (X.24) vanish. In Lemma 2, we specify some conditions under which we may simplify the y equation by choosing γ_{pq} so as to make the coefficients of $y^p\bar{y}^q$ vanish. However, we cannot *always* choose such a good γ_{pq}. It is most important that the reader understand what we do when we *cannot* choose γ_{pq} so as to make the terms in the brackets which multiply $y^p\bar{y}^q$ vanish.

Let us suppose that $\gamma_{pq}(t, \mu)$, $\mathbf{\Gamma}_{pq}(t, \mu)$ satisfying $\langle \mathbf{\Gamma}_{pq}, \boldsymbol{\zeta}^* \rangle = \langle \mathbf{\Gamma}_{pq}, \bar{\boldsymbol{\zeta}}^* \rangle = 0$, are known when $p + q \le k - 1$. Then b_{pq} and $\mathbf{B}_{pq}$ in (X.23) and (X.24) are known when $p + q = k$ and, since $(X.18)_2$ is linear in $\mathbf{\Gamma}$, $\mathbf{B}_1$, $\hat{\mathbf{B}}_{pq}$ and $\mathbf{\Gamma}_{pq}$ are orthogonal to $\boldsymbol{\zeta}^*$ and $\bar{\boldsymbol{\zeta}}^*$, $\langle \mathbf{B}_{pq}, \boldsymbol{\zeta}^* \rangle = \langle \mathbf{B}_{pq}, \bar{\boldsymbol{\zeta}}^* \rangle = 0$. Lemma 1, below, says that we can always find $\mathbf{\Gamma}_{pq}$, orthogonal to $\boldsymbol{\zeta}^*$ and $\bar{\boldsymbol{\zeta}}^*$, such that

$$\dot{\mathbf{\Gamma}}_{pq} - \mathbf{f}_u(t, \mu \,|\, \mathbf{\Gamma}_{pq}) + (p\sigma + q\bar{\sigma})\mathbf{\Gamma}_{pq} + \mathbf{B}_{pq} = 0.$$

In this case (X.24) becomes

$$\dot{\mathbf{Y}} = \mathbf{f}_u(t, \mu \,|\, \mathbf{Y}) + \tilde{\mathbf{B}}_1(t, \mu, y, \bar{y}, \mathbf{Y}). \tag{X.25}$$

Lemma 1. *The problem is to find T-periodic vectors* $\mathbf{\Gamma}$ *satisfying*

$$\dot{\mathbf{\Gamma}} - \mathbf{f}_u(t, \mu \,|\, \mathbf{\Gamma}) + (p\sigma + q\bar{\sigma})\mathbf{\Gamma} + \mathbf{B} = 0 \qquad (X.26)_1$$

where $\mathbf{B}(t) = \mathbf{B}(t + T)$ *and* $\langle \mathbf{B}(t), \boldsymbol{\zeta}^*(t) \rangle = \langle \mathbf{B}(t), \bar{\boldsymbol{\zeta}}^*(t) \rangle = 0$. *This problem admits solutions (of class* C^{k+1}, *if* $\mathbf{B}$ *is of class* C^k*) and there is one and only one solution such that*

$$\langle \mathbf{\Gamma}(t), \boldsymbol{\zeta}^*(t) \rangle = \langle \mathbf{\Gamma}(t), \bar{\boldsymbol{\zeta}}^*(t) \rangle = 0. \qquad (X.26)_2$$

PROOF. To prove lemma 1 we first note that the eigenvalues of

$$-\frac{d}{dt} + \mathbf{f}_u(t, \mu \,|\, \cdot) - (p\sigma + q\bar{\sigma}) \qquad (X.26)_3$$

which are close to the imaginary axis are

$$\sigma(1 - p) - q\bar{\sigma} + \frac{2k\pi i}{T}$$

and

$$-p\sigma + (1 - q)\bar{\sigma} + \frac{2k'\pi i}{T}$$

where $k, k' \in \mathbb{Z}$. The other eigenvalues of $(X.26)_3$ are "far" from the imaginary axis. We want to know when zero is an eigenvalue of $(X.26)_3$ with $\mu = 0$; that is, we want to know the values of parameters under which

$$i\omega_0(q + 1 - p) + \frac{2k\pi i}{T} = 0$$

or

$$i\omega_0(q - 1 - p) + \frac{2k'\pi i}{T} = 0.$$

Case 1. $r = \omega_0 T/2\pi$ is irrational. Then $r(q + 1 - p) + k = 0$ if and only if $p = q + 1$ and $k = 0$, and $r(q - 1 - p) + k' = 0$ if and only if $q = 1 + p$ and $k' = 0$. It follows that if $p \ne q + 1$ and $q \ne 1 + p$ then zero is not an eigenvalue of $(X.26)_3$ at criticality and $(X.26)_3$ is invertible for $\mu = 0$ and for μ close to 0.

If $p = 1 + q$, then the eigenvector belonging to the eigenvalue near to zero is $\zeta(t) \in \mathbb{P}_T$ satisfying $-\dot{\zeta} + \mathbf{f}_u(t, \mu|\zeta) = \sigma\zeta$ also satisfies

$$\left[-\frac{d}{dt} + \mathbf{f}_u(t, \mu|) - p\sigma - q\bar{\sigma}\right]\zeta = [\sigma(1 - p) - q\bar{\sigma}]\zeta$$

$$= -q(\sigma + \bar{\sigma})\zeta = -2q\xi(\mu)\zeta$$

where $\xi(0) = 0$. In the same way we find that if $q = 1 + p$ then

$$\left[-\frac{d}{dt} + \mathbf{f}_u(t, \mu|) - p\sigma - q\bar{\sigma}\right]\bar{\zeta} = [-p\sigma + (1 - q)\bar{\sigma})\bar{\zeta} = -2p\xi(\mu)\bar{\zeta}.$$

It follows that when r is irrational we can invert $(X.26)_1$ for $\Gamma \in \mathbb{P}_T$ provided that

$$[\mathbf{B}, \zeta^*]_T = [\mathbf{B}, \bar{\zeta}^*]_T = 0$$

where $[\cdot, \cdot]_T$ is the scalar product defined by (IX.16). The required orthogonality is automatically satisfied by vectors $\mathbf{B}(t)$ for which $\langle \mathbf{B}(t), \zeta^*(t)\rangle = \langle \mathbf{B}(t), \bar{\zeta}^*(t)\rangle = 0$, independent of t.

Case 2. $r = \omega_0 T/2\pi = m/n$ is a positive rational fraction less than one and $n \geq 3$. Then zero is an eigenvalue of $(X.26)_3$ at criticality for k, k' satisfying

$$\frac{m}{n}(q + 1 - p) + k = 0$$

or

$$\frac{m}{n}(q - 1 - p) + k' = 0.$$

Since m and n have no common divisor, these equations can be satisfied only if there is $l, l' \in \mathbb{Z}$ such that $p = q + 1 + ln$ or $q = p + 1 + l'n$. If there are no such values l, l', then $(X.26)_3$ is invertible for $\mu = 0$ and for μ close to zero.

Suppose now that there is an l such that $p = q + 1 + ln$. Then $q - p - 1 - l'n = -2 - n(l + l')$ cannot vanish because $n \geq 3$. Moreover, at criticality $(q + 1 + ln)\sigma + q\bar{\sigma} = (1 + ln)i\omega_0$ and the eigenvector of $(X.26)_3$ at criticality is $\zeta_0 e^{-2\pi imlt/T}$ where $\omega_0 = 2\pi m/nT$

and

$$\left[-\frac{d}{dt} + \mathbf{f}_u(t, 0|) - (1 + ln)i\omega_0\right]\zeta_0 e^{-2\pi imlt/T} = 0$$

It is easy to verify that

$$\left[-\frac{d}{dt} + \mathbf{f}_u(t, \mu|) - (p\sigma + q\bar{\sigma})\right]\zeta e^{-2\pi i m l t/T}$$
$$= \left[-(p-1)\sigma - q\bar{\sigma} + \frac{2\pi i l m}{T}\right]\zeta e^{-2\pi i m l t/T}$$
$$= \left[-q(\sigma + \bar{\sigma}) - ln\sigma + \frac{2\pi i l m}{T}\right]\zeta e^{-2\pi i m l t/T}$$
$$= \left[-2\xi(\mu)q + l\left(\frac{2\pi i m}{T} - \sigma n\right)\right]\zeta e^{-2\pi i m l t/T}$$

vanishes when $\mu = 0$ and (X.26), is invertible at $\mu = 0$ and near $\mu = 0$ when

$$[\mathbf{B}(t), \zeta^*(t)e^{-2\pi i m l t/T}]_T = [\mathbf{B}(t), \bar{\zeta}^*(t)e^{2\pi i m l t/T}]_T = 0$$

where ζ^* is adjoint to ζ. These orthogonality conditions are automatically satisfied by vectors $\mathbf{B}(t)$ satisfying $\langle \mathbf{B}(t), \zeta^*(t)\rangle = \langle \mathbf{B}(t), \bar{\zeta}^*(t)\rangle$, independent of t.

We next remark that every solution $\mathbf{\Gamma}(t) = \mathbf{\Gamma}(t + T)$ of $(X.26)_1$ with $\langle \mathbf{B}(t), \zeta^*(t)\rangle = \langle \mathbf{B}(t), \bar{\zeta}^*(t)\rangle = 0$ is defined up to functions on the null space of the operator $(X.26)_3$, and is unique among $\mathbf{\Gamma}$'s which also satisfy

$$\langle \mathbf{\Gamma}(t), \zeta^*(t)\rangle = \langle \mathbf{\Gamma}(t), \bar{\zeta}^*(t)\rangle = 0.$$

In fact, when μ is close to zero,

$$-\frac{d}{dt}\langle \mathbf{\Gamma}, \zeta^*\rangle = [(p-1)\sigma + \bar{\sigma}q]\langle \mathbf{\Gamma}, \zeta^*\rangle$$

and

$$-\frac{d}{dt}\langle \mathbf{\Gamma}, \bar{\zeta}^*\rangle = [p\sigma + (q-1)\bar{\sigma}]\langle \mathbf{\Gamma}, \zeta^*\rangle.$$

The only periodic solutions of these equations are constants, and the constants vanish unless the coefficients on the right vanish. The reader may wish to verify that the $\mathbf{B}_{pq}$ in (X.24) differs from the $\hat{\mathbf{B}}_{pq}$ in $(X.18)_2$ by terms linear in $\mathbf{\Gamma}_{pq}$ which vanish in projection.

Lemma 2. *The problem is to find T-periodic functions γ satisfying the equation*

$$\dot{\gamma} + [\sigma(p-1) + \bar{\sigma}q]\gamma + b = 0 \tag{X.27}$$

where b is a given T-periodic function of class C^k.

i. *If $(p - q - 1)\omega_0 T/2\pi$ is not an integer (positive, negative, or zero) there is a unique solution γ of class C^{k+1} of the problem.*

ii. *If $(p - q - 1)\omega_0 T/2\pi = -l_0$ is an integer (positive, negative, or zero)* *there is a solution γ, smooth in μ, only if the Fourier coefficient*

$$b_{l_0} = \frac{1}{T}\int_0^T b(t)e^{-2\pi i l_0 t/T}\,dt = 0.$$

Moreover, if we require that $\gamma_{l_0} = 0$, the solution γ of class C^{k+1}, smooth in μ, is unique.

PROOF. We set

$$[\gamma(t,\mu), b(t,\mu)] = \sum_{l\in\mathbb{Z}} [\gamma_l(\mu), b_l(\mu)]e^{2i\pi lt/T}$$

and find that

$$\left[\frac{2i\pi l}{T} + (p-1)\sigma + q\bar{\sigma}\right]\gamma_l + b_l = 0, \qquad p \geq 0, \quad q \geq 0, \quad p+q \geq 2. \tag{X.28}$$

If $[(p-1)\sigma + q\bar{\sigma}]T/2i\pi$ is not an integer, the coefficient of γ_l does not vanish and we may solve for γ. Since we are interested in solutions for μ near to zero, we may solve for γ if the coefficient of γ_l does not vanish when $\mu = 0$. If $(p-q-1)\omega_0 T/2\pi = -l_0 \neq 0$ and if the necessary condition $b_{l_0} = 0$ for the solvability of (X.28) is satisfied, we choose $\gamma_{l_0}(\mu) = 0$ when μ is small. We omit the proof of regularity in t of T-periodic solutions $\gamma(t)$ of (X.27) which is needed to complete the proof.

We next apply Lemma 2 to simplify (X.23). We first expand the coefficient γ_{pq} of $y^p\bar{y}^q$ into Fourier series, using (X.27), and choose the coefficients $\gamma_{pql}(\mu)$ at criticality ($\mu = 0$) so that*

$$\gamma_{pql}(0) = \frac{-b_{pql}(0)}{i[(2\pi l/T) + (p-1-q)\omega_0]}. \tag{X.29}$$

The main consequence of Lemma 2 then takes form in the following reduction:

$$\begin{aligned}
&\dot{\gamma}_{pq} + [p-1-q]i\omega_0\gamma_{pq} + b_{pq}\\
&\quad = \sum_{l\in\mathbb{Z}}\left\{\left[\frac{2\pi l}{T} + (p-1-q)\omega_0\right]i\gamma_{pql} + b_{pql}\right\}e^{2\pi i l t/T}\\
&\quad = \sum_{\mathrm{ES}} b_{pql}e^{2\pi i l t/T},
\end{aligned} \tag{X.30}$$

where ES stands for "exceptional set" defined by

$$\mathrm{ES} = \left\{l \text{ such that there are numbers } p, q, \omega_0 \text{ satisfying } \omega_0 = \frac{2\pi r}{T},\right.$$
$$\left. p \geq 0, q \geq 0, p+q \geq 2, 0 < r < 1, r \neq \frac{1}{2}, l + r(p-q-1) = 0\right\} \tag{X.31}$$

* There are small divisors for some $(p, q, 1)$ when $\omega_0 T/2\pi$ is irrational (see Exercise X.5).

and of course,

$$b_{pql} = \frac{1}{T}\int_0^T b_{pq}(s, 0)e^{-2\pi ils/T}\,ds.$$

All possible types of subharmonic bifurcation arising when a complex-conjugate pair of simple Floquet multipliers pass out of the unit disk are included in set ES. The special cases of T-periodic ($r = 0$) and $2T$-periodic ($r = \frac{1}{2}$) bifurcation, which are associated with $\lambda_0 = 1$ and $\lambda_0 = -1$, are excluded (see §IX.12). If l is in set ES we *cannot* choose $\gamma_{pql}(0)$ as in (X.29). So we merely set $\gamma_{pql}(\mu) = 0$ and find that $\gamma_{pq}(\mu, t)$ satisfies

$$\dot{\gamma}_{pq} + [(p-1)\sigma + q\bar{\sigma}]\gamma_{pq} + b_{pq} = \sum_{\text{ES}} b_{pql}(\mu)e^{2\pi ilt/T},$$

where, of course,

$$b_{pql}(\mu) = \frac{1}{T}\int_0^T b_{pq}(s, \mu)e^{-2\pi ils/T}\,ds.$$

It follows now from Lemmas 1 and 2 that (X.23) and (X.24) may be reduced to

$$\dot{y} = \sigma y + \sum_{p+q\geq 2}^{N} y^p\bar{y}^q \sum_{\text{ES}} b_{pql}(\mu)e^{2\pi ilt/T} + \tilde{b}_1(t, \mu, y, \bar{y}, \mathbf{Y}) \qquad (\text{X.32})_1$$

and

$$\dot{\mathbf{Y}} = \mathbf{f}_u(t, \mu|\mathbf{Y}) + \tilde{\mathbf{B}}_1(t, \mu, y, \bar{y}, \mathbf{Y}) \qquad (\text{X.32})_2$$

It is convenient to decompose the exceptional set into two subsets:

I. The *mean set*:

$$(p, q, l, r) = (q+1, q, 0, r), \qquad r \stackrel{\text{def}}{=} \frac{\omega_0 T}{2\pi}$$

II. The *resonant* set ($r = m/n$, $m < n$, $l \stackrel{\text{def}}{=} -km$, $kn = p - q - 1$):

$$\left(p, q, l, \frac{m}{n}\right) = \left(q + 1 + nk, q, -km, \frac{m}{n}\right).$$

When r is irrational, only terms in the mean set can arise, and (X.32) may be reduced to

$$\dot{y} = \sigma y + \sum_{q\geq 1}^{2q+1\leq N} y^{q+1}\bar{y}^q b_{q+1,q,0}(\mu) + \tilde{b}_1(t, \mu, y, \bar{y}, \mathbf{Y}), \qquad (\text{X.33})$$

where $\mathbf{Y}$ satisfies $(\text{X}.32)_2$. When there is an $n \geq 1$ such that $\lambda_0^n = 1$ we get terms from the mean set and terms from the resonant set

$$\begin{aligned}\dot{y} = \sigma y &+ \sum_{q \geq 1}^{2q+1 \leq N} y^{q+1}\bar{y}^q b_{q+1,q,0}(\mu)\\ &+ \sum_{k>0} \sum_{q \geq 0}^{2q-1+kn \leq N} \{y^{q+1+kn}\bar{y}^q b_{q+1+nk,\,q,\,-km}(\mu)e^{-2\pi ikmt/T}\\ &+ y^q\bar{y}^{q-1+kn} b_{q,\,q-1+kn,\,km}(\mu)e^{2\pi ikmt/T}\}\\ &+ \tilde{b}_1(t, \mu, y, \bar{y}, \mathbf{Y})\end{aligned} \tag{X.34}$$

It is important in the analysis to follow that equation (X.34) *with* $\tilde{b}_1$ *neglected* can be transformed into an *autonomous* equation for $x = e^{-i\omega_0 t}y$, where $\omega_0 = 2\pi r/T$, $r = m/n$. To simplify the writing of this autonomous equation and other equations in what follows, we define

$$\begin{aligned}a_{q,0}(\mu) &\overset{\text{def}}{=} b_{q+1,q,0}(\mu) = a_q(\mu)\\ a_{q,k}(\mu) &\overset{\text{def}}{=} b_{q+1+kn,q,-km}(\mu)\\ a_{q,-k}(\mu) &\overset{\text{def}}{=} b_{q,q-1+kn,km}(\mu).\end{aligned}$$

The autonomous equation for $x = e^{-i\omega_0 t}y$ which arises from (X.34) when $\tilde{b}_1$ is neglected is then in the form

$$\begin{aligned}\dot{x} = \mu[\hat{\xi}(\mu) + i\hat{\omega}(\mu)]x &+ \sum_{q \geq 1}^{2q+1 \leq N} x|x|^{2q}a_q\\ &+ \sum_{k>0} \sum_{q \geq 0}^{2q-1+kn \leq N} |x|^{2q}\{x^{1+kn}a_{qk} + \bar{x}^{kn-1}a_{q,-k}\},\end{aligned} \tag{X.35}$$

where we have set

$$\sigma = i\omega_0 + \mu[\hat{\xi}(\mu) + i\hat{\omega}(\mu)], \tag{X.36}$$

where $\hat{\xi}(0) > 0$ because we have assumed that the loss of stability of the null solution is strict. Many of the properties of bifurcation of solutions of (X.1) may be determined from (X.35), or even from the lowest-order truncation of (X.35):

$$\begin{aligned}\dot{x} = \mu[\hat{\xi}(\mu) + i\hat{\omega}(\mu)]x &+ |x|^2 x a_1 + |x|^4 x a_2 + \cdots\\ &+ \bar{x}^{n-1}a_{0,-1} + \cdots.\end{aligned} \tag{X.37}$$

X.4 The Normal Equations in Polar Coordinates

Our goal now is to derive analytical expressions for the torus shown in Figure X.1, whose cross section is given by $\rho = \rho(\theta)$, where (ρ, θ) are polar coordinates related to x and y by

$$y = e^{i\omega_0 t}x, \qquad x = \rho e^{i\theta}. \tag{X.38}$$

To convert the equation (X.33) governing the evolution of y in the case in which r is irrational and the equation (X.34) governing the evolution of y in the case in which $r = m/n$ is rational into polar coordinates, we introduce (X.38) and the relation

$$\dot{y} = [\dot{\rho} + i\omega_0 \rho + i\dot{\theta}\rho]e^{i(\theta + \omega_0 t)} \tag{X.39}$$

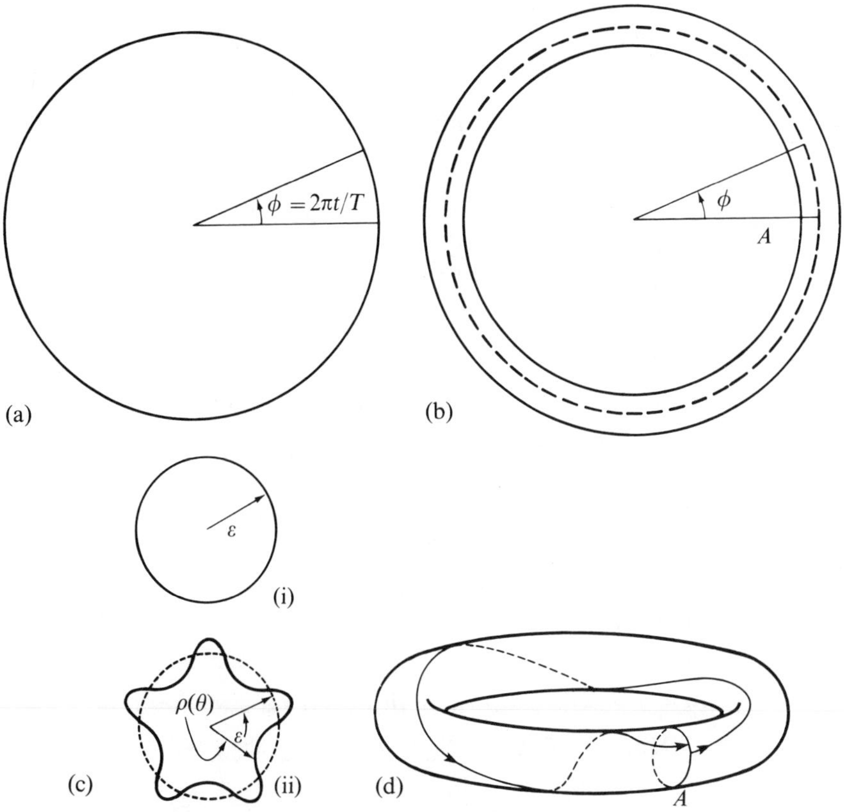

Figure X.1 Schematic sketch of the bifurcating torus T^2. (a) The torus T^1 (a limit cycle corresponding to the forced T-periodic solution $\mathbf{U}(t) = \mathbf{U}(t + T)$). (b) The torus T^2 (in top view) may be visualized as a tube around the limit cycle. A cross section of this torus is in $\mathbb{R}^n$. (c) Cross section of the torus T^2 of mean radius ε for small ε when (i) the Floquet multiplier is not a root of unity and (ii) the Floquet multiplier is a root of unity with $n = 5$. (d) The two-dimensional torus T^2 and a period segment of a trajectory which winds repeatedly around the torus

Then, *in the rational case*, we get

$$\begin{aligned}\dot{\rho} = \rho\Bigg\{&\mu\hat{\xi}(\mu) + \sum_{q\geq 1}^{2q+1\leq N} \operatorname{Re}\,[a_{q,0}(\mu)\rho^{2q}] \\ &+ \sum_{k>0}\sum_{q\geq 0}^{2q+1+kn\leq N} \rho^{2q+kn}\operatorname{Re}\,[a_{q,k}(\mu)e^{ikn\theta}] \\ &+ \sum_{k>0}\sum_{q\geq 0}^{2q-1+kn\leq N} \rho^{2q-2+kn}\operatorname{Re}\,[a_{q,-k}(\mu)e^{-ikn\theta}]\Bigg\} \\ &+ R_1(t, \mu, \rho, \theta, \mathbf{Y})\end{aligned} \tag{X.40}$$

and

$$\begin{aligned}\rho\dot{\theta} = \rho\Bigg\{&\mu\hat{\omega}(\mu) + \sum_{q\geq 1}^{2q+1\leq N} \operatorname{Im}\,[a_{q,0}(\mu)\rho^{2q}] \\ &+ \sum_{k>0}\sum_{q\geq 0}^{2q+1+kn\leq N} \rho^{2q+kn}\operatorname{Im}\,[a_{q,k}(\mu)e^{ikn\theta}] \\ &+ \sum_{k>0}\sum_{q\geq 0}^{2q-1+kn\leq N} \rho^{2q-2+kn}\operatorname{Im}\,[a_{q,-k}(\mu)e^{-ikn\theta}]\Bigg\} \\ &+ R_2(t, \mu, \rho, \theta, \mathbf{Y}),\end{aligned} \tag{X.41}$$

where

$$R_1 + iR_2 = e^{-i(\theta+\omega_0 t)}\tilde{b}_1(t, \mu, y, \bar{y}, \mathbf{Y}), \tag{X.42}$$

with y and $\bar{y}$ given by (X.38) and $\mathbf{Y}$ satisfying (X.25) with

$$\mathbf{B}_2(t, \mu, \rho, \theta, \mathbf{Y}) \stackrel{\text{def}}{=} \tilde{\mathbf{B}}_1(t, \mu, y, \bar{y}, \mathbf{Y}). \tag{X.43}$$

When r *is irrational* we put

$$a_{q,k} = a_{q,-k} = 0, \tag{X.44}$$

so that the terms in the double summations on the right-hand side of (X.40) and (X.41) vanish. In both cases of rational and irrational r we have the estimates

$$|R_l(t, \mu, \rho, \theta, \mathbf{Y})| = O(\rho^{N+1} + \rho\|\mathbf{Y}\| + \|\mathbf{Y}\|^2), \qquad l = 1, 2, \tag{X.45}$$

$$|\mathbf{B}_2(t, \mu, \rho, \theta, \mathbf{Y})| = O(\rho^{N+1} + \rho\|\mathbf{Y}\| + \|\mathbf{Y}\|^2). \tag{X.46}$$

X.5 The Torus and Trajectories on the Torus in the Irrational Case

The analysis is least complicated when r is irrational, so we use this case to introduce the notion of a torus and of an asymptotically quasi-periodic solution on the torus. We observe that the systems (X.40, 41) are *autonomous* up to the terms R_j which may be neglected in the first approximation. The **Y** equation (X.25) has a linear contracting part $J(\mu) = -(d/dt) + \mathbf{f}_u(t, \mu|\cdot)$ because all the Floquet exponents $\sigma = \xi + i\eta$ are of negative real part in the subspace of **Y**'s that is among **Y**'s satisfying $\langle \mathbf{Y}, \boldsymbol{\zeta}^* \rangle = 0$. For autonomous problems in $\mathbb{R}^2$ solutions of permanent form are fixed points or closed curves corresponding to limit cycles. In fact, we will find a closed curve $\rho(\theta)$ for the system approximating (X.40, 41, 44, 25) when R_1, R_2, and $\mathbf{B}_2$ are set to zero. And we show that the approximating problem gives rise to a periodic solution with a frequency depending on the amplitude, where the amplitude is the mean radius of the closed curve $\rho(\theta)$. The solution of the original problem, more precisely, the approximation to that solution through terms of order ρ^N, is a composition of T-periodic functions and of the aforementioned periodic solutions with a period τ which depends on the average value ε of $\rho(\theta)$.

To visualize what is meant by a two-dimensional torus, it is convenient to proceed in steps. We first recall that the basic problem (X.1) has been reduced to local form; the solution $\mathbf{u} = 0$ of (X.1) corresponds to a forced T-periodic solution $\mathbf{U}(t) = \mathbf{U}(t + T)$. This solution may be represented by a circle in the $n + 1$-dimensional phase space whose coordinates are the components of $\mathbf{u}$ and the time t. The "circle" is identified by $\mathbf{u} = 0$ in $\mathbb{R}^n$ and the set of numbers $t \in [0, T)$, where $t + T$ is identified with the point $t \in [0, T)$ (because $\mathbf{U}(t + T) = \mathbf{U}(t)$). The interval $[0, T)$ and identification rule for $t + T$ is called a *one-dimensional torus* T^1. *Identification* means that we "join" the ends of the interval $[0, T)$ and form a circle $T^1 \times 0$ in $\mathbb{R}^{n+1}$. This circle is a limit cycle for the forced periodic problem which may be visualized on a true circle of any radius with angle $\phi = 2\pi t/T$, where $\phi \in [0, 2\pi)$ and $\phi + 2\pi$ is identified with ϕ, as indicated in Figure X.1.

A two-dimensional torus T^2 of mean radius ε bifurcates from T^1 at $\varepsilon = 0$. The radius of the cross section of the torus is $\rho(\theta) = \rho(\theta + 2\pi)$. A *trajectory* is a curve

$$\theta = \theta(t), \qquad \rho = \rho(\theta(t)), \qquad \phi = \frac{2\pi t}{T} \tag{X.47}$$

on the torus. Trajectories wind around the torus repeatedly. They cut a particular cross section, say A in Figure X.1, each time t is advanced through a period T (and ϕ through an angle 2π). If, after a certain number of circuits, say n, in which the angle ϕ increases by $2\pi n$, the angle θ is also periodically repeated,

$$\theta(t + nT) = \theta(t), \tag{X.48}$$

the solution on the torus is *periodic*. If the solution is *quasi-periodic*, it has two rationally independent frequencies $2\pi/T$ and $\omega(\varepsilon)$, with $\theta = \omega(\varepsilon)t$. In the quasi-periodic case the solution trajectories are dense on the surface of the torus T^2 and are said to be *ergodic*.

The curious reader may now ask why we speak of an approximating problem through terms of order ρ^N when N is arbitrary? Why not pass to the limit? The answer here is that the solutions we obtain are asymptotic (for $\varepsilon \to 0$) and, in general, diverge (see remarks closing this section). It is frequently the case with divergent approximations that there is an optimal $N(\varepsilon)$ depending on ε such that approximations for a certain ε become better and better as $N < N(\varepsilon)$ is increased and worse and worse as $N > N(\varepsilon)$ is increased. So we are well advised to think of an approximation of a fixed, but arbitrary, order.

Now we shall derive the form of the function $\rho(\theta)$, the trajectories $\theta(t)$, and show that the approximate solution

$$\mathbf{u}^{(N)}(t) = Z^{(N)}(t, \varepsilon)\zeta(\mu^{(N)}, t) + \bar{Z}^{(N)}(t, \varepsilon)\bar{\zeta}(\mu^{(N)}, t) + \mathbf{W}^{(N)}(t, \varepsilon) \quad \text{(X.49)}$$

is quasi-periodic, with two frequencies when $r = \omega_0 T/2\pi$ is irrational, $0 < r < 1$. In this case we find that $\rho^{(N)}(\theta) = \varepsilon$ is a constant independent of θ. To show this we solve (X.40) in the approximation with $R_1 = 0$ for steady solutions

$$\mu\hat{\xi}(\mu) + \sum_{q \geq 1}^{2q+1 \leq N} \operatorname{Re}\,(a_q(\mu)\rho^{2q}) = 0. \quad \text{(X.50)}$$

It is generally convenient to solve for μ in powers of ρ^2. For steady solutions we put $\rho = \varepsilon$ and develop (X.50) in powers of ε, assuming that $a_q(\mu)$ and $\hat{\xi}(\mu)$ can be developed in powers of μ. After identifying the coefficients of the independent powers of ε, we find that $\mu = \mu^{(N)}(\varepsilon)$,

$$\begin{aligned}\mu^{(N)}(\varepsilon) &= -\left(\frac{(\operatorname{Re}\,a_1(0))\varepsilon^2}{\hat{\xi}(0)}\right) + O(\varepsilon^4)\\ &= \mu^{(N)}(-\varepsilon), \qquad\qquad \rho^{(N)} = \varepsilon. \end{aligned} \quad \text{(X.51)}$$

Moreover, in the same approximation, neglecting R_2 we find that the solution (X.41) is in the form

$$\theta^{(N)} = \varepsilon^2\hat{\theta}^{(N)}(\varepsilon^2)t, \quad \text{(X.52)}$$

where

$$\varepsilon^2\hat{\theta}^{(N)}(\varepsilon^2) = \mu^{(N)}(\varepsilon)\hat{\omega}[\mu^{(N)}(\varepsilon)] + \sum_{q \geq 1}^{2q+1 \leq N} \operatorname{Im}\,(a_q[\mu^{(N)}(\varepsilon)]\varepsilon^{2q}) \quad \text{(X.53)}$$

After tracing back through the changes of variables, we find that the approximate solution, up to N (arbitrary) terms, of (X.1) is (X.49), with

$$\begin{aligned} Z^{(N)}(t, \varepsilon) &= \varepsilon \exp i(\omega_0 + \varepsilon^2\hat{\theta}^{(N)}(\varepsilon^2))t \\ &\quad + \sum_{p+q \geq 2}^{N} \gamma'_{pq}(t, \mu^{(N)}(\varepsilon))\varepsilon^{p+q} \exp i(p - q)[\omega_0 + \varepsilon^2\hat{\theta}^{(N)}(\varepsilon^2)]t \end{aligned}$$

$$\text{(X.54)}_1$$

and

$$\mathbf{W}^{(N)}(t, \varepsilon) = \sum_{p+q\geq 2}^{N} \mathbf{\Gamma}'_{pq}(t, \mu^{(N)}(\varepsilon))\varepsilon^{p+q} \exp (p - q)i[\omega_0 + \varepsilon^2\hat{\theta}^{(N)}(\varepsilon^2)]t, \tag{X.54$_2$}$$

where $\mu^{(N)}(\varepsilon)$ and $\hat{\theta}^{(N)}(\varepsilon^2)$ are defined by (X.51) and (X.53). We assert without proof that solutions on the torus satisfy

$$\begin{aligned} \mu(\varepsilon) - \mu^{(N)}(\varepsilon) &= O(\varepsilon^{N+2}) \\ \mathbf{W}(t, \mu) - \tilde{\mathbf{W}}^{(N)}(t, \varepsilon) &= O(\varepsilon^{N+1}) \\ Z(t, \mu) - \tilde{Z}^{(N)}(t, \varepsilon) &= O(\varepsilon^{N+1}) \\ \boldsymbol{\zeta}(\mu, t) - \boldsymbol{\zeta}(\mu^{(N)}, t) &= O(\varepsilon^{N+2}), \end{aligned}$$

where $\tilde{\mathbf{W}}^{(N)}$ and $\tilde{Z}^{(N)}$ are obtained by replacing $\varepsilon^2\hat{\theta}^{(N)}(\varepsilon^2)t$ by $\theta(t)$ in the expressions (X.54) and

$$\theta(t) - \varepsilon^2\hat{\theta}^{(N)}(\varepsilon^2)t = \chi(t, \varepsilon).$$

These estimates are uniformly valid in t, even if $\chi(t, \varepsilon)$ contains secular terms which, like terms which are linear in t, are unbounded. However, $|\dot{\chi}(t, \varepsilon)| = O(\varepsilon^N)$. The functions $\boldsymbol{\zeta}(\mu^{(N)}, t)$, $\gamma'_{pq}(t, \mu^{(N)})$, and $\mathbf{\Gamma}'_{pq}(t, \mu^{(N)})$ are all T-periodic.

We now claim that the vector

$$\mathbf{u}^{(N)}(t) = \mathcal{U}^{(N)}(\tau_1(t), \tau_2(t)) \tag{X.55$_1$}$$

given by (X.49) is a doubly periodic, vector-valued function with

$$\tau_1(t) = t \quad \text{and} \quad \tau_2(t) = [\omega_0 + \varepsilon^2\hat{\theta}^{(N)}(\varepsilon^2)]t \tag{X.55$_2$}$$

and that

$$\mathcal{U}^{(N)}(\tau_1, \tau_2) = \mathcal{U}^{(N)}(\tau_1 + T, \tau_2) = \mathcal{U}^{(N)}(\tau_1, \tau_2 + 2\pi).$$

So we say that the flow is *asymptotically quasi-periodic* with two fundamental frequencies

$$\omega_1 = \frac{2\pi}{T}, \qquad \omega_2 = \omega_0 + \varepsilon^2\hat{\theta}^{(N)}(\varepsilon^2).$$

The second frequency is a polynomial in ε^2, but the series for it which arises formally as $N \to \infty$ is divergent in general. There is as yet no direct proof of divergence, but convergence would contradict certain slightly exotic mathematical theorems which are outside the scope of an elementary book (see G. Iooss, *Bifurcation of Maps and Applications* (Amsterdam: North-Holland, 1979)). Note that $\mathcal{U}^{(N)}$ is quasi-periodic for values of ε such that $T/2\pi(\omega_0 + \varepsilon^2\hat{\theta}(\varepsilon^2))$ is irrational. In the rational case it is periodic; but the strong fact is that the linear dependence in (X.55)$_2$ of τ_1, τ_2 on t is not in general true of the true solution $\mathbf{u}(t)$ for all values of ε.

X.6 The Torus and Trajectories on the Torus When $\omega_0 T/2\pi$ is a Rational Point of Higher Order $(n \geq 5)$

The interesting fact is that when there is an $n \geq 5$ such that $\lambda_0^n = 1$, we get a torus and the solutions on it are asymptotically quasi-periodic. So the basic physical results implied by analysis of bifurcation of periodic solutions are qualitatively independent of whether r is rational or irrational. But the analysis is more delicate in the rational case and the formulas for the torus and the trajectories on it are different.

Suppose now that $r = m/n$ is an irreducible fraction and $n \geq 5$. Our first goal is to determine an approximation $\rho_N(\theta)$ to the cross section of the torus $\rho(\theta)$. The equation governing this approximation can be obtained by dropping R_1 and R_2 in (X.40) and (X.41):

$$\frac{d\rho}{dt} = \rho\left[\mu\hat{\xi} + \sum_{q\geq 1}^{2q+1\leq N} \alpha_q \rho^{2q} + \sum_{k>0} \sum_{q\geq 0}^{2q-1+kn\leq N} (\alpha_{qk} e^{ikn\theta} + \bar{\alpha}_{qk} e^{-ikn\theta})\rho^{2q-2+kn}\right], \qquad (\text{X.56})_1$$

and

$$\frac{d\theta}{dt} = \mu\hat{\omega} + \sum_{q\geq 1}^{2q+1\leq N} \beta_q \rho^{2q} + \sum_{k>0} \sum_{q\geq 0}^{2q-1+kn\leq N} (\beta_{qk} e^{ikn\theta} + \bar{\beta}_{pk} e^{-ikn\theta})\rho^{2q-2+kn}, \qquad (\text{X.56})_2$$

where all coefficients $\hat{\xi}$, $\hat{\omega}$, α_q, β_q, α_{qk}, β_{qk} are functions of μ, as smooth as we wish,

$$\begin{aligned}
\hat{\xi}(\mu) &= \hat{\xi}_0 + \mu\hat{\xi}_1 + \mu^2\hat{\xi}_2 + \cdots, \qquad \hat{\xi}_0 > 0\\
\hat{\omega}(\mu) &= \hat{\omega}_0 + \mu\hat{\omega}_1 + \mu^2\hat{\omega}_2 + \cdots\\
\alpha_q(\mu) &= \alpha_{q0} + \mu\alpha_{q1} + \mu^2\alpha_{q2} + \cdots\\
\alpha_{qk}(\mu) &= \alpha_{qk0} + \mu\alpha_{qk1} + \mu^2\alpha_{qk2} + \cdots\\
\beta_q(\mu) &= \beta_{q0} + \mu\beta_{q1} + \mu^2\beta_{q2} + \cdots\\
\beta_{qk}(\mu) &= \beta_{qk0} + \mu\beta_{qk1} + \mu^2\beta_{qk2} + \cdots,
\end{aligned}$$

and where by construction

$$\begin{aligned}
\alpha_q + i\beta_q &= a_{q,0}\\
\alpha_{qk} &= \tfrac{1}{2}(a_{q-1,k} + \bar{a}_{q,-k}), \qquad a_{q-1,k} \overset{\text{def}}{=} 0 \ \text{ if } \ q = 0 \qquad (\text{X.57})\\
\beta_{qk} &= \frac{1}{2i}(a_{q-1,k} - \bar{a}_{q,-k}).
\end{aligned}$$

To solve (X.56), we introduce an amplitude ε, defined as the mean radius of the cross section of the torus, as in Figure X.1:

$$\varepsilon \stackrel{\text{def}}{=} \frac{1}{2\pi}\int_0^{2\pi} \rho(\theta, \mu)\, d\theta \stackrel{\text{def}}{=} \bar{\bar{\rho}} \tag{X.58}$$

The equation governing $\rho(\theta, \mu)$ can be deduced from the relation

$$\frac{d\rho}{dt} = \frac{d\rho}{d\theta}\frac{d\theta}{dt}, \tag{X.59}$$

where $d\rho/dt$ and $d\theta/dt$ are given by (X.56). To solve (X.59) we develop μ and ρ in powers of ε:

$$\mu = \sum_{p=1}^{N} \mu_p \varepsilon^p + O(\varepsilon^{N+1}),$$

$$\rho = \sum_{p=1}^{N} \rho_p(\theta)\varepsilon^p + O(\varepsilon^{N+1}) \tag{X.60}$$

$$\bar{\bar{\rho}}_1 = 1, \qquad \bar{\bar{\rho}}_p = 0 \quad \text{for} \quad p \geq 2.$$

Identification of the coefficient of ε^2 in (X.59) gives

$$\hat{\xi}_0 \mu_1 \rho_1(\theta) = \mu_1 \hat{\omega}_0 \rho_1'(\theta). \tag{X.61}$$

Taking the mean value on $(0, 2\pi)$, we find that $\hat{\xi}_0 \mu_1 = 0$; hence

$$\mu_1 = 0.$$

Identification of the coefficient of ε^3 in (X.59) gives:

$$\rho_1(\theta)[\hat{\xi}_0 \mu_2 + \alpha_{10}\rho_1^2(\theta)] = \rho_1'(\theta)[\mu_2 \hat{\omega}_0 + \beta_{10}\rho_1^2(\theta)]. \tag{X.62}$$

Taking the mean value of (X.62) on $(0, 2\pi)$, we find that

$$\hat{\xi}_0 \mu_2 + \alpha_{10}\bar{\bar{\rho}}_1^3 = 0. \tag{X.63}$$

Now it is not hard to show from (X.62) and (X.63) that any periodic solution of mean value 1 must satisfy $\overline{\overline{\rho_1^{2+\nu}}} = \overline{\overline{\rho_1^{\nu}}}\,\overline{\overline{\rho_1^{3}}}$ for all integers $\nu \geq 0$. Hence, for any integer $p \geq 1$,

$$\left[\frac{1}{2\pi}\int_0^{2\pi} |\rho_1|^{2p}\, d\theta\right]^{1/p} = \frac{1}{2\pi}\int_0^{2\pi} |\rho_1|^2\, d\theta = \overline{\overline{\rho_1^2}},$$

and since ρ_1 is continuous,

$$\left[\frac{1}{2\pi}\int_0^{2\pi} |\rho_1|^{2p}\, d\theta\right]^{1/p} \xrightarrow[p\to\infty]{} \underset{\theta\in[0,2\pi)}{\text{l.u.b.}}\ |\rho_1(\theta)|^2.$$

So $\overline{\overline{\rho_1^2}} = \text{l.u.b.}\ |\rho_1(\theta)|$; that is, $|\rho_1(\theta)| = 1$ and

$$\rho_1(\theta) \equiv 1 \tag{X.64}$$

$$\mu_2 = -\frac{\alpha_{10}}{\hat{\xi}_0}. \tag{X.65}$$

We stop the general analysis here. Further results depend on the value of $n \geq 5$ for which $\lambda_0^n = 1$.

X.7 The Form of the Torus in the Case $n = 5$

We now suppose that $\lambda_0^5 = 1$. Identifying the coefficient of ε^4 in (X.59) we find that

$$\begin{aligned}\hat{\xi}_0(\mu_3 + \mu_2\rho_2(\theta)) + 3\alpha_{10}\rho_2(\theta) + \alpha_{010}e^{5i\theta} + \bar{\alpha}_{010}e^{-5i\theta} \\ = \rho_2'(\theta)[\mu_2\hat{\omega}_0 + \beta_{10}].\end{aligned} \tag{X.66}$$

After taking the mean value of (X.66) we find that $\hat{\xi}_0\mu_3 = 0$, hence

$$\mu_3 = 0 \tag*{(X.67)$_1$}$$

and

$$\rho_2(\theta) = g_1 e^{5i\theta} + \bar{g}_1 e^{-5i\theta}, \tag*{(X.67)$_2$}$$

where g_1 is a complex constant satisfying

$$g_1[2\alpha_{10} - 5i(\mu_2\hat{\omega}_0 + \beta_{10})] + \alpha_{010} = 0. \tag{X.68}$$

We can compute g_1 from (X.68) provided that the coefficient of g_1 does not vanish. Since (X.65) shows that $\mu_2 = 0$ when $\alpha_{10} = 0$, we conclude that (X.68) may be solved for g_1 except for the exceptional case in which $\alpha_{10} = \beta_{10} = 0$. In this exceptional case bifurcation into an invariant torus need not occur. We shall not consider such exceptional cases.

Proceeding as before we identify the coefficient of ε^5 in (X.59) and find that

$$\begin{aligned}\hat{\xi}_0[\mu_4 + \mu_2\rho_3] + \hat{\xi}_1\mu_2^2 + 3\alpha_{10}\rho_2^2 + 3\alpha_{10}\rho_3 + \alpha_{11}\mu_2 \\ + \alpha_{20} + 4\rho_2(\alpha_{010}e^{5i\theta} + \bar{\alpha}_{010}e^{-5i\theta}) \\ = \rho_3'(\mu_2\hat{\omega}_0 + \beta_{10}) + \rho_2'(2\beta_{10}\rho_2 + \beta_{010}e^{5i\theta} + \bar{\beta}_{010}e^{-5i\theta}).\end{aligned} \tag{X.69}$$

The mean value of (X.69) is

$$\begin{aligned}\hat{\xi}_0\mu_4 = (5i\bar{g}_1\beta_{010} - 5ig_1\bar{\beta}_{010}) - 4(\alpha_{010}\bar{g}_1 + \bar{\alpha}_{010}g_1) \\ - \mu_2^2\hat{\xi}_1 - 6\alpha_{10}|g_1|^2 - \alpha_{11}\mu_2 - \alpha_{20},\end{aligned} \tag{X.70}$$

and (X.69) and (X.70) imply

$$\rho_3(\theta) = g_2 e^{10i\theta} + \bar{g}_2 e^{-10i\theta},$$

where g_2 may be computed as g_1 if α_{10} and β_{10} are not both zero.

Turning next to the coefficient of ε^6 we find that

$$\hat{\xi}_0 \mu_5 + 2\alpha_{10}\rho_4 + F_3 e^{15i\theta} + \bar{F}_3 e^{-15i\theta} + F_1 e^{5i\theta} + \bar{F}_1 e^{-5i\theta}$$
$$= \rho_4'(\mu_2 \hat{\omega}_0 + \beta_{10}). \tag{X.71}$$

Hence

$$\begin{aligned} \mu_5 &= 0 \\ \rho_4(\theta) &= g_{30} e^{15i\theta} + \bar{g}_{30} e^{-15i\theta} + g_{31} e^{5i\theta} + \bar{g}_{31} e^{-5i\theta}, \end{aligned} \tag{X.72}$$

where g_{30}, g_{31} are determined by identification in (X.71) and F_3, F_1 may be computed easily in terms of known coefficients.

More generally, it can be shown by mathematical induction that

$$\begin{aligned} \mu_{2p+1} &= 0 \\ \rho_{p+1}(\theta) &= \sum_{q \geq 0}^{Q_p} g_{pq} e^{5(p-2q)i\theta} + \bar{g}_{pq} e^{5(2q-p)i\theta}, \end{aligned} \tag{X.73}$$

where $Q_p = (p-1)/2$ if p is odd and $Q_p = (p/2) - 1$ if p is even. All the numbers g_{pq}, like g_1 and g_2, may be determined by identification.

X.8 Trajectories on the Torus When $n = 5$

We next turn to the problem of trajectories on the torus. In particular, we seek $\theta = \theta(t, \varepsilon)$ solving $(X.56)_2$. To solve this problem, we define

$$\tilde{\theta} = \theta + \sum_{l=1}^{N-1} \varepsilon^l h_l(\theta) \tag{X.74}$$

and construct periodic functions $h_l(\theta) = h_l(\theta + 2\pi)$ of mean value zero, $\overline{\overline{h_l}} = 0$, in such a way that $\tilde{\theta}(t)$ is constant up to order ε^N. It turns out that these functions $h_l(\theta)$ are $2\pi/5$-periodic; that is

$$h_l(\theta) = h_l\left(\theta + \frac{2\pi}{5}\right), \qquad \overline{\overline{h_l}} = 0. \tag{X.75}$$

The differential equation satisfied by $\tilde{\theta}(t)$ is

$$\frac{d\tilde{\theta}}{dt} = \left\{1 + \sum_{l=1}^{N-1} \varepsilon^l h_l'(\theta)\right\} \frac{d\theta}{dt}, \tag{X.76}$$

where $d\theta/dt$ is given by $(X.56)_2$ with $n = 5$. After expanding the right-hand side of $(X.56)_2$ in powers of ε,

$$\begin{aligned} \mu &= \mu_2 \varepsilon^2 + \mu_4 \varepsilon^4 + \mu_6 \varepsilon^6 + \cdots \\ \rho(\theta) &= \varepsilon + \varepsilon^2 \rho_2(\theta) + \varepsilon^3 \rho_3(\theta) + \varepsilon^4 \rho_4(\theta) + \cdots, \end{aligned} \tag{X.77}$$

we find that

$$\frac{d\theta}{dt} = \Omega_0 \varepsilon^2 + \Theta_1(\theta)\varepsilon^3 + \Theta_2(\theta)\varepsilon^4 + \Theta_3(\theta)\varepsilon^5 + \cdots, \tag{X.78}$$

where

$$\begin{aligned}
\Omega_0 &= \mu_2 \hat{\omega}_0 + \beta_{10} \\
\Theta_1(\theta) &= 2\beta_{10}\rho_2(\theta) + \beta_{010}e^{5i\theta} + \bar{\beta}_{010}e^{-5i\theta} \\
\Theta_2(\theta) &= \mu_4\hat{\omega}_0 + \mu_2^2\hat{\omega}_1 + \mu_2\beta_{11} + 2\beta_{10}\rho_3(\theta) + \beta_{10}\rho_2^2(\theta) + \beta_{20} \\
&\quad + 3\rho_2(\theta)(\beta_{010}e^{5i\theta} + \bar{\beta}_{010}e^{-5i\theta}) \\
\Theta_3(\theta) &= 2\beta_{10}\rho_2(\theta)\rho_3(\theta) + 2\mu_2\beta_{11}\rho_2(\theta) + 4\beta_{20}\rho_2(\theta) \\
&\quad + 3[\rho_2^2(\theta) + \rho_3(\theta)][\beta_{010}e^{5i\theta} + \bar{\beta}_{010}e^{-5i\theta}] \\
&\quad + 2\beta_{11}\rho_4(\theta) + [\beta_{110}e^{5i\theta} + \bar{\beta}_{110}e^{-5i\theta}] \\
&\quad + \mu_2[\beta_{011}e^{5i\theta} + \bar{\beta}_{011}e^{-5i\theta}],
\end{aligned}$$

and so on. Here, and in general

$$\bar{\bar{\Theta}}_{2l} \neq 0 \quad \text{and} \quad \bar{\bar{\Theta}}_{2l+1} = 0, \qquad l \geq 1 \tag{X.79}$$

and

$$\Theta_l(\theta) = \Theta_l\left(\theta + \left(\frac{2\pi}{5}\right)\right). \tag{X.80}$$

Equations (X.77) and (X.78) imply that

$$\begin{aligned}
\frac{d\tilde{\theta}}{dt} &= [1 + \varepsilon h_1'(\theta) + \varepsilon^2 h_2'(\theta) + \varepsilon^3 h_3'(\theta) + \cdots][\Omega_0\varepsilon^2 + \Theta_1(\theta)\varepsilon^3 \\
&\quad + \Theta_2(\theta)\varepsilon^4 + \Theta_3(\theta)\varepsilon^5 + \cdots] + O(\varepsilon^N)
\end{aligned} \tag{X.81}$$

Now we shall construct periodic functions $h_l(\theta)$ to simplify (X.81). We seek

$$h_l(\theta) = h_l\left(\theta + \left(\frac{2\pi}{5}\right)\right) \tag{X.82}$$

with

$$\bar{h}_l = 0 \tag{X.83}$$

for all $l \geq 1$ such that

$$\frac{d\tilde{\theta}}{dt} = \varepsilon^2\Omega(\varepsilon^2) + O(\varepsilon^N), \tag{X.84}$$

where $\Omega(\varepsilon^2)$ is a polynomial independent of t and θ. Our method of selection is as follows. First we arrange the right-hand side of (X.81) in powers of ε:

$$\begin{aligned}\frac{1}{\varepsilon^2}\frac{d\tilde{\theta}}{dt} &= \Omega_0 + (\Omega_0 h_1' + \Theta_1)\varepsilon + (\Omega_0 h_2' + \Theta_2 + \Theta_1 h_1')\varepsilon^2 \\ &\quad + (\Omega_0 h_3' + \Theta_3 + \Theta_2 h_1' + \Theta_1 h_2')\varepsilon^3 + \cdots \\ &\quad + (\Omega_0 h_l' + \Theta_l + \Theta_{l-1} h_1' + \cdots + \Theta_1 h_{l-1}')\varepsilon^l \\ &\quad + \cdots + O(\varepsilon^{N-2}). \end{aligned} \tag{X.85}$$

We are assuming that $\Omega_0 \neq 0$. Then we choose $h_n(\theta)$ sequentially so that each coefficient is replaced by its average value. For the first coefficient we put

$$\Omega_0 h_1' + \Theta_1 = \bar{\bar{\Theta}}_1,$$

where

$$\Theta_1 = \Theta_{10} e^{5i\theta} + \bar{\Theta}_{10} e^{-5i\theta}, \qquad \bar{\bar{\Theta}}_1 = 0.$$

So

$$h_1 = -\frac{\Theta_{10}}{5i\Omega_0} e^{5i\theta} + \frac{\bar{\Theta}_{10}}{5i\Omega_0} e^{-5i\theta}.$$

For the second coefficient we find that

$$\Omega_0 h_2' + \Theta_2 + \Theta_1 h_1' = \bar{\bar{\Theta}}_2 + \overline{\overline{\Theta_1 h_1'}} \neq 0. \tag{X.86}$$

We easily calculate $h_2(\theta)$ satisfying (X.86), (X.82), and (X.83). For the third coefficient we have

$$\Omega_0 h_3' + \Theta_3 + \Theta_2 h_1' + \Theta_1 h_2' = \bar{\bar{\Theta}}_3 + \overline{\overline{\Theta_2 h_1'}} + \overline{\overline{\Theta_1 h_2'}} = 0,$$

and so on. The average values of the coefficients of odd orders vanish and

$$\begin{aligned}\frac{1}{\varepsilon^2}\frac{d\tilde{\theta}}{dt} &= \Omega_0 + [\bar{\bar{\Theta}}_2 + \overline{\overline{\Theta_1 h_1'}}]\varepsilon^2 + [\bar{\bar{\Theta}}_4 + \overline{\overline{\Theta_3 h_1'}} + \overline{\overline{\Theta_2 h_2'}} + \overline{\overline{\Theta_1 h_3'}}]\varepsilon^4 \\ &\quad + \cdots + O(\varepsilon^{N-2}) \\ &\stackrel{\text{def}}{=} \Omega(\varepsilon^2) + O(\varepsilon^{N-2}). \end{aligned} \tag{X.87}$$

When $\lambda_0^5 = 1$, the trajectories on the torus are given in general by an asymptotic expression of the form

$$\varepsilon^2\Omega(\varepsilon^2)t = \theta + \varepsilon h_1(\theta) + \varepsilon^2 h_2(\theta) + \cdots + \varepsilon^{N-1} h_{N-1}(\theta) + \chi(t, \varepsilon), \tag{X.88}$$

where $h_l(\theta)$ is $2\pi/5$-periodic in θ, of mean value zero, N is unrestricted, and $\dot{\chi}(t, \varepsilon) = O(\varepsilon^N)$.

X.9 The Form of the Torus When $n > 5$

We return now to the rational case with $n > 5$ and consider (X.59) with

$$\rho = \varepsilon R(\theta, \varepsilon)$$
$$\mu = \varepsilon^2 \tilde{\mu}(\varepsilon), \tag{X.89}$$

where

$$\begin{bmatrix} R(\theta, \varepsilon) \\ \tilde{\mu}(\varepsilon) \end{bmatrix} = \sum_{l=0} \begin{bmatrix} R_l(\theta) \\ \tilde{\mu}_l \end{bmatrix} \varepsilon^l, \qquad \tilde{\mu}_l = \mu_{l+2}, \qquad R_l = \rho_{l+1}, \tag{X.90}$$

$$R_0 = 1, \qquad \tilde{\mu}_0 = -\frac{\alpha_{10}}{\hat{\xi}_0}.$$

We find that the approximation (X.54) of the solution satisfies

$$\tilde{\mu} L(\mu) R + \sum_{q \geq 0}^{2q+3 \leq N} \varepsilon^{2q} L^{\langle 2q+3 \rangle}(\mu) R^{2q+3}$$
$$+ \sum_{k > 0} \sum_{q \geq 0}^{2q-1+kn \leq N} \varepsilon^{2q-4+kn} L^{\langle 2q, k \rangle}(\mu, \theta) R^{2q-1+kn} = 0, \tag{X.91}$$

where

$$L(\mu) = \hat{\xi}(\mu) - \hat{\omega}(\mu) \frac{d}{d\theta}$$

$$L^{2q+3}(\mu) = \alpha_{q+1}(\mu) - \frac{\beta_{q+1}(\mu) d}{2q + 3\, d\theta}$$

and

$$L^{\langle 2q, k \rangle}(\mu, \theta) = \alpha_{qk}(\mu) e^{ikn\theta} + \bar{\alpha}_{qk}(\mu) e^{-ikn\theta}$$
$$- \frac{1}{2q - 1 + kn} [\beta_{qk}(\mu) e^{ikn\theta} + \bar{\beta}_{qk}(\mu) e^{-ikn\theta}] \frac{d}{d\theta}.$$

The first nonzero term in the last summation of (X.91) is the one for which $q = 0$, $k = 1$ and

$$\varepsilon^{2q-4+kn} = \varepsilon^{n-4}.$$

So we may identify the coefficients of successive powers ε^l, $l < n - 4$, without considering the last summation of (X.91). These coefficients may be computed by writing

$$R^m(\theta, \varepsilon) = \sum_{p=0} [R^m(\theta)]_p \varepsilon^p$$
$$\tilde{\mu}(\varepsilon) L(\varepsilon^2 \tilde{\mu}(\varepsilon)) = \sum_{p=0} [\tilde{\mu} L(\mu)]_p \varepsilon^p, \tag{X.92}$$
$$\varepsilon^{2q} L^{\langle 2q+3 \rangle}(\varepsilon^2 \tilde{\mu}) = \sum_{p=0} \varepsilon^{2q+p} [L^{\langle 2q+3 \rangle}(\mu)]_p.$$

After identification, we find that

$$\sum_{\nu=l+p} [\tilde{\mu}L(\mu)]_l R_p + \sum_{\nu=2q+p+l} [L^{\langle 2q+3\rangle}(\mu)][R^{2q+3}(\theta)]_p = 0 \quad \text{(X.93)}$$

for $\nu = 0, 1, \ldots, n-5, q \geq 0, p \geq 0, l \geq 0$. This problem is in fact *identical* to (X.50) through terms of $O(\varepsilon^{n-5})$. Hence $R_0 = 1$,

$$R_l(\theta) = 0, \qquad 0 < l < n-4 \quad \text{(X.94)}$$

and

$$\begin{aligned} \tilde{\mu}_{2l-1} &= 0, \qquad 2l-1 \leq n-5 \\ \mu_{2l-1} &= 0, \qquad 2l-1 \leq n-3. \end{aligned} \quad \text{(X.95)}$$

Now we demonstrate that

$$R_{n-4}(\theta) = \rho_{n-3}(\theta) = g_{10}e^{in\theta} + g_{10}e^{-in\theta}, \quad \text{(X.96)}$$

where g_{10} is a constant depending on the resonance number n. To prove (X.96) we identify the coefficient of ε^ν with $\nu \geq n-4$ in (X.91) and find that

$$\begin{aligned} &\sum_{\nu=l+p} [\tilde{\mu}L(\mu)]_l R_p + \sum_{\nu=2g+p+l} [L^{\langle 2q+3\rangle}(\mu)]_l [R^{2q+3}(\theta)]_p \\ &\quad + \sum_{2q+kn-4+l+p=\nu} [L^{\langle 2q,k\rangle}(\mu,\theta)]_l [R^{2q+kn-1}]_p = 0. \end{aligned} \quad \text{(X.97)}$$

There are two cases to consider.

(i) n is even and $\nu = n-4$. Then, using (X.94), we may write (X.97) as

$$\begin{aligned} &[\tilde{\mu}L(\mu)]_{n-4}R_0 + [\tilde{\mu}L(\mu)]_0 R_{n-4} + [L^{\langle 3\rangle}(\mu)]_0[R^3]_{n-4} \\ &\quad + \sum_{n-4=2q+l} [L^{\langle 2q+3\rangle}(\mu)]_l[R^{2q+3}]_0 + [L^{\langle 0,1\rangle}(\mu,\theta)]_0[R^{n-1}]_0 = 0, \end{aligned} \quad \text{(X.98)}$$

where

$$[\tilde{\mu}L(\mu)]_{n-4} = \tilde{\mu}_{n-4}L_0 + \text{l.o.t.}$$

$$R_0 = [R^{n-1}]_0 = 1, \qquad [\tilde{\mu}L(\mu)]_0 = \tilde{\mu}_0\left(\hat{\xi}_0 - \hat{\omega}_0 \frac{d}{d\theta}\right)$$

$$[L^{\langle 3\rangle}(\mu)]_0[R^3]_{n-4} = 3\alpha_{10}R_{n-4} - \beta_{10}R'_{n-4}$$

$$[L^{\langle 0,1\rangle}(\mu,\theta)][R^{n-1}]_0 = L^{\langle 0,1\rangle}(0,\theta) = \alpha_{010}e^{in\theta} + \bar{\alpha}_{010}e^{-in\theta}$$

and

$$\tilde{\mu}_l = \mu_{l+2}.$$

Hence

$$\begin{aligned} &\mu_{n-2}\hat{\xi}_0 + \text{l.o.t.} + (\mu_2\hat{\xi}_0 + 3\alpha_{10})R_{n-4} - (\beta_{10} + \mu_2\hat{\omega}_0)R'_{n-4} \\ &\quad + \sum_{n=2q+l+4} [L^{\langle 2q+3\rangle}(\mu)]_l + \alpha_{010}e^{in\theta} + \alpha_{010}e^{-in\theta} = 0 \end{aligned} \quad \text{(X.99)}$$

where the operator $[L^{\langle 2q+3\rangle}(\mu)]_l$ acts on the constant unit function 1. The average value of (X.99) is

$$\mu_{n-2}\hat{\bar{\xi}}_0 + \text{l.o.t.} + \sum_{n=2q+l+4} [L^{\langle 2q+3\rangle}(\mu)]_l = 0. \tag{X.100}$$

This determines μ_{n-2} in terms of lower order. It then follows from (X.99) that $R_{n-4}(\theta)$ is in the form given by (X.96).

(ii) n is odd and $v = n - 4$. Here the second two terms of (X.100) vanish because $\mu_{2l-1} = 0$ for $2l - 1 < n - 2$, so that μ_{n-2} also vanishes.

We next establish that when n is even ($n \geq 6$)

$$R_{2m+1}(\theta) = \rho_{2m}(\theta) = 0 \tag{X.101}$$

for all m such that $2m < N$ (recall that N is unrestricted so that the even ε derivatives of $\rho(\theta, \varepsilon)$ all vanish in every approximation). This follows from the fact that when n is even only even powers of ε appear in (X.91). The same observation establishes that

$$\mu_{2m-1} = 0 \tag{X.102}$$

when n is even.

Equation (X.102) also holds when n is odd ($n \geq 5$). Assume $\mu_l = 0$ when $l < v$ and l is odd. Then all odd-order derivatives with respect to ε of functions of μ must vanish and the average of (X.97) may be written as

$$\begin{aligned}\tilde{\mu}_v\hat{\bar{\xi}}_0 &+ \sum_{v=2q+wl+2p+1} [L^{\langle 2q+3\rangle}(\mu)]_{2l}[\overline{\overline{R^{2q+3}}}]_{2p+1} \\ &+ \sum_{2q+kn-4+2l+p=v} \overline{\overline{[L^{\langle 2q,k\rangle}(\mu,\theta)]_{2l}[R^{2q+kn-1}]_p}} = 0.\end{aligned} \tag{X.103}$$

Now $R_l(\theta)$, $l > 0$ is an even (odd) polynomial in harmonics of $e^{in\theta}$ if l is even (odd) and $\overline{\overline{R}}_l = 0$. Then $[R^m(\theta)]_l (m \geq 1)$ is also a polynomial in $e^{in\theta}$ and $\overline{\overline{R^m}} = 0$ if l is odd. Similarly $[L^{\langle 2q,k\rangle}(\mu, \theta)]_{2l}[R^{2q+kn-1}]_p$ is a polynomial in the harmonics of $e^{in\theta}$ of mean value zero when $k + p$ is odd. Since v and n are odd $kn + p$ is odd when $k + p$ is. It follows that the average terms in (X.103) vanish and $\tilde{\mu}_v = 0$ when v is odd.

In general, we have

$$\mu = \sum_{p=1}^{2p \leq N} \mu_{2p}\varepsilon^{2p} + O(\varepsilon^{N+1}) \tag{X.104}$$

and, when $\lambda_0^n = 1$, $n \geq 5$, and n is odd,

$$\begin{aligned}\rho(\theta, \varepsilon) &= \varepsilon + \varepsilon^{n-3}(g_{10}e^{in\theta} + \bar{g}_{10}\bar{e}^{in\theta}) + \varepsilon^{n-2}(g_{20}e^{2in\theta} + \bar{g}_{20}e^{-2in\theta}) \\ &\quad + \varepsilon^{n-1}(g_{30}e^{3in\theta} + \bar{g}_{30}e^{-3in\theta} + g_{31}e^{in\theta} + \bar{g}_{31}e^{-in\theta}) + O(\varepsilon^n) \\ &= \varepsilon + \sum_{k=1}^{k \leq N+4-n} \varepsilon^{n-4+k} \sum_{q=0}^{k-2q>0} [g_{kq} \exp n(k - 2q)i\theta \\ &\quad + \bar{g}_{kq} \exp(-n(k - 2q)i\theta)] + O(\varepsilon^{N+1}).\end{aligned} \tag{X.105}$$

When $n = 2\nu$ is even, we have

$$\begin{aligned}\rho(\theta, \varepsilon) &= \varepsilon + \varepsilon^{2\nu-3}(g_{00}e^{2\nu i\theta} + \bar{g}_{00}e^{-2i\nu\theta}) \\ &\quad + \varepsilon^{2\nu-1}(g_{11}e^{2\nu i\theta} + \bar{g}_{11}e^{-2i\nu\theta} + g_{10}e^{4\nu i\theta} + \bar{g}_{10}e^{-4i\nu\theta}) \\ &\quad + \varepsilon^{2\nu+1}(g_{22}e^{2i\nu\theta} + \bar{g}_{22}e^{-2i\nu\theta} + g_{21}e^{4i\nu\theta} + \bar{g}_{21}e^{-4i\nu\theta} \\ &\quad + g_{20}e^{6i\nu\theta} + \bar{g}_{20}e^{-6i\nu\theta}) + O(\varepsilon^{2\nu+3}) \\ &= \varepsilon + \sum_{k=0}^{2(k+\nu)\leq N+3} \varepsilon^{2\nu-3+2k} \sum_{q=0}^{k} (g_{kq} \exp 2\nu(k+1-q)i\theta \\ &\quad + \bar{g}_{kq} \exp(-2\nu(k+1-q)i\theta)) + O(\varepsilon^{N+1}). \qquad \text{(X.106)}\end{aligned}$$

The verification of the forms (X.105) and (X.106) is left as an exercise for the reader.

X.10 Trajectories on the Torus When $n \geq 5$

The procedure we use to find the trajectories on the torus is exactly the one used in §X.8 to study the case in which $n = 5$.

We have first to express $\rho(\theta, \varepsilon)$ in (X.56) with the explicit asymptotic expressions (X.105) and (X.106). We find that when n is odd

$$\begin{aligned}\frac{1}{\varepsilon^2}\frac{d}{dt} &= \Omega_0 + \bar{\psi}_2\varepsilon^2 + \bar{\psi}_4 e^4 + \cdots + \bar{\psi}_{2\nu}\varepsilon^{2\nu} + \cdots + \varepsilon^{n-4}\psi^*_{n-4}(\theta) \\ &\quad + \varepsilon^{n-3}(\bar{\psi}_{n-3} + \psi^*_{n-3}(\theta)) + \cdots + O(\varepsilon^{N-2}), \qquad \text{(X.107)}\end{aligned}$$

where $\Omega_0 = \mu_2\hat{\omega}_0 + \beta_{10}$ is assumed to be nonvanishing, and

$$\begin{aligned}&\psi_l \overset{\text{def}}{=} \bar{\psi}_l + \psi^*_l(\theta), \qquad \bar{\psi}^*_l \overset{\text{def}}{=} 0 \\ &\bar{\psi}_{2l+1} = 0 \\ &\psi^*_l(\theta) = 0 \quad \text{for } l < n-4 \\ &\psi^*_{n-5+l}(\theta) = \sum_{q=0}^{l-2q>0} [\theta_{lq}e^{n(l-2q)i\theta} + \bar{\theta}_{lq}e^{-n(l-2q)i\theta}]\end{aligned}$$

and θ_{lq} are all constants. (For example, $\psi^*_{n-4}(\theta) = \theta_{10}e^{in\theta} + \bar{\theta}_{10}e^{-in\theta}$ where $\theta_{10} = 2\beta_{10}g_{10} + \beta_{010}$.)

When $n = 2\nu$ is even

$$\begin{aligned}\frac{1}{\varepsilon^2}\frac{d\theta}{dt} &= \Omega_0 + \bar{\psi}_2\varepsilon^2 + \bar{\psi}_4\varepsilon^4 + \cdots \\ &\quad + \varepsilon^{2\nu-4}(\bar{\psi}_{2\nu-4} + \psi^*_{2\nu-4}\theta)) + \cdots + O(\varepsilon^{N-2}), \qquad \text{(X.108)}\end{aligned}$$

where $\psi_l = \bar{\psi}_l + \psi_l^*(\theta)$, as before, and

$$\psi_{2l+1}(\theta) = 0 \quad \text{for all } l > 0,$$
$$\psi_{2l}(\theta) = 0 \quad \text{for } 2l < 2\nu - 4$$

and

$$\psi^*_{2\nu-4+2l}(\theta) = \sum_{q=0}^{l} [\theta_{lq} \exp 2\nu(l+1-q)i\theta + \bar{\theta}_{lq} \exp(-2\nu(l+1-q)i\theta)].$$

To solve (X.107) and (X.108) we proceed as in §X.8 and introduce

$$\tilde{\theta} = \theta + \varepsilon^{n-4}h_{n-4}(\theta) + \varepsilon^{n-3}h_{n-3}(\theta) + \cdots + \varepsilon^{N-1}h_{N-1}(\theta), \quad \text{(X.109)}$$

where $h_l(\theta)$ is a to-be-determined function satisfying

$$h_l(\theta) = h_l\left(\theta + \frac{2\pi}{n}\right), \qquad \bar{\bar{h}}_l(\theta) = 0. \quad \text{(X.110)}$$

It follows that

$$\frac{1}{\varepsilon^2}\frac{d\tilde{\theta}}{dt} = \{1 + \varepsilon^{n-4}h'_{n-4}(\theta) + \varepsilon^{n-3}h'_{n-3}(\theta) + \cdots + O(\varepsilon^N)\}\frac{1}{\varepsilon^2}\frac{d\theta}{dt} \quad \text{(X.111)}$$

where $d\theta/dt$ is given by (X.107) when n is odd and by (X.108) when $n = 2\nu$ is even. Let $C(\varepsilon^2)$ be all of the mean terms in (X.107) and (X.108). Then, in either case,

$$\frac{1}{\varepsilon^2}\frac{d\theta}{dt} = C(\varepsilon^2) + \varepsilon^{n-4}\Phi^*(\theta, \varepsilon) + O(\varepsilon^N) \quad \text{(X.112)}$$

where $\bar{\bar{\Phi}}^*(\theta, \varepsilon) = 0$ and $C(0) = \Omega_0$. Combining (X.111) and (X.112) we can generate an ordered sequence of equations for $h_l(\theta)$ satisfying (X.110) by identifying the independent coefficients of ε in

$$\begin{aligned} C(\varepsilon^2)\{h'_{n-4}(\theta) + \varepsilon h'_{n-3}(\theta) + \cdots\} &+ \Phi^*(\theta, \varepsilon) \\ &+ \varepsilon^{n-4}\Phi^*(\theta, \varepsilon)\{h'_{n-4}(\theta) + \varepsilon h'_{n-3}(\theta) + \cdots\} \\ &= \varepsilon^{n-4}\overline{\overline{\Phi^*(\theta, \varepsilon)\{h'_{n-4}(\theta) + \varepsilon h'_{n-3}(\theta) + \cdots\}}} \\ &= \varepsilon^{n-4}\{\overline{\overline{\psi_{n-4}(\theta)h'_{n-4}}}(\theta) + \cdots\}. \end{aligned} \quad \text{(X.113)}$$

The $h_l(\theta)$ are given by

$$h_l = 0 \quad \text{for } l \le n - 5$$

$$h_{n-5+p}(\theta) = \sum_{l=0}^{p-2l>0} [\nu_{pl}e^{in\theta(p-2l)} + \bar{\nu}_{pl}e^{-in\theta(p-2l)}], \qquad p \ge 1 \quad \text{(X.114)}_1$$

when n is odd, and

$$\begin{aligned} h_l(\theta) &= 0 \quad \text{for } l \le 2v - 5 \\ h_{2l+1}(\theta) &= 0 \\ h_{2v-4+2p} &= \sum_{l=0}^{p} [v_{pl} \exp i2v(p + 1 - l)\theta \\ &\quad + \bar{v}_{pl} \exp(-i2v(p + 1 - l)\theta)], \quad p \ge 0 \end{aligned} \tag{X.114$_2$}$$

when $n = 2v$ is even.

Using (X.113), we may reduce (X.111) to

$$\begin{aligned} \frac{1}{\varepsilon^2}\frac{d\tilde{\theta}}{dt} &= C(\varepsilon^2) + \varepsilon^{2n-8}\overline{\overline{\psi^*_{n-4}h'_{n-4}}} + \cdots + O(\varepsilon^{N-2}) \\ &= \Omega(\varepsilon^2) + O(\varepsilon^{N-2}) \end{aligned} \tag{X.115}$$

and, as a consequence of (X.115), (X.109) may be written as

$$\varepsilon^2\Omega(\varepsilon^2)t = \theta + \varepsilon^{n-4}h_{n-4}(\theta) + \varepsilon^{n-3}h_{n-3}(\theta) + \cdots + \chi(t, \varepsilon)$$

where $|\dot{\chi}(t, \varepsilon)| = O(\varepsilon^N)$ and the $h_l(\theta)$ are given by (X.114) and are such that

$$h_l(\theta) = h_l\left(\theta + \frac{2\pi}{n}\right), \qquad \bar{\bar{h}}_l(\theta) = 0.$$

X.11 Asymptotically Quasi-Periodic Solutions

We summarize the results given in this chapter up to now. The solution is decomposed into a biorthogonal sum

$$\mathbf{u}(t) = Z(t)\boldsymbol{\zeta}(\mu, t) + \bar{Z}(t)\bar{\boldsymbol{\zeta}}(\mu, t) + \mathbf{W}(t), \tag{X.116}$$

where $\boldsymbol{\zeta}(\mu, t)$ is the eigenfunction belonging to eigenvalue $\sigma(\mu)$ of largest real part of the operator $-d/dt + \mathbf{f}_u(t, \mu|\cdot)$,

$$\begin{aligned} Z(t) &= \langle \mathbf{u}(t), \boldsymbol{\zeta}^*(t)\rangle \\ &= \rho(t) \exp i[\omega_0 t + \theta(t)] \\ &\quad + \sum_{p+q=2}^{N} \gamma'_{pq}(t, \mu)[\rho(t)]^{p+q} \exp i(p - q)[\omega_0 t + \theta(t)] + O(\varepsilon^{N+1}) \end{aligned} \tag{X.117}$$

and

$$\mathbf{W}(t) = \sum_{p+q=2}^{N} \boldsymbol{\Gamma}'_{qp}(t, \mu)[\rho(t)]^{p+q} \exp(p - q)i[\omega_0 t + \theta(t)] + O(\varepsilon^{N+1}), \tag{X.118}$$

where $\langle \mathbf{W}(t), \zeta^* \rangle = 0$, the truncation number N is unrestricted, and $\rho(t)$, $\theta(t)$, and μ are parameterized by

$$\varepsilon = \bar{\bar{\rho}} = \frac{1}{2\pi} \int_0^{2\pi} \rho(\theta)\, d\theta, \tag{X.119}$$

the mean radius of the torus. In all cases $\mu_{2l+1} = 0$,

$$\mu = \mu_2 \varepsilon^2 + \mu_4 \varepsilon^4 + \mu_6 \varepsilon^6 + \cdots + O(\varepsilon^{N+1}), \tag{X.120}$$

and

$$\rho(t) = \varepsilon + \varepsilon^{n-3} \rho_{n-3}(\theta) + \varepsilon^{n-2} \rho_{n-2}(\theta) + \cdots + O(\varepsilon^{N+1}), \tag{X.121}$$

where the number n is the one for which $\lambda_0^n = 1$, the $\rho_l(\theta)$ are defined by (X.105) when n is odd and (X.106) when $n = 2\nu$ is even, and $\bar{\bar{\rho}}_k = 0$ when $k \geq 1$. For $\theta(t)$ we have the relation

$$\begin{aligned} &\theta(t) + \varepsilon^{n-4} h_{n-4}(\theta(t)) + \varepsilon^{n-3} h_{n-3}(\theta(t)) + \cdots \\ &\quad = \varepsilon^2 \Omega(\varepsilon^2) t + \chi(t, \varepsilon), \qquad |\dot{\chi}| = O(\varepsilon^N), \end{aligned} \tag{X.122}$$

where the $h_l(\theta)$ are given by (X.114) and satisfy $h_l(\theta + (2\pi/n)) = h_l(\theta)$ and $\bar{\bar{h}}_l = 0$.

The formulas given in the previous paragraph hold when the ratio $\omega_0/(2\pi/T)$ of frequencies at criticality is irrational. We may obtain the results for the irrational case by letting $n \to \infty$, or more simply, merely by setting all terms involving n to zero.

Since the approximate solution through terms or order ε^N (N unrestricted) is a composition of T-periodic functions ($\zeta(\mu, t)$, $\gamma'_{pq}(t, \mu)$, $\mathbf{\Gamma}'_{pq}(t, \mu)$) and polynomials of harmonics of $e^{i\tau(t)}$, $\tau(t) = \omega_0 t + \theta(t)$, the solution is in the form

$$\mathbf{u}(t) \approx \mathcal{U}(t, \tau(t))$$

with $\tau(t) = \omega_0 t + \theta(t)$. The function $\mathcal{U}(\cdot, \cdot)$ is T-periodic in its first argument and 2π-periodic in its second argument. In fact, it is not hard to show that $\tau(t) \approx F(t, \omega_0 t + \varepsilon^2 \Omega(\varepsilon^2) t)$, $F(t + T, t') = F(t, t')$, $F(t, t' + 2\pi) = 2\pi + F(t, t')$, where F is the function solving the following functional equation in $\omega_0 t + \theta$:

$$\begin{aligned} &\omega_0 t + \theta + \varepsilon^{n-4} H_{n-4}(t, \omega_0 t + \theta) + \varepsilon^{n-3} H_{n-3}(t, \omega_0 t + \theta) + \cdots \\ &\quad = (\omega_0 + \varepsilon^2 \Omega(\varepsilon^2)) t, \end{aligned}$$

and $H_{n-4}(t, \omega_0 t + \theta) = H_{n-4}(t + T, \omega_0 t + \theta) = h_{n-4}(\theta)$ where $h_{n-4}(\theta)$ is a polynomial in exponentials and

$$\begin{aligned} \exp ikn\theta &= \exp ik[n(\omega_0 t + \theta) - n\omega_0 t] \\ &= \exp\left(-\frac{2\pi imkt}{T}\right) \exp ik[n(\omega_0 t + \theta)] \end{aligned}$$

because $\omega_0 = 2\pi m/nT$. So

$$\mathbf{u}(t) \approx \mathcal{U}(t, \tau(t)) \approx \mathbf{V}(t, [\omega_0 + \varepsilon^2\Omega(\varepsilon^2)]t),$$

where $\mathbf{V}$, like $\mathcal{U}$, is T-periodic in its first argument and 2π-periodic in its second argument. We have therefore shown that each and every approximation (every N) of the solution is a doubly periodic function, quasi-periodic when $[\omega_0 + \varepsilon^2\Omega(\varepsilon^2)]T/2\pi$ is irrational.

X.12 Stability of the Bifurcated Torus

It is necessary to draw attention to the fact that in the present case the bifurcating object is not a unique trajectory, but is a one-parameter family of trajectories lying on a torus in the phase space. Our understanding of stability here is the attracting or repelling property of the torus itself, instead of the stability of a single trajectory on it.

The decomposition (X.2) shows that the trajectories are contracted exponentially in the $\mathbf{W}$ directions of the phase space because the Floquet exponents corresponding to these directions stay on the left-hand side of the complex plane. This suggests that the two-dimensional projection $Z\zeta + \overline{Z\zeta}$ or, equivalently, the image of this projection in $\mathbb{R}^2$ with coordinates (ρ, θ) defined by (X.38), controls the stability of the whole solution. This suggestion is a provable fact which can be established as a consequence of the center manifold theorem (see O. Lanford III, Bifurcation of periodic solutions into invariant tori: the work of Ruelle and Takens, in *Nonlinear Problems in the Physical Sciences and Biology*, Lecture Notes in Mathematics No. 322, (New York–Heidelberg–Berlin: Springer-Verlag, 1973), pp. 159–192 or G. Iooss, *Bifurcation of Maps and Applications*, op. cit.).

So our differential equation may be reduced to the two given by (X.56). These two equations are asymptotically valid through terms $O(\varepsilon^N)$, N unrestricted, and they are satisfied by the flows on the bifurcated torus

$$\rho(t) = \varepsilon R(\theta(t), \varepsilon),$$

where R may be computed up to terms of $O(\varepsilon^N)$ as in §§X.5–10:

$$R(\theta, \varepsilon) = 1 + \varepsilon R_1(\theta) + \cdots.$$

To study stability, we perturb the torus and set

$$\rho = \varepsilon R(\theta(t), \varepsilon) + \rho', \tag{X.123}$$

where $\theta(t) \in [0, 2\pi)$ is any one of the solutions of (X.56) on the torus. Combining (X.123) with the ρ equation (X.56), we find that ρ' satisfies

$$\dot{\rho}' = \varepsilon^2(\mu_2\hat{\xi}_0 + 3\alpha_{10})\rho' + O(\varepsilon^3)\rho' + O(|\rho'|^2). \tag{X.124}$$

We recall that $\mu_2\hat{\xi}_0 + 3\alpha_{10} = -2\mu_2\hat{\xi}_0$, and $\hat{\xi}_0 > 0$ by virtue of the assumption of strict loss of stability of $\mathbf{u} = 0$. It then follows that $|\rho'(t)| \to 0$

as $t \to \infty$ if $\mu_2 > 0$ and $|\rho(0)|$ is small enough; that is, we get stability if the torus bifurcates supercritically. And if $\mu_2 < 0$ the torus is unstable. Small perturbations of the torus are attracted to the supercritical torus and are repelled by the subcritical torus.

X.13 Subharmonic Solutions on the Torus

To understand what happens to trajectories on and near the supercritical torus it is necessary to consider the properties of subharmonic solutions on the torus which arise as a result of frequency locking. A brief discussion of this is given in §X.15. For now, it will suffice to develop the properties of subharmonic solutions *on the torus* which bifurcates at criticality when the Floquet exponent is a rational point.

Assume that $\omega_0 = 2\pi m/nT$, $n \geq 3$. If x is a steady solution of (X.35), $y(t) = e^{i\omega_0 t}x$ is nT-periodic and $Z(t) = y(t) - \gamma(t, \mu, Z(t), \bar{Z}(t))$, and $\mathbf{u}(t) = Z(t)\zeta(t) + \bar{Z}(t)\bar{\zeta}(t) - \mathbf{\Gamma}(t, \mu, Z(t), \bar{Z}(t))$ are the compositions of T-periodic and nT-periodic functions. So we get an approximation to subharmonic solutions of $O(|x|^{N+1})$ from steady solutions of (X.35).

Consider the cases $n = 3$ and $n = 4$ of strong resonance:

$$n = 3: \qquad \dot{x} = \mu\hat{\sigma}x + x|x|^2 a_1 + \bar{x}^2 a_{0,-1} + O(|x|^4)$$

$$n = 4: \qquad \dot{x} = \mu\hat{\sigma}x + x|x|^2 a_1 + \bar{x}^3 a_{0,-1} + O(|x|^5).$$

We find steady solutions x in the form given by (IX.68) and (IX.80). Following the *ansatz* used in Chapter IX, we define an amplitude δ (formerly ε which here is defined as the mean radius of $\rho(\theta)$) and set $x = \delta e^{i\phi(\delta)}$ and $\mu = \mu^{(1)}\delta + \mu^{(2)}\delta^2 + \mu^{(3)}\delta^3 + \cdots$. When $n = 3$, we find the leading balance

$$\mu^{(1)}\hat{\sigma}_0 e^{i\phi_0} + e^{-2i\phi_0}a_{0,-1} = 0 \tag{X.125}$$

corresponding to (IX.68). One $3T$-periodic solution $\mathbf{u}$ of (X.1) bifurcates on both sides of criticality and both solutions are unstable when δ is small (§IX.14).

And when $n = 4$ we get $\mu^{(1)} = 0$ and

$$(\mu^{(2)}\hat{\sigma}_0 + a_1)e^{i\phi_0} + e^{-3i\phi_0}a_{0,-1} = 0 \tag{X.126}$$

corresponding to (IX.80). We find that two $4T$-periodic solutions $\mathbf{u}$ of (X.1) bifurcate, provided that a certain inequality implied by (X.126) (see (IX.83)) is satisfied.

When $n \geq 5$ we enter into the case of *weak resonance* and find that subharmonic solutions are possible only when exceptional conditions hold. At order δ we find that $\mu^{(1)} = 0$ and, at order δ^2,

$$\mu^{(2)}\hat{\sigma}_0 + a_1 = 0. \tag{X.127}$$

It is not possible to solve (X.127) for a real-valued $\mu^{(2)}$ unless $a_1/\hat{\sigma}_0$ is real-valued. This is the first exceptional condition; it is the same as (IX.101) and it holds for all $n \geq 5$. When $n = 5$ we have

$$\dot{x} = \mu\hat{\sigma}x + x|x|^2 a_1 + \bar{x}^4 a_{0,-1} + O(|x|^5). \tag{X.128}$$

At order δ^3 we get

$$\mu^{(3)}\hat{\sigma}_0 + e^{-5i\phi_0}a_{0,-1} = 0. \tag{X.129}$$

And when $n = 6$,

$$\dot{x} = \mu\hat{\sigma}x + x|x|^2 a_1 + x|x|^4 a_2 + \bar{x}^5 a_{0,-1} + O(|x|^7), \tag{X.130}$$

and $\mu^{(1)} = \mu^{(3)} = \mu^{(2n+1)} = 0$, $\mu^{(2)}\hat{\sigma}_0 + a_1 = 0$, and

$$\mu^{(4)}\hat{\sigma}_0 + a_2 + e^{-6i\phi_0}a_{0,-1} = 0. \tag{X.131}$$

When $n > 6$, we get $\mu^{(1)} = \mu^{(3)} = 0$ and, besides (X.127), we have a second exceptional condition arising from the equation

$$\mu^{(4)}\hat{\sigma}_0 + a_2 = 0. \tag{X.132}$$

Supposing now that both exceptional conditions are satisfied; then we get

$$\mu^{(5)}\hat{\sigma}_0 + e^{-7i\phi_0}a_{0,-1} = 0$$

when $n = 7$, and when $n = 8$, $\mu^{(5)} = 0$ and

$$\mu^{(6)}\hat{\sigma}_0 + a_2 + e^{8i\phi_0}a_{0,-1} = 0.$$

In deriving the equations for $n \geq 6$ we have assumed for simplicity that $\hat{\sigma}, a_1, a_{0,-1}$ are independent of μ. The analysis shows that the results for subharmonic bifurcation require exceptional conditions; a new condition is added at each odd value of n, starting with $n = 5$. When $n \geq 5$ is odd, the computation of bifurcation is like that given when $n = 3$ with the following differences. Since $\mu^{(1)} = 0$, $\mu^{(2)} \neq 0$ and $\mu^{(n-2)} \neq 0$ when $n \geq 5$ is odd, the bifurcation is one-sided but $\mu(\delta)$ is not even, so that there are two solutions with the same μ but different amplitudes δ. (It is perhaps necessary here to caution the reader against confusing the amplitude δ with the amplitude ε, used earlier.)

When $n \geq 6$ is even, the computation of bifurcation is like that given when $n = 4$ and two nT-periodic solutions bifurcate when a certain additional inequality which guarantees solvability is satisfied.

Now we shall show that the subharmonic solutions which bifurcate when $n \geq 5$ lie on the torus. We first note that we may always compute the cross section $\rho(\theta)$ unless conditions even more exceptional than the ones required for subharmonic bifurcation with $n \geq 5$ are satisfied. In fact, $\mu^{(2)} \neq 0$ is sufficient for the existence of $\rho(\theta)$. For subharmonic bifurcation we must have $\dot{\rho} = \dot{\theta} = 0$. Under these conditions the equation $\dot{\rho} = \rho'(\theta)\dot{\theta}$ is satisfied identically and the subharmonic solutions lie on the torus

$$\rho(\theta, \varepsilon) = \sum_{p=1}^{N} \rho_p(\theta)\varepsilon^p + O(\varepsilon^{N+1}),$$

where ρ_p satisfies (X.105) or (X.106). Piercing points of the periodic solution on the closed curve $\rho(\theta, \varepsilon)$ are determined by roots for which $\dot{\theta} = 0$; that is for roots θ such that (X.107) or (X.108) vanishes. Since there are no disposable parameters left, the equation $\dot{\theta} = 0$ determines $2n$ piercing points $\theta(\varepsilon)$ on the closed curve $\rho(\theta, \varepsilon)$.

It is useful to show how this calculation proceeds at the lowest significant order when $n = 5$. We first note that

$$\rho(\theta) = \varepsilon + \varepsilon^2\rho_2(\theta), \qquad \bar{\bar{\rho}}_2 = 0, \qquad \mu_2\hat{\xi}_0 + \alpha_{10} = 0, \tag{X.133}$$

where $\rho_2(\theta)$ is given by $(X.67)_2$ and (X.68). Now $\dot{\theta} = 0$ implies that

$$\Omega_0 = \mu_2\hat{\omega}_0 + \beta_{10} = 0. \tag{X.134}$$

Then

$$\begin{aligned} \rho_2(\theta) &= -\frac{\alpha_{010}}{2\alpha_{10}} e^{5i\theta} - \frac{\bar{\alpha}_{010}}{2\alpha_{10}} e^{-5i\theta} \\ &= \frac{|\alpha_{010}|}{\mu_2\hat{\xi}_0} \cos(5\theta + \arg \alpha_{010}). \end{aligned} \tag{X.135}$$

The first approximation $\theta(0) = \theta_0$ of $\theta(\varepsilon)$ is obtained from $\dot{\theta} = 0$ by requiring that $\dot{\theta} = 0$ through $O(\varepsilon^3)$:

$$\Theta_1(\theta_0) = 2\beta_{10}\rho_2(\theta_0) + \beta_{010}e^{5i\theta_0} + \bar{\beta}_{010}e^{-5i\theta_0} = 0. \tag{X.136}$$

We now show that (X.136) determines the first approximation to the ten piercing points $\theta(\varepsilon)$ on the closed curve $\rho(\theta, \varepsilon)$. We note first that (X.57) says that $\alpha_{0k} + i\beta_{0k} = 0$; hence, $\beta_{010} = i\alpha_{010}$ and

$$\begin{aligned} \Theta_1(\theta) &= \frac{1}{\hat{\xi}_0}\left(i\hat{\sigma}_0\alpha_{010}e^{5i\theta} - \overline{i\hat{\sigma}_0\alpha_{010}e^{-5i\theta}}\right) \\ &= -\frac{2}{\hat{\xi}_0}|\hat{\sigma}_0\alpha_{010}| \sin(5\theta + \arg \alpha_{010} + \arg \hat{\sigma}_0). \end{aligned} \tag{X.137}$$

Hence

$$5\theta_0 + \arg \alpha_{010} + \arg \hat{\sigma}_0 = k\pi, \qquad k = 0, 1, 2, \ldots, 9 \tag{X.138}$$

and there are ten values of θ_0. Returning now to (X.135) with (X.138) we find that there are two values of

$$\begin{aligned} \rho_2(\theta_0) &= \frac{|\alpha_{010}|}{\mu_2\hat{\xi}_0} \cos(k\pi - \arg \hat{\sigma}_0) \\ &= \frac{|\alpha_{010}|}{\mu_2\hat{\xi}_0} \times \begin{cases} \cos(\arg \hat{\sigma}_0), & \text{for } k = 0, 2, 4, 6, 8 \\ -\cos(\arg \hat{\sigma}_0), & \text{for } k = 1, 3, 5, 7, 9. \end{cases} \end{aligned} \tag{X.139}$$

Since $\hat{\xi}_0 = \xi_\mu > 0, \hat{\omega}_0 = \omega_\mu, \hat{\sigma}_0 = \sigma_\mu$ at $\mu = 0$ we have $-\pi/2 < \arg \hat{\sigma}_0 < \pi/2$, and since $\mu_2 > 0$ (we are considering the stable, that is, supercritical torus), $\rho_2(\theta_0) > 0$ when k is even and $\rho_2(\theta_0) < 0$ when k is odd. The largest and smallest values of $\rho_2(\theta)$ are attained when $\hat{\omega}_0 = \arg \hat{\sigma}_0 = 0$. So the position of the piercing points on the closed curve $\rho(\theta) \approx \varepsilon + \varepsilon^2\rho_2(\theta)$, $\varepsilon > 0$ are rotated through an angle $\arg \hat{\sigma}_0$ from troughs and crests.

Finally we note that the ten piercing points on the closed curve in the cross section of the torus are exactly the same as the ones determined by (X.129). The relation between the amplitude $\delta > 0$ used in (X.129) and ε can be determined at lowest order from the relation

$$x = \delta e^{i\phi_0} = [\varepsilon + \varepsilon^2\rho_2(\theta_0)]e^{i\theta_0}.$$

It follows that there are two values δ, δ_1, and δ_2 of

$$\delta = \varepsilon + \varepsilon^2\rho_2(\theta_0), \qquad \phi_0 = \theta_0$$

corresponding to the two values of $\rho_2(\theta_0)$ and

$$\begin{aligned}\mu(\delta(\varepsilon)) &= \mu^{(2)}\delta^2 + \mu^{(3)}\delta^3 + \cdots \\ &= \mu^{(2)}\varepsilon^2 + [2\mu^{(2)}\rho_2(\theta_0) + \mu^{(3)}]\varepsilon^3 + O(\varepsilon^4).\end{aligned}$$

The odd powers ε in the expansion $\mu(\delta(\varepsilon)) = \mu(\varepsilon) = \mu_2\varepsilon^2 + \mu_4\varepsilon^4 + \cdots$, vanish. For example, by identification we find that $\mu^{(2)} = \mu_2$ and (X.129) shows the $\mu^{(3)} = -2\mu^{(2)}\rho_2(\theta_0)$.

X.14 Stability of Subharmonic Solutions on the Torus

The supercritical torus ($\mu_2 > 0$) is stable when ε is small. But one of the two subharmonic solutions on the torus is unstable. To study the stability of these solutions we set $\rho = \rho(\theta, \varepsilon) = \varepsilon + \varepsilon^2\rho_2(\theta) + O(\varepsilon^3)$, $\mu = \mu_2\varepsilon^2 + O(\varepsilon^4)$ in (X.78) and find that

$$\dot{\theta} = \varepsilon^3\Theta_1(\theta) + O(\varepsilon^4)$$

where $\Theta_1(\theta)$ is given by (X.137). Now we perturb θ_0, $\theta = \theta_0 + \theta'$ and linearize, using (X.136), to get

$$\dot{\theta}' = [\varepsilon^3\Theta_1'(\theta_0) + O(\varepsilon^4)]\theta' + O(|\theta'|^2), \tag{X.140}$$

where θ_0 is given by (X.138), $\beta_{10}/\alpha_{10} = \hat{\omega}_0/\hat{\xi}_0$, $\beta_{010} = i\alpha_{010}$ and

$$2\beta_{10}\rho_2'(\theta_0) = 5i\frac{\beta_{10}}{\alpha_{10}}[-\alpha_{010}e^{5i\theta_0} + \overline{\alpha_{010}}e^{-5i\theta_0}].$$

After some easy manipulations, using the relations just cited, we reduce (X.140) to

$$\dot{\theta}' = -\left[\frac{5\varepsilon^3}{\xi_0}|\hat{\sigma}_0||\alpha_{010}|\cos k\pi\right] + [O(\varepsilon^4)]\theta' + O(|\theta'|^2).$$

So the $5T$-periodic solution with the 5 piercing points ($k = 0, 2, 4, 6, 8$) nearer to the crests is stable and the other $5T$-periodic solution with piercing points near troughs ($k = 1, 3, 5, 7, 9$) is unstable (see Figure X.2).

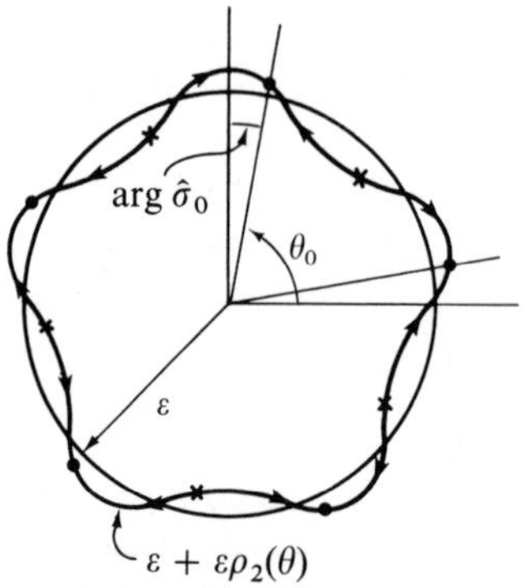

Figure X.2 Bifurcation and stability of $5T$-periodic solutions on the torus. There are two $5T$-periodic solutions, each with 5 piercing points in the cross section. The solution with positive values of $\rho_2(\theta_0)$ is unstable and the one with negative values is stable. If arg $\hat{\sigma}_0 = 0$ then $\hat{\omega}_0 = \omega_\mu(0) = 0$ and the Floquet multiplier $\lambda = e^{\sigma T} = e^{i\omega_0 T}e^{\mu\hat{\xi}_0 T}$ crosses the unit circle along a ray from the origin. In this case the stable solutions are the 5 points on the crests where $\rho'(\theta_0) = 0$ and the unstable solutions are the 5 points in the trough where $\rho'(\theta_0) = 0$

The stability results for subharmonic solutions with $n > 5$ are like the one just given. There are two distinct periodic solutions on the torus, each with n piercing points; half of these are unstable, the other half are stable, and the stable and unstable solutions separate each other. (Details for these stability calculations can be found in G. Iooss, *Bifurcation of Maps and Applications*, op. cit.)

X.15 Frequency Locking

Frequency locking may be said to occur in a dynamical system when oscillations with two independent frequencies influence one another in such a way as to produce synchronization of the two oscillations into a periodic oscillation with a common longer period (a subharmonic oscillation). This phenomenon is ubiquitous and very complicated.

The phenomenon of phase locking on the torus T^2 occurs when all the trajectories on the torus are captured by a periodic one as μ increases. To understand the phenomenon of capture it is useful to introduce the Poincaré map and the rotation number. The Poincaré map (first return map) is defined by a monotone function $f(\cdot)$:

$$\theta \mapsto f(\theta), \qquad 0 \le \theta < 2\pi$$

where θ and $f(\theta)$ are real numbers, f is such that

$$f(\theta + 2\pi) = f(\theta) + 2\pi,$$

and f maps the starting point of a trajectory on the curve $\rho = \varepsilon R(\theta, \varepsilon)$ on the torus into the intersection of the trajectory with this curve after time T, the curve being parameterized by θ. So we may suppose that the trajectory starts at the place $\theta = \theta_0$ on the closed curve $\rho = \varepsilon R(\theta, \varepsilon)$. The first return pierces the closed curve at $\theta = \theta_1$; that is $\theta_1 = f(\theta_0)$. The trajectory winds around the torus again and after an increase in time of T it hits the closed curve at $\theta = \theta_2 = f(\theta_1) = f^2(\theta_0)$, and so on. The angular increment between successive hits is given by $f(\theta)$. So we get the sequence $\theta_0, f(\theta_0) = \theta_1$, $f^2(\theta_0) = f(\theta_1) = \theta_2, \ldots, f^n(\theta_0) = f^{n-1}(\theta_1) = \cdots = f(\theta_{n-1}) = \theta_n$. Suppose that $\theta = \theta_0 + \omega t$. Then $f(\theta_0) = \theta_0 + \omega T = \theta_1$, $f^2(\theta_0) = \theta_0 + 2\omega T, \ldots$, $f^n(\theta_0) = \theta_0 + n\omega T$. We note that if $\omega = 2\pi m/nT$, then the trajectory on the torus will be nT-periodic.

We next introduce the rotation number $\hat{\rho}(f)$ of f:

$$\hat{\rho}(f) = \lim_{\nu \to \infty} \frac{1}{2\pi\nu} [f^\nu(\theta) - \theta]. \tag{X.141}$$

Poincaré, who first introduced this number, proved that this limit exists and is independent of θ. If the rotation number $\hat{\rho}(f)$ is an *irrational number r*, then it may be shown that the solutions on the torus are quasi-periodic and that a change of variable in θ leads to $f(\theta) = \theta + \omega T$, $\omega = 2\pi r/T$, $0 < r < 1$, which is just a rotation on the closed curve (Denjoy, Bohl). Since $f^\nu(\theta) - \theta = 2\pi\nu r$ we get $\hat{\rho}(f) = r$. In the irrational case each iteration of the map produces a new point on the curve $\rho = \varepsilon R(\theta, \varepsilon)$ and no point is ever repeated, so that piercing points of any trajectory are dense on the curve $\rho = \varepsilon R(\theta, \varepsilon)$ in the cross section of the torus. Hence any trajectory on the torus eventually fills up the entire torus.

For quasi-periodic solutions on a torus the rotation number $\hat{\rho} = r = \omega/(2\pi/T)$ is the ratio of frequencies. If the rotation $\hat{\rho}(f)$ is a *rational number*, $r = m/n$, then there is a θ_0 such that $f^n(\theta_0) = \theta_0$ modulo 2π, and the corresponding trajectory is nT-periodic. In this case there are, in general, two nT-periodic trajectories, one being attractive (stable) and the other repelling (unstable), as, for example, in Figure X.2. Trajectories near to the attracting torus will eventually be trapped by the stable trajectory on the torus.

The approximate doubly periodic solutions which are asymptotic to true solutions up to terms of order ε^N, N arbitrary, are of the form

$$\mathbf{u}(t) = \mathbf{V}(t, [\omega_0 + \varepsilon^2\Omega(\varepsilon^2)]t) \tag{X.142}$$

with $\mathbf{V}(t + T, t') = \mathbf{V}(t, t' + 2\pi) = \mathbf{V}(t, t')$ (see the remarks concluding Appendix X.3). This type of behavior corresponds to a rotation number

$$\hat{\rho}(\varepsilon) = \frac{[\omega_0 + \varepsilon^2\Omega(\varepsilon^2)]T}{2\pi} \tag{X.143}$$

which is a polynomial in ε. For this rotation number, most of the values of ε correspond to irrational values of $\hat{\rho}(\varepsilon)$ whenever $\Omega(\varepsilon^2) \not\equiv 0$.

It is necessary to distinguish between the rotation number $\hat{\rho}(\varepsilon)$ of (X.143) for the asymptotic representation of the flow on the torus and the true rotation number $\hat{\rho}(f_\varepsilon)$, where

$$f_\varepsilon(\theta) = \theta + \omega_0 + \varepsilon^2\Omega(\varepsilon^2) + \varepsilon^N h(\theta, \varepsilon),$$

where N is arbitrary and $h(\theta, \varepsilon)$ is not known. The qth iterate of the true map is

$$f_\varepsilon^q(\theta) = \theta + q(\omega_0 + \varepsilon^2\Omega(\varepsilon)) + \varepsilon^N h_q(\theta, \varepsilon). \tag{X.144}$$

The function $\hat{\rho}(\varepsilon)$ is analytic. Unlike the true rotation number, discussed below, it cannot have the steps required when the solution locks frequencies (see Figure X.3).

The true map f_ε need not be analytic in ε; even if the map f_ε is analytic, the rotation number $\hat{\rho}(f_\varepsilon)$ need not be smooth in ε, though Poincaré has shown that $\hat{\rho}(f_\varepsilon)$ is at least continuous in ε. In fact, the following argument might be interpreted as suggesting that the function $\hat{\rho}(f_\varepsilon)$ is not smooth but,

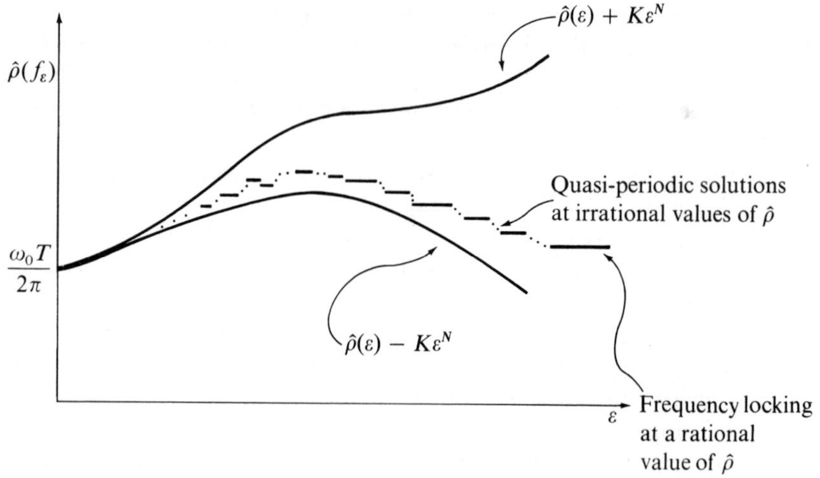

Figure X.3 Rotation number of the Poincaré map. The rotation number seems smooth as $\varepsilon \to 0$

instead, takes on constant values on an interval of ε at the rational points $\hat{\rho}(f_\varepsilon) = p/q$. This leads to a continuous curve containing steps as in Figure X.3. Suppose that $\hat{\rho} = p/q$ when $\varepsilon = \varepsilon_0$ and that θ_0 is a fixed point of order q of the map $\theta \mapsto f_\varepsilon(\theta)$ when $\varepsilon = \varepsilon_0$, that is,

$$f^q_{\varepsilon_0}(\theta_0) - \theta_0 = 0. \tag{X.145}$$

This fixed point corresponds to a periodic solution of period qT of our original problem and the ratio $0 < p/q < 1$ plays the role of m/n of Chapter IX. To prove that the rotation number $\hat{\rho}(f_\varepsilon)$ remains constant on an interval containing ε_0, it is enough to show that

$$g_\varepsilon(\theta) \overset{\text{def}}{=} f^q_\varepsilon(\theta) - \theta = 0 \tag{X.146}$$

holds for (ε, θ) close to $(\varepsilon_0, \theta_0)$. Here q is the integer multiple corresponding to the qT-periodic solution and p/q is determined by the continuity of $\hat{\rho}(f_\varepsilon)$. The implicit function theorem guarantees that (X.146) holds if (X.145) holds and

$$\frac{\partial g_\varepsilon}{\partial \theta} = \varepsilon_0^N \frac{\partial h_q}{\partial \theta} \neq 0 \tag{X.147}$$

when $(\varepsilon, \theta) = (\varepsilon_0, \theta_0)$. Then there is an interval of ε containing ε_0 for which there is a solution $\theta(\varepsilon)$ of (X.146) with rotation number $\hat{\rho}(f_\varepsilon) = p/q$. This leads to the flat segments, the steps, shown in Figure X.3. The proof of the implicit function theorem shows that a lower bound for the size of steps is of order ε_0^N.

We have shown that for every N the flow on the torus is at least approximated by a quasi-periodic flow. For all such, $h_q \equiv 0$ and the implicit function theorem does not give frequency locking. The function $\hat{\rho}(\varepsilon)$ is analytic for all these approximations.

The implication of the fact that the truncation number N is arbitrary is as follows. The norm of the difference of truncated approximation $\mathbf{u}^{(N)}(\varepsilon)$ and the true solution $\mathbf{u}(\varepsilon)$ is of the form

$$\|\mathbf{u}(\varepsilon) - \mathbf{u}^{(N)}(\varepsilon)\| = \varepsilon^N \delta_N(\varepsilon),$$

where in general a finite $N = N(\varepsilon)$ may give a smaller error $\varepsilon^N \delta_{N(\varepsilon)}(\varepsilon)$ than a larger N and the best $N = N(\varepsilon)$ is such that

$$\lim_{\varepsilon \to 0} N(\varepsilon) \to \infty.$$

So at the very least we may assert that the *lengths of the intervals on which* $\hat{\rho}(f_\varepsilon)$ *is constant must tend to zero faster than any power of* ε.

M. Herman has shown (Mesure de Lebesgue et nombre de rotation, in *Geometry and Topology*, Lecture Notes in Mathematics No. 597 (New York–Heidelberg–Berlin: Springer-Verlag, 1977), pp. 271–293) that if $\hat{\rho}(f_\varepsilon)$ is not identically constant then the set of points ε for which $\hat{\rho}(f_\varepsilon)$ is irrational has a positive measure. The set of ε's corresponding to frequency-locked solutions is of positive measure if there are points $(\varepsilon_0, \theta_0)$ satisfying (X.147), but intervals of ε around ε_0 are small when ε is small, so that it might be difficult to observe frequency-locked subharmonic solutions when ε is small. But for larger values of ε, asymptotically quasi-periodic solutions and frequency-locked subharmonic periodic solutions are expected and observed in applications.

Recent experiments suggest that the bifurcation of periodic solutions into invariant tori is common in fluid mechanics. The quasi-periodic solutions are detected in experiments by examination of a Fourier analysis of some time-dependent observable in the flow, for example, a velocity component. In the analysis of the spectrum of a quasi-periodic motion there are a large number of spikes corresponding to periodic components in the oscillation plus smaller-amplitude noise. If the motion is quasi-periodic with two frequencies it is possible to identify all the sharp spectral features as given by the sums and differences of harmonics of the two frequencies. The ratio of the two frequencies gives the rotation number just mentioned. In experiments this ratio appears to be a smooth function of μ near the point of bifurcation. For larger values of μ the solutions can lock into a subharmonic one in which the ratio of the two frequencies is constant and rational on intervals. (See Figure X.3.)

Appendix X.1 Computation of Asymptotically Quasi-Periodic Solutions Which Bifurcate at Rational Points of Higher Order ($n \geq 5$) by the Method of Power Series Using the Fredholm Alternative

The analysis of asymptotic solutions on the bifurcating torus can be determined from the autonomous Equations (X.35) by the method of power series used in Chapter VIII. The structure of the solutions which we compute is a composition of T-periodic functions and $y = e^{i\omega_0 t}x(s)$, where s is again a reduced time related to t by a mapping depending on the amplitude

$$\varepsilon^2\Omega(\varepsilon)t = s \tag{X.148}$$

which maps $2\pi/\varepsilon^2\Omega(\varepsilon)$ intervals of t into intervals of 2π in s.

It serves our purpose here to assume that the reduction to the autonomous equation (X.35) actually terminates at some N, or that $N = \infty$, but the right-hand side of (X.35) is analytic in x and $\bar{x}$ when μ is small. Given any one of an equivalent class of definitions of the amplitude ε of x we could justify the formal construction given below; that is, given the assumption of the analyticity of (X.35) in x and $\bar{x}$, we could use the implicit function theorem to prove that the series (X.154) in powers of ε converges to a unique solution of (X.35) when ε is small.

Let $a(\cdot) \in \mathbb{P}_{2\pi}$, $b(\cdot) \in \mathbb{P}_{2\pi}$. Then

$$[a, b]_{2\pi} \stackrel{\text{def}}{=} \frac{1}{2\pi} \int_0^{2\pi} a(s)\bar{b}(s)\, ds.$$

The amplitude of the bifurcating solution may be defined by

$$\varepsilon = [x, e^{is}]_{2\pi}. \tag{X.149}$$

We find $\mu(\varepsilon)$, $\Omega(\varepsilon)$, and $x(s, \varepsilon) \in \mathbb{P}_{2\pi}$ satisfying (X.35) in the form

$$\mu = \varepsilon^2 \tilde{\mu}(\varepsilon), \qquad \Omega = \Omega(\varepsilon), \qquad x = \varepsilon \tilde{\chi}(s, \varepsilon), \tag{X.150}$$

where

$$\begin{aligned} \Omega \frac{d\tilde{\chi}}{ds} = \tilde{\mu}\hat{\sigma}\tilde{\chi} &+ \sum_{q \geq 1} \tilde{\chi}|\tilde{\chi}|^{2q} a_q \varepsilon^{2q-2} \\ &+ \sum_{k>0} \sum_{q \geq 0} |\tilde{\chi}|^{2q} \{a_{q,k} \tilde{\chi}^{1+kn} \varepsilon^{2q+kn-2} \\ &+ a_{q,-k} \bar{\tilde{\chi}}^{kn-1} \varepsilon^{2q+kn-4}\}. \end{aligned} \tag{X.151}$$

$\tilde{\chi}(s, \varepsilon)$ satisfying (X.151) is translationally invariant to shifts in the origin of s and is rotationally invariant to rotations of $\tilde{\chi}(s, \varepsilon)$ through angles of $2\pi/n$; that is, if $\tilde{\chi}(s, \varepsilon)$ solves (X.151) so does $\tilde{\chi}(s + \phi, \varepsilon)$ for any ϕ and so does $\tilde{\chi}(s, \varepsilon)\, e^{2\pi i/n}$.

The previous analysis of (X.35) shows that

$$x = \rho e^{i\theta} = \varepsilon e^{is}\{1 + \varepsilon \chi_1(s) + \cdots\},$$

where $\chi_1(\cdot) \in \mathbb{P}_{2\pi}$. We therefore set

$$\tilde{\chi}(s, \varepsilon) = e^{is}\chi(s, \varepsilon),$$

where, from (X.149),

$$\frac{1}{2\pi} \int_0^{2\pi} \chi(s, \varepsilon)\, ds = 1. \tag{X.152}$$

In fact, we already know and are going to demonstrate again that $\chi(s, \varepsilon)$ is not only 2π-periodic but is $2\pi/n$-periodic. For the present we note that the solution $\chi(\cdot, \varepsilon) \in \mathbb{P}_{2\pi}$, $\mu(\varepsilon)$ and $\Omega(\varepsilon)$ of

$$\begin{aligned} (i\Omega - \tilde{\mu}\hat{\sigma})\chi + \Omega \frac{d\chi}{ds} = &\sum_{q \geq 1} \chi|\chi|^{2q} a_q \varepsilon^{2q-2} \\ &+ \sum_{k>0} \sum_{q \geq 0} |\chi|^{2q} \{a_{q,k} e^{ikns} \chi^{1+kn} \varepsilon^{2q+kn-2} \\ &+ a_{q,-k} e^{-ikns} \bar{\chi}^{kn-1} \varepsilon^{2q+kn-4}\} \end{aligned} \tag{X.153}$$

is analytic in ε when ε is small; it is unique and may be constructed as a power series:

$$\begin{aligned} \chi(s, \varepsilon) &= \chi_0(s) + \varepsilon\chi_1(s) + \varepsilon^2\chi_2(s) + \cdots \\ \tilde{\mu}(\varepsilon) &= \tilde{\mu}_0 + \varepsilon\tilde{\mu}_1 + \varepsilon^2\tilde{\mu}_2 + \cdots \\ \Omega(\varepsilon) &= \Omega_0 + \varepsilon\Omega_1 + \varepsilon^2\Omega_2 + \cdots, \end{aligned} \tag{X.154}$$

where the coefficients in the series are determined by the perturbation equations arising from identification of independent powers of ε in (X.152) and (X.153). To simplify the writing we have assumed that $\hat{\sigma}$, a_l, and a_{lk} are independent of μ. We find that

$$\frac{1}{2\pi}\int_0^{2\pi} \chi_0(s)\, ds = 1 \tag{X.155}$$

$$(i\Omega_0 - \tilde{\mu}_0\hat{\sigma})\chi_0 + \Omega_0\frac{d\chi_0}{ds} = |\chi_0|^2\chi_0 a_1. \tag{X.156}$$

Hence $\chi_0 = 1$ and

$$i\Omega_0 = \tilde{\mu}_0\hat{\sigma} + a_1, \tag{X.157}$$

where, by assumption, the loss of stability of $\mathbf{u} = 0$ is strict ($\text{Re}\, \hat{\sigma} \neq 0$) and weakly resonant subharmonic bifurcation does not occur ($\Omega_0 \neq 0$). Under these assumptions we may solve (X.157) for $\tilde{\mu}_0$ and Ω_0.

The coefficient of ε in (X.152) and (X.153) will vanish if

$$\frac{1}{2\pi}\int_0^{2\pi} \chi_1(s)\, ds = 0 \tag{X.158}$$

and

$$(i\Omega_1 - \tilde{\mu}_1\hat{\sigma}) + (i\Omega_0 - \tilde{\mu}_0\hat{\sigma})\chi_1 + \Omega_0\frac{d\chi_1}{ds} = (\bar{\chi}_1 + 2\chi_1)a_1 + g_1, \tag{X.159}$$

or, using (X.157),

$$\Omega_0\frac{d\chi_1}{ds} = (\chi_1 + \bar{\chi}_1)a_1 + g_1 - (i\Omega_1 - \tilde{\mu}_1\hat{\sigma}), \tag{X.160}$$

where

$$g_1 = a_{0,-1}e^{-5is} \quad \text{if } n = 5 \tag*{(X.161)$_1$}$$

or

$$g_1 = 0 \qquad \text{if } n > 5. \tag*{(X.161)$_2$}$$

The following remarks specify the procedure to be used in solving (X.160) and (X.161). The linear problem

$$\Omega_0\frac{dy}{ds} - (y + \bar{y})a_1 = \hat{g} \in \mathbb{P}_{2\pi}, \tag{X.162}$$

where $\hat{g}(s)$ has a zero mean value:

$$\frac{1}{2\pi}\int_0^{2\pi} \hat{g}(s)\,ds = 0, \tag{X.163}$$

has a unique 2π-periodic solution $y(s)$ of zero mean value. If $\hat{g}(s) \in \mathbb{P}_{2\pi/n}$, then $y(s) \in \mathbb{P}_{2\pi/n}$. To prove this we note that

$$\Omega_0 \frac{d(y + \bar{y})}{ds} - 2\,\mathrm{Re}\,(a_1)(y + \bar{y}) = 2\,\mathrm{Re}\,\hat{g}(s). \tag{X.164}$$

Since (X.164) has no 2π-periodic solutions $(y + \bar{y})(s)$ when $\hat{g}(s) = 0$, and the mean value of $\mathrm{Re}\,\hat{g}(s) = 0$, its solution, $y(s) + \bar{y}(s)$, must be unique also have a zero mean value. Then y solving (X.162) is uniquely determined and of zero mean value.

Returning now to (X.160) we may construct $\hat{g} = g_1 - (i\Omega_1 - \tilde{\mu}_1\hat{\sigma})$ with a zero mean value if and only if $\tilde{\mu}_1 = \Omega_1 = 0$. Then, if $n = 5$

$$\chi_1(s) = Ae^{5is} + Be^{-5is},$$

where

$$A = \frac{a_1\bar{a}_{0,-1}}{5i\Omega_0(5i\Omega_0 - a_1 - \bar{a}_1)}$$

and

$$B = \frac{-a_{0,-1}(5i\Omega_0 + \bar{a}_1)}{5i\Omega_0(5i\Omega_0 + a_1 + \bar{a}_1)}.$$

Higher-order solutions are expressible in terms of polynomials in $e^{\pm i5s}$.

The equations governing $\chi_l(s)$ for $l > 1$ are of the form (X.160). We get $\tilde{\mu}_n$ and Ω_n by choosing them so that the mean value of $\chi_l(s)$ vanishes. If $n > 5$ there are no inhomogeneous terms with zero mean values for $0 \le l \le n - 5$. For these values of l, $\chi_l(s) = 0$, the first nonzero $\chi_l(s)$ for $l > 0$ arises at order $l = n - 4$.

The construction of $\chi(s, \varepsilon)$ in this appendix has already shown that $\chi(s, \varepsilon)$ is $2\pi/n$-periodic in s. This reduced (from 2π) periodicity comes from the translational and rotational invariance of $\tilde{\chi}(s, \varepsilon)$ satisfying (X.151). Since the solution $\tilde{\chi}(\cdot, \varepsilon)$ is unique up to a translation in s and invariant under a $2\pi/n$ rotation of $\tilde{\chi}$, there is a ϕ such that

$$\tilde{\chi}(s + \phi, \varepsilon) = e^{i(s+\phi)}\chi(s + \phi, \varepsilon) = e^{is}e^{2\pi i/n}\chi(s, \varepsilon),$$

where

$$\chi(\cdot, \varepsilon) = 1 + \varepsilon\chi_1(\cdot) + \varepsilon^2\chi_2(\cdot) + \cdots.$$

So

$$e^{i(s+\phi)}\{1 + \varepsilon\chi_1(s + \phi) + \cdots\} = e^{i(s+(2\pi/n))}\{1 + \varepsilon\chi_1(s) + \cdots\}.$$

Hence $\phi = 2\pi/n$ and

$$\chi_l\left(s + \left(\frac{2\pi}{n}\right)\right) = \chi_l(s) \quad \text{for } l \geq 1.$$

Appendix X.2 Direct Computation of Asymptotically Quasi-Periodic Solutions Which Bifurcate at Irrational Points Using the Method of Two Times, Power Series, and the Fredholm Alternative

Now we shall solve (X.1) when the ratio of frequencies at criticality $\omega_0 T/2\pi$ is irrational. We seek a doubly periodic solution $\mathbf{u}(t, s, \varepsilon)$

$$\mathbf{u}(\cdot, \cdot, \varepsilon) \in \mathbb{P}_{T, 2\pi} \tag{X.165}$$

which is T-periodic in t and 2π-periodic in s and such that

$$s = \omega(\varepsilon)t, \qquad \omega(0) \stackrel{\text{def}}{=} \omega_0, \qquad \mathbf{u}(t, s, 0) = 0.$$

The amplitude ε of $\mathbf{u}$ is defined by the projection (X.180).

First we expand the solution

$$\begin{bmatrix} \mathbf{u}(t, s, \varepsilon) \\ \omega(\varepsilon) - \omega_0 \\ \mu(\varepsilon) \end{bmatrix} = \sum_{p=1}^{\infty} \frac{\varepsilon^p}{p!} \begin{bmatrix} \mathbf{u}_p(t, s) \\ \omega_p \\ \mu_p \end{bmatrix} \tag{X.166}$$

and identify independent powers of ε. To simplify the writing of the perturbation equations which arise from identification, we note that

$$\mu(\varepsilon) \text{ and } \omega(\varepsilon) \text{ are even functions.}$$

We can prove this again using the present method at a cost of a longer, but not more difficult analysis (see D. D. Joseph, Remarks about bifurcation and stability of quasi-periodic solutions which bifurcate from periodic solutions of the Navier–Stokes equations in *Nonlinear Problems in the Physical Sciences and Biology*, Lecture Notes in Mathematics No. 322 (New York–Heidelberg–Berlin: Springer-Verlag, 1973)). Noting now that the two times imply that

$$\frac{d\mathbf{u}}{dt} = \frac{\partial \mathbf{u}}{\partial t} + \omega \frac{\partial \mathbf{u}}{\partial s}$$

we identify the perturbation equations. These equations are nearly the same as (IX.50), (IX.51), and (IX.52), except that odd derivatives of μ and ω are zero, the perturbation of the product term must be taken into account, and an operator

$$\mathscr{L}_0 = -\omega_0 \frac{\partial}{\partial s} - \frac{\partial}{\partial t} + \mathbf{f}_u(t|\cdot) = -\omega_0 \frac{\partial}{\partial s} + J_0, \tag{X.167}$$

whose domain is doubly periodic functions of t and s (dom $\mathscr{L}_0 = \mathbb{P}_{T,2\pi}$) replaces $\mathbb{J}$. The operator J_0 is exactly as defined in §IX.4, except that we write $\partial/\partial t$ in recognition of the fact that when J_0 operates on functions in $\mathbb{P}_{T,2\pi}$ the second variable s is held constant. In particular

$$J_0 \zeta = i\omega_0 \zeta,$$

where $i\omega_0$ is a simple eigenvalue so that

$$\mathbf{z} = e^{is}\zeta \quad \text{and} \quad \bar{\mathbf{z}} \tag{X.168}$$

are the only eigenfunctions on the null space of $\mathscr{L}_0$; that is, $\mathscr{L}_0 \mathbf{z} = \mathscr{L}_0 \bar{\mathbf{z}} = 0$. To show this we express the null vectors of $\mathscr{L}_0$ as Fourier series and identify the coefficients of e^{iks}. Then, on taking account of the fact that $\omega_0 T/2\pi$ is irrational, we get

$$\pm i\omega_0 + \frac{2\pi i k}{T} \neq i\omega_0 k' \quad \text{for any } k \neq 0 \text{ and } k'.$$

The adjoint of $\mathscr{L}_0$ in $\mathbb{P}_{T,2\pi}$ is computed using the scalar product

$$[\![\cdot,\cdot]\!] \overset{\text{def}}{=} \frac{1}{2\pi}\int_0^{2\pi} [\cdot,\cdot]_T\, ds, \tag{X.169}$$

where $[\cdot,\cdot]_T$ is an integral as in Chapter IX. We have

$$[\![\mathscr{L}_0 \mathbf{a}, \mathbf{b}]\!] = [\![\mathbf{a}, \mathscr{L}_0^* \mathbf{b}]\!]$$

for all $\mathbf{a}$, $\mathbf{b}$ in $\mathbb{P}_{T,2\pi}$ and

$$\mathscr{L}_0^* = \omega_0 \frac{\partial}{\partial s} + \frac{\partial}{\partial t} + \mathbf{f}_u^*(t|\quad) = \omega_0 \frac{\partial}{\partial s} + J_0^*, \tag{X.170}$$

where J_0^* is as in §§IX.2–4.

We next define $\mathbf{z}^*(t, s) = e^{is}\zeta^*(t)$ and suppose that

$$\mathscr{L}_0 \mathbf{y}(t, s) = \mathbf{h}(t, s), \qquad h(\cdot,\cdot) \in \mathbb{P}_{T,2\pi}. \tag*{(X.171)$_1$}$$

Then there exists a solution $\mathbf{y}(t, s)$, $\mathbf{y}(\cdot,\cdot) \in \mathbb{P}_{T,2\pi}$ only if

$$[\![\mathbf{h}, \mathbf{z}^*]\!] = [\![\mathbf{h}, \bar{\mathbf{z}}^*]\!] = 0. \tag*{(X.171)$_2$}$$

(Equations (X.171)$_2$ are also sufficient for solvability when $\mathbf{h}(t, s)$ is expressible in a finite Fourier series in s. In the general case there is a "problem of small divisors" in (X.171)$_1$ and solvability can be guaranteed when stronger conditions are required for $\mathbf{h}(\cdot,\cdot)$ and ω_0 satisfies a diophantine condition. Equations (X.171) form a Fredholm alternative for the solvability of the perturbation equations which are listed below. If $\mathbf{h}(t, s)$ has real values, then one of the equations $(X.171)_2$ implies the other.

Identification of powers of ε in (X.1) leads to

$$0 = \mathscr{L}_0 \mathbf{u}_1 \tag{X.172}$$

$$0 = \mathscr{L}_0 \mathbf{u}_2 + \mathbf{f}_{uu}(t|\mathbf{u}_1|\mathbf{u}_1) \tag{X.173}$$

$$3\omega_2 \frac{\partial \mathbf{u}_1}{\partial s} = \mathscr{L}_0 \mathbf{u}_3 + 3\mu_2 \mathbf{f}_{u\mu}(t|\mathbf{u}_1) + 3\mathbf{f}_{uu}(t|\mathbf{u}_1|\mathbf{u}_2) + \mathbf{f}_{uuu}(t|\mathbf{u}_1|\mathbf{u}_1|\mathbf{u}_1) \tag{X.174}$$

and, for $p > 3$,

$$p\omega_{p-1} \frac{\partial \mathbf{u}_1}{\partial s} = \mathscr{L}_0 \mathbf{u}_p + p\mu_{p-1} \mathbf{f}_{u\mu}(t|\mathbf{u}_1) + p\mathbf{f}_{uu}(t|\mathbf{u}_1|\mathbf{u}_{p-1}) + \mathbf{g}_p, \tag{X.175}$$

where $\mathbf{g}_p(t, s)$ depends on terms of order lower than $p - 1$. We solve each equation for $\mathbf{u}_p(t, s)$, $\mathbf{u}_p(\cdot, \cdot) \in \mathbb{P}_{T, 2\pi}$, ω_{p-1}, and μ_{p-1}.

The Fredholm alternative $(X.171)_2$ for $\mathscr{L}_0$ implies that (X.173) is solvable if

$$[\![\mathbf{f}_{uu}(t|\mathbf{u}_1|\mathbf{u}_1), \mathbf{z}^*]\!] = 0 \tag{X.176}$$

We recall that $[\![\cdot, \cdot]\!]$ is given by (X.169) as an integral over t and s. Equation (X.174) is solvable if

$$\begin{aligned} 3\omega_2 \left[\!\!\left[\frac{\partial \mathbf{u}_1}{\partial s}, \mathbf{z}^* \right]\!\!\right] &= 3\mu_2 [\![\mathbf{f}_{u\mu}(t|\mathbf{u}_1), \mathbf{z}^*]\!] \\ &\quad + [\![\{3\mathbf{f}_{uu}(t|\mathbf{u}_1|\mathbf{u}_2) + \mathbf{f}_{uuu}(t|\mathbf{u}_1|\mathbf{u}_1|\mathbf{u}_1)\}, \mathbf{z}^*]\!] \end{aligned} \tag{X.177}$$

and (X.175) is solvable if

$$p\omega_{p-1} \left[\!\!\left[\frac{\partial \mathbf{u}_1}{\partial s}, \mathbf{z}^* \right]\!\!\right] = p\mu_{p-1} [\![\mathbf{f}_{u\mu}(t|\mathbf{u}_1), \mathbf{z}^*]\!] + [\![\{p\mathbf{f}_{uu}(t|\mathbf{u}_1|\mathbf{u}_{p-1}) + \mathbf{g}_p\}, \mathbf{z}^*]\!]. \tag{X.178}$$

Now we shall show that

$$\begin{gathered} \left[\!\!\left[\frac{\partial \mathbf{u}_1}{\partial s}, \mathbf{z}^* \right]\!\!\right] = i, \\ [\![\mathbf{f}_{u\mu}(t|\mathbf{u}_1), \mathbf{z}^*]\!] = \sigma_\mu(0), \qquad \operatorname{Re} \sigma_\mu(0) \neq 0, \end{gathered} \tag{X.179}$$

and that $\mathbf{u}_1, \mathbf{u}_2, \mathbf{u}_3, \ldots, \mu_2, \omega_2, \ldots, \mu_9, \omega_9, \ldots$ may be computed sequentially. In fact (X.179) shows that (X.177) and (X.178) can be solved for μ_{2l} and ω_{2l} if the $\mathbf{u}_n$'s can be computed.

We shall again determine $\mathbf{u}(t, s, \varepsilon)$ in a decomposition

$$\mathbf{u}(t, s, \varepsilon) = \varepsilon[\mathbf{z}(t, s) + \bar{\mathbf{z}}(t, s)] + \mathbf{w}(t, s, \varepsilon), \tag*{(X.180)$_1$}$$

where

$$\varepsilon = [\![\mathbf{u}, \mathbf{z}^*]\!], \qquad [\![\mathbf{w}, \mathbf{z}^*]\!] = [\![\mathbf{w}, \bar{\mathbf{z}}^*]\!] = 0, \tag{X.180$_2$}$$

and $\mathbf{w} = (\varepsilon^2/2)\mathbf{w}_2 + (\varepsilon^3/3!)\mathbf{w}_3 + \cdots$. We may choose ε to be real because if the coefficient of $\mathbf{z}$ in (X.180) were complex we could redefine $\mathbf{z}$ given by (X.168) by translation of the origin of s so that the coefficient of the new $\mathbf{z}$ in (X.180) would be real. The decomposition (X.180) reduces (X.179) and (X.172) to identities and (X.173) may be written

$$\mathscr{L}_0 \mathbf{w}_2 + \mathbf{f}_{uu}(t|\mathbf{u}_1|\mathbf{u}_1) = 0. \tag{X.181}$$

It is easy to verify that when $\mathbf{u}_1 = \mathbf{z} + \bar{\mathbf{z}}$ the solvability equation (X.176) is satisfied and we may find $\mathbf{w}_2$. In fact, the function $\mathbf{w}_2(\cdot, \cdot) \in \mathbb{P}_{T,2\pi}$ may be decomposed into Fourier series

$$\mathbf{w}_2(t, s) = \sum_{k \in \mathbb{Z}} \mathbf{w}_{2k}(t) e^{iks},$$

which together with (X.181) and $\mathbf{u}_1 = e^{is}\zeta + e^{-is}\bar{\zeta}$ gives

$$\begin{aligned} (J_0 - 2i\omega_0)\mathbf{w}_{2,2} + \mathbf{f}_{uu}(t|\zeta|\zeta) &= 0 \\ (J_0 + 2i\omega_0)\mathbf{w}_{2,-2} + \mathbf{f}_{uu}(t|\bar{\zeta}|\bar{\zeta}) &= 0 \\ J_0 \mathbf{w}_{20} + 2\mathbf{f}_{uu}(t|\zeta|\bar{\zeta}) &= 0 \end{aligned} \tag{X.182}$$

and $\mathbf{w}_{2k}(t) = 0$ for all $k \in \mathbb{Z}$, $k \neq 0, \pm 2$.

Turning next to the equation (X.177), using (X.179) and (X.180), we find that

$$3\mu_2 \sigma_\mu(0) - 3i\omega_2 + [\![3\mathbf{f}_{uu}(t|\mathbf{u}_1|\mathbf{w}_2) + \mathbf{f}_{uuu}(t|\mathbf{u}_1|\mathbf{u}_1|\mathbf{u}_1), \mathbf{z}^*]\!] = 0. \tag{X.183}$$

Hence

$$\mu_2 \sigma_\mu(0) - i\omega_2 + [\mathbf{f}_{uu}(t|\zeta|\mathbf{w}_{2,0}) + \mathbf{f}_{uu}(t|\bar{\zeta}|\mathbf{w}_{2,2}) + \mathbf{f}_{uuu}(t|\zeta|\zeta|\bar{\zeta}), \zeta^*]_T = 0, \tag{X.184}$$

and since $\text{Re}\, \sigma_\mu(0) \neq 0$ we may solve the complex-valued equation (X.184) for μ_2 and ω_2.

When μ_2 and ω_2 are given we may solve (X.174) for $\mathbf{w}_3$ and we find that

$$\mathbf{w}_3(t, s) = \mathbf{w}_{3,3}(t)e^{3is} + \mathbf{w}_{3,1}(t)e^{is} + \bar{\mathbf{w}}_{3,1}(t)e^{-is} + \bar{\mathbf{w}}_{3,3}(t)e^{-3is}, \tag{X.185}$$

where

$$[\mathbf{w}_{3,1}, \zeta^*]_T = 0.$$

We leave the computation of $\mathbf{w}_{3,k}$ and higher-order terms as an exercise for the reader.

Appendix X.3 Direct Computation of Asymptotically Quasi-Periodic Solutions Which Bifurcate at Rational Points of Higher Order Using the Method of Two Times

Now we shall solve (X.1) when the ratio of the frequencies at criticality $\omega_0 T/2\pi = m/n$ is rational and $n \geq 5$. We saw in §X.11 that the solutions on the bifurcated torus have the form

$$\begin{aligned}\mathbf{u}(t) = \sum_{p+q\geq 1} & R(\varepsilon, t, [\omega_0 + \varepsilon^2\Theta(\varepsilon^2)]t)^{p+q} \\ & \times \exp\left(i(p-q)([\omega_0 + \varepsilon^2\Theta(\varepsilon^2)]t + H(\varepsilon, t, [\omega_0 + \varepsilon^2\Theta(\varepsilon^2)]t))\right) \\ & \times \mathbf{u}_{pq}(t, \mu(\varepsilon^2)) \end{aligned} \tag{X.186}$$

up to terms of higher order. Here $R(\varepsilon, t, s)$ and $H(\varepsilon, t, s)$ are T-periodic in t, $2\pi/n$-periodic in s, while $\mathbf{u}_{pq}$ is T-periodic in t and $\exp(i(p-q)s)$ is 2π-periodic in s. This solution suggests that we can find solutions of the form $\mathbf{u}(t, s)$ where $\mathbf{u}$ is doubly periodic, T-periodic in t, 2π-periodic in s, and where $s = (\omega_0 + \varepsilon^2\Theta(\varepsilon^2))t$. In the notation of (X.186) we have

$$\begin{aligned} R(\varepsilon, t, s) &= \varepsilon + \varepsilon^{n-3}R_{n-3}(t, s) + \cdots \\ H(\varepsilon, t, s) &= \varepsilon^{n-4}H_{n-4}(t, s) + \cdots \\ \Theta(\varepsilon^2) &= \Omega_0 + O(\varepsilon^2) \\ \mu(\varepsilon^2) &= \frac{\mu_2}{2}\varepsilon^2 + O(\varepsilon^4) \\ \mathbf{u}_{10}[t, \mu(\varepsilon^2)] &= \zeta(t) + O(\varepsilon^2) \\ \mathbf{u}_{01} &= \bar{\mathbf{u}}_{10}. \end{aligned} \tag{X.187}$$

We remark that any doubly periodic function $\mathbf{u}$ in $\mathbb{P}_{T,2\pi}$ may be written as

$$u(t, s) = \tilde{u}(t', s') = u(t', s' + \omega_0 t'). \tag{X.188}$$

Hence $\tilde{u}$ is in $\mathbb{P}_{nT,2\pi}$. But, on the contrary, functions in $\mathbb{P}_{nT,2\pi}$ are not necessarily in $\mathbb{P}_{T,2\pi}$ even after a change of variables. This little remark is nevertheless useful in the construction of the Fredholm alternative in $\mathbb{P}_{T,2\pi}$.

Following notation used in Appendix X.2, we define the operator

$$\mathscr{L}_0 = -\omega_0 \frac{\partial}{\partial s} + J_0 \tag{X.189}$$

in the space $\mathbb{P}_{T,2\pi}$. In the present case the kernel of $\mathscr{L}_0$ is infinite-dimensional. To see this we decompose $\mathbf{u}(t, s)$ into a Fourier series in s:

$$\mathbf{u}(t, s) = \sum_{k\in\mathbb{Z}} \mathbf{u}_k(t)e^{kis}.$$

Then $\mathscr{L}_0\mathbf{u} = 0$ implies that

$$(J_0 - ki\omega_0)\mathbf{u}_k = 0,$$

and $\mathbf{u}_k$ can differ from zero only if $k = \pm 1 + ln$, $l \in \mathbb{Z}$. Hence $\mathbf{u}$ has the form

$$\mathbf{u}(t, s) = \sum_l u_{1+ln} \exp((1 + ln)is - iln\omega_0 t)\ \zeta(t)$$
$$+ \sum_l u_{-1+ln} \exp((-1 + ln)is - iln\omega_0 t)\ \bar{\zeta}(t).$$

So the general form of the kernel (or null space) of $\mathscr{L}_0$ is

$$\mathbf{u}(t, s) = e^{is}\alpha(s - \omega_0 t)\zeta(t) + e^{-is}\beta(s - \omega_0 t)\bar{\zeta}(t), \tag{X.190}$$

where α and β are arbitrary $2\pi/n$-periodic functions of s.

To prepare the Fredholm alternative for $\mathscr{L}_0$ in $\mathbb{P}_{T,2\pi}$, we introduce the new variables

$$t' = t, \qquad s' = s - \omega_0 t$$

and write $\mathbf{u}(t, s) = \tilde{\mathbf{u}}(t', s')$, where now $\tilde{\mathbf{u}}$ *is in* $\mathbb{P}_{nT,2\pi}$. We wish to solve

$$\mathscr{L}_0\mathbf{u} = \mathbf{h} \in \mathbb{P}_{T,2\pi}, \tag{X.191}$$

which may now be written as

$$\mathbb{J}\tilde{\mathbf{u}} = \tilde{\mathbf{h}} \in \mathbb{P}_{nT,2\pi} \tag{X.192}$$

because

$$-\frac{\partial}{\partial t'} = -\omega_0\frac{\partial}{\partial s} - \frac{\partial}{\partial t}.$$

The linear operator $\mathbb{J}$ is the same as the one used in Chapter IX, except that s' appears as a parameter in (X.192). Hence the compatibility conditions are:

$$[\tilde{\mathbf{h}}(\cdot, s'), \mathbf{Z}^*]_{nT} = [\tilde{\mathbf{h}}(\cdot, s'), \bar{\mathbf{Z}}^*]_{nT} = 0, \tag{X.193}$$

where we recall that

$$\mathbf{Z}(t) = e^{i\omega_0 t}\zeta(t), \qquad \mathbf{Z}^*(t) = e^{i\omega_0 t}\zeta^*(t)$$

are null vectors of $\mathbb{J}$, $\mathbb{J}^*$ in $\mathbb{P}_{nT}$. We know that (X.192) has solutions $\tilde{\mathbf{u}} \in \mathbb{P}_{nT,2\pi}$ which may be made unique by imposing supplementary conditions of the form

$$[\tilde{\mathbf{u}}(\cdot, s'), \mathbf{Z}^*]_{nT} = [\tilde{\mathbf{u}}(\cdot, s'), \bar{\mathbf{Z}}^*]_{nT} = 0. \tag{X.194}$$

The conditions (X.193) imply that the coefficients of the Fourier series

$$\mathbf{h}(t, s) = \sum_{k\in\mathbb{Z}} \mathbf{h}_k(t)e^{kis}$$

satisfy orthogonality conditions of the form

$$[\mathbf{h}_k, \zeta^*]_T = 0 \quad \text{for } k = 1 + ln, l \in \mathbb{Z}$$

and

$$[\mathbf{h}_k, \bar{\zeta}^*]_T = 0 \quad \text{for } k = -1 + ln, l \in \mathbb{Z}.$$

Now we must verify that the solution $\tilde{\mathbf{u}}$ of (X.192) is such that $\tilde{\mathbf{u}}(t', s') = \tilde{\mathbf{u}}(t, s - \omega_0 t)$ is *T-periodic in t*. In this case, (X.191) will be solved.

In fact it is easy to see that $\tilde{\mathbf{u}}(t' + T, s' - \omega_0 T)$ is a solution of (X.192) with the same $\tilde{\mathbf{h}}$ because

$$\mathbf{h}(t, s) = \mathbf{h}(t + T, s) = \tilde{\mathbf{h}}(t' + T, s' - \omega_0 T) = \tilde{\mathbf{h}}(t', s'),$$

and because $\mathbb{J}$ has T-periodic coefficients. Moreover, since $\tilde{\mathbf{u}}(t' + T, s' - \omega_0 T)$ satisfies (X.194) uniquely,

$$\tilde{\mathbf{u}}(t' + T, s' - \omega_0 T) = \tilde{\mathbf{u}}(t', s')$$

and $\mathbf{u}(t, s - \omega_0 t)$ is in $\mathbb{P}_{T, 2\pi}$.

We now seek a solution of (X.1) in the form

$$\mathbf{u}(t, s, \varepsilon) = \varepsilon \mathbf{u}_1(t, s) + \frac{\varepsilon^2}{2!} \mathbf{u}_2(t, s) + \frac{\varepsilon^3}{3!} \mathbf{u}_3(t, s) + \cdots$$

$$\mu(\varepsilon) = \frac{\mu_2}{2!} \varepsilon^2 + \frac{\mu_4}{4!} \varepsilon^4 + \cdots \tag{X.195}$$

$$\tilde{\omega}(\varepsilon) = \omega_0 + \frac{\varepsilon^2}{2!} \tilde{\omega}_2 + \frac{\varepsilon^4}{4!} \tilde{\omega}_4 + \cdots,$$

where $\mathbf{u}_k \in \mathbb{P}_{T, 2\pi}$, μ, and $\tilde{\omega}$ are even in ε and $\tilde{\omega}_2/2 = \Omega_0$ is assumed to be nonvanishing to avoid subharmonic bifurcation, as in §X.13. To save writing, we have asserted that the odd coefficients vanish in the expansion of $\mu(\varepsilon)$ and $\tilde{\omega}(\varepsilon)$. The assertion is easy to prove. To solve (X.1) we set $s = \tilde{\omega}(\varepsilon)t$ in $\mathbf{u}(t, s, \varepsilon)$ and obtain the function

$$t \mapsto \mathbf{u}(t, \tilde{\omega}(\varepsilon)t, \varepsilon)$$

solving (X.1) in the form (X.186). Identification of powers of ε in the equation

$$\frac{\partial \mathbf{u}}{\partial t} + \tilde{\omega} \frac{\partial \mathbf{u}}{\partial s} = \mathbf{f}(t, \mu, \mathbf{u}) \tag{X.196}$$

then leads to

$$\mathscr{L}_0 \mathbf{u}_1 = 0 \tag{X.197}$$

$$\mathscr{L}_0 \mathbf{u}_2 + \mathbf{f}_{uu}(t|\mathbf{u}_1|\mathbf{u}_1) = 0 \tag{X.198}$$

$$\mathscr{L}_0 \mathbf{u}_3 - 3\tilde{\omega}_2 \frac{\partial \mathbf{u}_1}{\partial s} + 3\mu_2 \mathbf{f}_{u\mu}(t|\mathbf{u}_1) + 3\mathbf{f}_{uu}(t|\mathbf{u}_1|\mathbf{u}_2) + \mathbf{f}_{uuu}(t|\mathbf{u}_1|\mathbf{u}_1|\mathbf{u}_1) = 0 \tag{X.199}$$

$$\begin{aligned} \mathscr{L}_0 \mathbf{u}_4 &- 6\tilde{\omega}_2 \frac{\partial \mathbf{u}_2}{\partial s} + 6\mu_2 \mathbf{f}_{u\mu}(t|\mathbf{u}_2) + 4\mathbf{f}_{uu}(t|\mathbf{u}_1|\mathbf{u}_3) \\ &+ 3\mathbf{f}_{uu}(t|\mathbf{u}_2|\mathbf{u}_2) + 6\mathbf{f}_{uuu}(t|\mathbf{u}_1|\mathbf{u}_1|\mathbf{u}_2) \\ &+ 6\mu_2 \mathbf{f}_{uu\mu}(t|\mathbf{u}_1|\mathbf{u}_1) = 0 \end{aligned} \tag{X.200}$$

and, for $p > 4$,

$$\begin{aligned}\mathscr{L}_0 \mathbf{u}_p - p\tilde{\omega}_{p-1}\frac{\partial \mathbf{u}_1}{\partial s} - \frac{p(p-1)}{2}\left[\tilde{\omega}_2\frac{\partial \mathbf{u}_{p-2}}{\partial s} + \tilde{\omega}_{p-2}\frac{\partial \mathbf{u}_2}{\partial s}\right] \\ + p\mu_{p-1}\mathbf{f}_{u\mu}(t|\mathbf{u}_1) + p\mathbf{f}_{uu}(t|\mathbf{u}_1|\mathbf{u}_{p-1}) \\ + \frac{p(p-1)}{2}[\mu_2\mathbf{f}_{u\mu}(t|\mathbf{u}_{p-2}) + \mathbf{f}_{uuu}(t|\mathbf{u}_1|\mathbf{u}_1|\mathbf{u}_{p-2}) \\ + \mathbf{f}_{uu}(t|\mathbf{u}_2|\mathbf{u}_{p-2}) + \mu_{p-2}\mathbf{f}_{u\mu}(t|\mathbf{u}_2) \\ + \mu_{p-2}\mathbf{f}_{uu\mu}(t|\mathbf{u}_1|\mathbf{u}_1)] + \mathbf{g}_p = 0,\end{aligned} \qquad \text{(X.201)}$$

where $\mathbf{g}_p$ depends on terms of order lower than $p - 2$. We want to solve this system of equations in sequence for $\tilde{\omega}_p, \mu_p, \mathbf{u}_p \in \mathbb{P}_{T,2\pi}$. To show how this works, we begin by solving the first few equations.

The compatibility conditions (X.193) applied to (X.198–200) leads to:

$$[\mathbf{f}_{uu}(t'|\tilde{\mathbf{u}}_1|\tilde{\mathbf{u}}_1), \mathbf{Z}^*(t')]_{nT} = 0 \quad \text{(integrate on } t') \qquad \text{(X.202)}$$

$$\begin{aligned}3\tilde{\omega}_2\left[\frac{\partial \tilde{\mathbf{u}}_1}{\partial s'}, \mathbf{Z}^*(t')\right]_{nT} &= 3\mu_2[\mathbf{f}_{u\mu}(t'|\tilde{\mathbf{u}}_1), \mathbf{Z}^*(t')]_{nT} \\ &\quad + [3\mathbf{f}_{uu}(t'|\tilde{\mathbf{u}}_1|\tilde{\mathbf{u}}_2) + \mathbf{f}_{uuu}(t'|\tilde{\mathbf{u}}_1|\tilde{\mathbf{u}}_1|\tilde{\mathbf{u}}_1), \mathbf{Z}^*(t')]_{nT}\end{aligned} \qquad \text{(X.203)}$$

$$\begin{aligned}6\tilde{\omega}_2\left[\frac{\partial \tilde{\mathbf{u}}_2}{\partial s'}, \mathbf{Z}^*(t')\right]_{nT} &= 6\mu_2[\mathbf{f}_{u\mu}(t'|\tilde{\mathbf{u}}_2), \mathbf{Z}^*(t')]_{nT} \\ &\quad + [4\mathbf{f}_{uu}(t'|\tilde{\mathbf{u}}_1|\tilde{\mathbf{u}}_3) + 3\mathbf{f}_{uu}(t'|\tilde{\mathbf{u}}_2|\tilde{\mathbf{u}}_2) \\ &\quad + 6\mathbf{f}_{uuu}(t'|\tilde{\mathbf{u}}_1|\tilde{\mathbf{u}}_1|\tilde{\mathbf{u}}_2), \mathbf{Z}^*(t')]_{nT} \\ &\quad + 6\mu_2[\mathbf{f}_{uu\mu}(t'|\tilde{\mathbf{u}}_1|\tilde{\mathbf{u}}_1), \mathbf{Z}^*(t')]_{nT}.\end{aligned} \qquad \text{(X.204)}$$

In these equations we used the convention

$$\mathbf{u}(t, s) \stackrel{\text{def}}{=} \tilde{\mathbf{u}}(t', s')$$

with $t = t'$, $s = s' + \omega_0 t'$, for any $\mathbf{u}$ in $\mathbb{P}_{T,2\pi}$.

We determine $\mathbf{u}_p \in \mathbb{P}_{T,2\pi}$ in the following decomposition:

$$\begin{aligned}\mathbf{u}(t, s, \varepsilon) = \varepsilon[e^{is}\alpha(s - \omega_0 t, \varepsilon)\boldsymbol{\zeta}(t) + e^{-is}\bar{\alpha}(s - \omega_0 t, \varepsilon)\bar{\boldsymbol{\zeta}}(t)] \\ + \varepsilon^2\mathbf{w}(t, s, \varepsilon),\end{aligned} \qquad \text{(X.205)}$$

where α is $2\pi/n$-periodic in its argument, and where

$$[\overline{\mathbf{w}}(t', s', \varepsilon), \mathbf{Z}^*(t')_{nT}] = 0 \quad \text{(integrate on } t'). \qquad \text{(X.206)}$$

Note that

$$\tilde{\mathbf{u}}(t', s', \varepsilon) = \varepsilon(e^{is'}\alpha(s', \varepsilon)\mathbf{Z}(t') + e^{-is'}\bar{\alpha}(s', \varepsilon)\bar{\mathbf{Z}}(t')) + \varepsilon^2\tilde{\mathbf{w}}(t', s', \varepsilon), \qquad \text{(X.207)}$$

and that

$$\tilde{\mathbf{u}}_p(t', s') = p[e^{is'}\alpha_{p-1}(s')\mathbf{Z}(t') + e^{-is'}\bar{\alpha}_{p-1}(s')\overline{\mathbf{Z}}(t')] + p(p-1)\tilde{\mathbf{w}}_{p-2}(t', s'), \tag{X.208}$$

where all α_p are $2\pi/n$-periodic in s'.

The decomposition is made unique by requiring that

$$\varepsilon = \frac{1}{2\pi}\int_0^{2\pi} [\tilde{\mathbf{u}}(t', s', \varepsilon), \mathbf{Z}^*(t')]_{nT} e^{-is'}\, ds', \tag{X.209}$$

as is suggested by the form of the kernel of $\mathscr{L}_0^*$. This leads to

$$\frac{1}{2\pi}\int_0^{2\pi} \alpha(s', \varepsilon)\, ds' = 1.$$

Hence,

$$\frac{1}{2\pi}\int_0^{2\pi} \alpha_0(s')\, ds' = 1, \qquad \int_0^{2\pi} \alpha_p(s')\, ds' = 0, \qquad p \geq 1. \tag{X.210}$$

Returning to the systems (X.197–201) and (X.202–204) we find the solution of (X.197) in the form

$$\tilde{\mathbf{u}}_1(t', s') = \alpha_0(s')e^{is'}\mathbf{Z}(t') + \bar{\alpha}_0(s')e^{-is'}\overline{\mathbf{Z}}(t'), \tag{X.211}$$

where α_0 is of mean value 1 and $2\pi/n$-periodic and equation (X.202) is automatically satisfied because $n \neq 1, 3$ (see Chapter IX). Hence the Fredholm alternative guarantees a solution $\mathbf{u}_2 \in \mathbb{P}_{T, 2\pi}$ of (X.198), up to terms in the kernel of $\mathscr{L}_0$, i.e. $\mathbf{w}_0$ is determined. We have found that, in $\mathbb{P}_{nT, 2\pi}$,

$$\begin{aligned} \mathbb{J}2\tilde{\mathbf{w}}_0 &+ \alpha_0^2(s') \exp 2i(s' + \omega_0 t')\, \mathbf{f}_{uu}(t' | \zeta(t') | \zeta(t')) \\ &+ \bar{\alpha}_0^2(s') \exp(-2i(s' + \omega_0 t'))\, \mathbf{f}_{uu}(t' | \bar{\zeta}(t') | \bar{\zeta}(t')) \\ &+ 2|\alpha_0(s')|^2 \mathbf{f}_{uu}(t' | \zeta(t') | \bar{\zeta}(t')) = 0, \end{aligned} \tag{X.212}$$

so

$$\begin{aligned} 2\tilde{\mathbf{w}}_0 = {} & \alpha_0^2(s') \exp 2i(s' + \omega_0 t')\, \mathbf{w}_{01} \\ &+ \bar{\alpha}_0^2(s') \exp -2i(s' + \omega_0 t')\, \overline{\mathbf{w}}_{01} + |\alpha_0(s')|^2 \mathbf{w}_{02}, \end{aligned} \tag{X.213}$$

where $\mathbf{w}_{01}(t')$, $\mathbf{w}_{02}(t')$ are T-periodic,

$$\mathbb{J}(\mathbf{w}_{01}e^{2i\omega_0 t'}) + \mathbf{f}_{uu}(t' | \zeta | \zeta)e^{2i\omega_0 t'} = 0$$

$$\mathbb{J}(\mathbf{w}_{02}) + 2\mathbf{f}_{uu}(t' | \zeta | \bar{\zeta}) = 0,$$

and $\mathbf{w}_{01}$ and $\mathbf{w}_{02}$ are exactly the T-periodic functions which appear in (IX.79). We observe that $\mathbf{w}_0 \in \mathbb{P}_{T, 2\pi}$.

We turn next to (X.203). For the computation we shall need the following identities:

$$\left[\frac{\partial \tilde{\mathbf{u}}_p}{\partial s'}, \mathbf{Z}^*\right]_{nT} = p\frac{d}{ds'}(\alpha_{p-1}(s')e^{is'})$$

$$= pe^{is'}\left[\frac{d\alpha_{p-1}(s')}{ds'} + i\alpha_{p-1}(s')\right] \tag{X.214}$$

$$[\mathbf{f}_{u\mu}(t'|\tilde{\mathbf{u}}_p), \mathbf{Z}^*]_{nT} = p\sigma_\mu(0)\alpha_{p-1}(s')e^{is'} + p(p-1)[\mathbf{f}_{u\mu}(t'|\tilde{\mathbf{w}}_{p-2}), \mathbf{Z}^*]_{nT} \tag{X.215}$$

$$[\mathbf{f}_{uu}(t'|\tilde{\mathbf{u}}_1|\tilde{\mathbf{u}}_p), \mathbf{Z}^*]_{nT} = p(p-1)[\mathbf{f}_{uu}(t'|\tilde{\mathbf{u}}_1|\tilde{\mathbf{w}}_{p-2}), \mathbf{Z}^*]_{nT}. \tag{X.216}$$

Now (X.203) may be written as:

$$3\tilde{\omega}_2\left(\frac{d\alpha_0}{ds'} + i\alpha_0\right) = 3\mu_2\sigma_\mu(0)\alpha_0 + e^{-is'}[3\mathbf{f}_{uu}(t'|\tilde{\mathbf{u}}_1|2\tilde{\mathbf{w}}_0)$$

$$+ \mathbf{f}_{uuu}(t'|\tilde{\mathbf{u}}_1|\tilde{\mathbf{u}}_1|\tilde{\mathbf{u}}_1), \mathbf{Z}^*(t')]_{nT}.$$

We also have the identities

$$[\mathbf{f}_{uu}(t'|\beta\mathbf{Z} + \bar{\beta}\overline{\mathbf{Z}}|2\tilde{\mathbf{w}}_0), \mathbf{Z}^*]_{nT} = \beta|\alpha_0|^2[\mathbf{f}_{uu}(t'|\zeta|\tilde{\mathbf{w}}_{02}), \zeta^*(t')]_T$$

$$= \bar{\beta}\alpha_0^2 e^{2is'}[\mathbf{f}_{uu}(t'|\bar{\zeta}|\mathbf{w}_{01}), \zeta^*(t')]_T \tag{X.218}$$

$$[\mathbf{f}_{uuu}(t'|\tilde{\mathbf{u}}_1|\tilde{\mathbf{u}}_1|\beta\mathbf{Z} + \bar{\beta}\overline{\mathbf{Z}}), \mathbf{Z}^*]$$

$$= (2\beta|\alpha_0|^2 + \bar{\beta}\alpha_0^2 e^{2is'})[\mathbf{f}_{uuu}(t'|\zeta|\zeta|\bar{\zeta}), \zeta^*]_T, \tag{X.219}$$

and (X.217) leads to

$$\tilde{\omega}_2\left(\frac{d\alpha_0}{ds'} + i\alpha_0\right) = \mu_2\sigma_\mu(0)\alpha_0 + \Lambda_2\alpha_0|\alpha_0|^2, \tag{X.220}$$

where Λ_2 is the scalar product in $\mathbb{P}_T$ defined by (IX.80). The only possible periodic solution α_0 of mean value 1 of (X.220) is

$$\alpha_0 = 1, \tag{X.221}$$

and this implies that

$$i\tilde{\omega}_2 = \mu_2\sigma_\mu(0) + \Lambda_2, \tag{X.222}$$

which is exactly the condition (X.157), and it *determines* μ_2 *and* $\tilde{\omega}_2$.

Exercise

X.3 Multiply (X.220) by $\bar{\alpha}_0$ and add the conjugate equation to prove that $|\alpha_0|^2 =$ constant. Then integrate (X.220) over a period, to find a relationship between coefficients necessary to get a nonzero periodic solution. Then, conclude that $\alpha_0 = 1$ is the only possibility of mean value one.

Note that if we *impose* $\tilde{\omega}_2 = 0$, we cannot, in general, solve (X.222). In (IX.101) we used this fact to show that bifurcation into subharmonic solutions at rational points ($n \geq 5$) is not possible except under exceptional circumstances.

When (X.203) is satisfied we have

$$
\begin{aligned}
\tilde{\mathbf{u}}_1(t', s') &= e^{is'}\mathbf{Z}(t') + e^{-is'}\overline{\mathbf{Z}}(t') \\
\tilde{\mathbf{u}}_2(t', s') &= 2[e^{is'}\alpha_1(s')\mathbf{Z}(t') + e^{-is'}\bar{\alpha}_1(s')\overline{\mathbf{Z}}(t')] + 2\tilde{\mathbf{w}}_0(t', s') \\
2\tilde{\mathbf{w}}_0 &= \exp 2i(s' + \omega_0 t')\, \mathbf{w}_{01}(t') \\
&\quad + \exp(-2i(s' + \omega_0 t')\, \overline{\mathbf{w}}_{01}(t') + \mathbf{w}_{02}(t'),
\end{aligned}
\tag{X.223}
$$

where μ_2, $\tilde{\omega}_2$ are known, and $\mathbf{w}_{0j}$ are known T-periodic functions. α_1 is a to-be-determined $2\pi/n$-periodic function of zero mean value. Returning to (X.199), we obtain

$$
\mathbb{J}6\tilde{\mathbf{w}}_1 + 6\mathbf{f}_{uu}(t'|\tilde{\mathbf{u}}_1|e^{is'}\alpha_1(s')\mathbf{Z}(t') + e^{-is'}\bar{\alpha}_1(s')\overline{\mathbf{Z}}(t')) + \mathbf{R} = 0, \tag{X.224}
$$

with

$$
\mathbf{R} = 3\mu_2 \mathbf{f}_{u\mu}(t'|\tilde{\mathbf{u}}_1) + 3\tilde{\omega}_2 \frac{\partial \tilde{\mathbf{u}}_1}{\partial s'} + 3\mathbf{f}_{uu}(t'|\tilde{\mathbf{u}}_1|2\tilde{\mathbf{w}}_0) + \mathbf{f}_{uuu}(t'|\tilde{\mathbf{u}}_1|\tilde{\mathbf{u}}_1|\tilde{\mathbf{u}}_1).
$$

Hence

$$
\begin{aligned}
6\tilde{\mathbf{w}}_1 &= 6\alpha_1(s') \exp 2i(s' + \omega_0 t')\, \mathbf{w}_{01}(t') \\
&\quad + 6\bar{\alpha}_1(s') \exp(-2i(s' + \omega_0 t')\, \overline{\mathbf{w}}_{01}(t') \\
&\quad + 3[\alpha_1(s') + \bar{\alpha}_1(s')]\mathbf{w}_{02}(t') - \mathbb{J}^{-1}\mathbf{R}.
\end{aligned}
\tag{X.225}
$$

Now, the compatibility condition (X.204) allows us to determine $\alpha_1(s')$. For this computation we use the identities

$$
[\mathbf{f}_{uu\mu}(t'|\tilde{\mathbf{u}}_1|\tilde{\mathbf{u}}_1), \mathbf{Z}^*(t')]_{nT} = 0 \tag{X.226}
$$

$$
\begin{aligned}
&[4\mathbf{f}_{uu}(t'|\tilde{\mathbf{u}}_1|6\tilde{\mathbf{w}}_1) + 3\mathbf{f}_{uu}(t'|\tilde{\mathbf{u}}_2|\tilde{\mathbf{u}}_2) + 6\mathbf{f}_{uuu}(t'|\tilde{\mathbf{u}}_1|\tilde{\mathbf{u}}_1|\tilde{\mathbf{u}}_2), \mathbf{Z}^*(t')]_{nT} \\
&\quad = 12\Lambda_2 e^{is'}[2\alpha_1(s') + \bar{\alpha}_1(s')] - 4[\mathbf{f}_{uu\mu}(t'|\tilde{\mathbf{u}}_1|\mathbb{J}^{-1}\mathbf{R}), \mathbf{Z}^*(t')]_{nT} \\
&\quad\quad + 12[\mathbf{f}_{uu}(t'|\tilde{\mathbf{w}}_0|\tilde{\mathbf{w}}_0) + \mathbf{f}_{uuu}(t'|\tilde{\mathbf{u}}_1|\tilde{\mathbf{u}}_1|\tilde{\mathbf{w}}_0), \mathbf{Z}^*(t')]_{nT}.
\end{aligned}
\tag{X.227}
$$

Then (X.204) leads to

$$
\tilde{\omega}_2\left(\frac{d\alpha_1}{ds'} + i\alpha_1\right) = \mu_2 \sigma_\mu(0)\alpha_1 + \Lambda_2(2\alpha_1 + \bar{\alpha}_1) + P(s'), \tag{X.228}
$$

with

$$
\begin{aligned}
P(s') &= e^{-is'}[\mathbf{f}_{uu}(t'|\tilde{\mathbf{w}}_0|\tilde{\mathbf{w}}_0) + \mathbf{f}_{uuu}(t'|\tilde{\mathbf{u}}_1|\tilde{\mathbf{u}}_1|\tilde{\mathbf{w}}_0), \mathbf{Z}^*(t')]_{nT} \\
&\quad - \frac{e^{-is'}}{3}[\mathbf{f}_{uu}(t'|\tilde{\mathbf{u}}_1|\mathbb{J}^{-1}\mathbf{R}), \mathbf{Z}^*(t')]_{nT}.
\end{aligned}
$$

A careful examination of $P(s')$ using (X.223–225) shows that

$$P(s')\begin{cases} = \mathbf{P}_5 e^{-5is'} & \text{if } n = 5, \\ \equiv 0 & \text{if } n > 5, \end{cases} \tag{X.229}$$

that is, P is $2\pi/n$-periodic of mean value zero. Now (X.222) allows us to simplify (X.228), which becomes

$$\tilde{\omega}_2 \frac{d\alpha_1}{ds'} = \Lambda_2(\alpha_1 + \bar{\alpha}_1) + P(s'). \tag{X.230}$$

Exercises

X.4 Show that (X.230) has a unique solution α_1, $2\pi/n$-periodic in s', of zero mean value. (*Hint*: Appendix X.1. Deduce that $\mathbf{u}_2$ and $\mathbf{w}_1$ are therefore completely and uniquely determined and are in $\mathbb{P}_{T,2\pi}$. (*Hint*: See (X.225).) Prove that at each and every step in the sequential computation of $\mathbf{u}_p$, μ_p, $\tilde{\omega}_p$ it is necessary to solve a differential equation of the form (X.230) for α_p, $p \geq 1$, whose second member is $2\pi/n$-periodic with a zero mean value.

X.5 Suppose that $r = \omega_0 T/2\pi = m/n$ is rational. Show that

$$|\gamma_{pql}(0)| \leq \frac{nT|b_{pql}(0)|}{2\pi}.$$

Suppose now that r is irrational. Show that there is no number C, independent of p, q, and l, $p \neq q + 1$ such that

$$|\gamma_{pql}(0)| \leq C|b_{pql}(0)|.$$

Conclude that there are large coefficients $|\gamma_{pql}(0)|$ (small divisors) in the irrational case.

Notes

The results proved in this chapter describe the dynamics of problems in $\mathbb{R}^n$ and much of the observed behavior of continuum of solutions in infinite-dimensional spaces (Banach spaces), which are such that the dynamics really occur in two-dimensional spaces formed under projection. (Here, in fact, we work in a three-dimensional space where the time t is the third dimension.) Such problems arise, for example, in the fluid dynamics of small systems where the "small" serves to separate the eigenvalues in the spectrum of the governing linear operator. Some of these problems are reviewed in the volume on fluid mechanics edited by H. Swinney and J. Gollub (*Hydrodynamic Instabilities and the Transition to Turbulence*, Topics in Current Physics (New York–Heidelberg–Berlin: Springer-Verlag, 1980)). In general, we get sequences of bifurcations into steady symmetry-breaking solutions, into time-periodic solutions and into subharmonic and asymptotically quasi-periodic solutions on a torus. Frequency locking is also observed in some experiments involving fluid motions, as well as in classical experiments with tuning forks and electric circuits.

We acknowledge A. Chenciner for many valuable discussions about the nature of flow on T^2.

Historical note: It seems that J. Neimark was the first to announce the theorem about invariant two-dimensional tori which bifurcate from a periodic solution (or invariant circles which bifurcate from fixed points of maps, such as the Poincaré map). He gave no proof of his result and he gave no result about periodic solutions at points of strong resonance. He does exclude the points $n = 1, 2, 3, 4$ of strong resonance ($\lambda_0^n = 1$) by an assumption of weak attractivity of the origin at criticality. R. J. Sacker gave the first proof of the existence of the invariant tori under conditions clearly excluding the points of strong resonance. He also gave some partial indications that subharmonic solutions might be expected at such resonant points. Sacker's results were rediscovered by D. Ruelle and F. Takens, who mistakenly included $n = 5$ in the excluded set of points of strong resonance. The paper of Ruelle and Takens is best known for the basic idea that "turbulence" is a property of attracting sets which can already be associated with dynamics typical of differential equations in $\mathbb{R}^m$ with m small; for their paper $m = 4$. This idea is very important because it means that even after a few bifurcations one may see chaotic dynamics. The main results about bifurcating subharmonic solutions at points of strong resonance were proved in the formulation of Chapter IX by Iooss and Joseph (1977), *op. cit.* Poincaré treated the case of subharmonic bifurcation with $n = 1$. Y. H. Wan proved that a torus bifurcates when $\lambda_0^4 = 1$ and there is no $4T$-periodic bifurcation. All the resonant cases are treated in an original way by V. I. Arnold. Arnold introduces two parameters and develops some conjectures, based on the two-parameter analysis, to explain frequency locking.

V. I. Arnold, Loss of stability of self-oscillations close to resonance and versal deformations of equivariant vector fields, *Funk. Anal. Ego. Prilog.* **11**, 1–10 (1977).

J. Neimark, On some cases of periodic motions depending on parameters, *Dokl. Akad. Nauk. SSR*, 736–739 (1959).

H. Poincaré, *Les méthodes nouvelles de la mécanique céleste*, Gauthier–Villars, Paris 1892 (see §§37, 38).

D. Ruelle and F. Takens, On the nature of turbulence, *Com. Math. Phys.* **20**, 167–192 (1971).

R. J. Sacker, *On Invariant Surfaces and Bifurcation of Periodic Solutions of Ordinary Differential Equations*, New York Univ. IMM-NYU 333 (1964).

Y. H. Wan, Bifurcation into invariant tori at points of resonance, *Arch. Rational Mech. Anal.* **68**, 343–357 (1978).

CHAPTER XI

Secondary Subharmonic and Asymptotically Quasi-Periodic Bifurcation of Periodic Solutions (of Hopf's Type) in the Autonomous Case

In Chapters IX and X we considered the problems of stability and bifurcation of the solution $\mathbf{u} = 0$ of the evolution problem reduced to local form, $\dot{\mathbf{u}} = \mathbf{f}(t, \mu, \mathbf{u}) = \mathbf{f}(t + T, \mu, \mathbf{u})$. In §I.3 we showed how the reduced problem arises from the study of forced T-periodic solutions $\mathbf{U}(t) = \mathbf{U}(t + T)$ of evolution problems in the form

$$\dot{\mathbf{U}} = \mathbf{F}(t, \mu, \mathbf{U}) = \mathbf{F}(t + T, \mu, \mathbf{U}), \tag{XI.1$_1$}$$

where $\mathbf{U} = 0$ is *not* a solution because

$$\mathbf{F}(t, \mu, 0) = \mathbf{F}(t + T, \mu, 0) \not\equiv 0. \tag{XI.1$_2$}$$

In this type of problem the outside world communicates with the dynamical system governed by (XI.1)$_1$ through the imposed data (XI.1)$_2$. The dynamical system sees the outside world as precisely T-periodic and it must adjust its own evolution to fit this fact.

Now we want to consider the bifurcation of periodic solutions in a different class. We suppose that we have a $T(\varepsilon)$-periodic ($T(\varepsilon) = 2\pi/\omega(\varepsilon)$) bifurcating solution $\mathbf{U}(\omega(\varepsilon)t, \varepsilon) = \tilde{\mathbf{U}}(\mu(\varepsilon)) + \mathbf{u}(\omega(\varepsilon)t, \varepsilon) = \mathbf{U}(\omega(\varepsilon)t + 2\pi, \varepsilon)$, of an autonomous problem

$$\frac{d\mathbf{u}}{dt} = \mathbf{F}(\mu, \tilde{\mathbf{U}} + \mathbf{u}) = \mathbf{f}(\mu, \mathbf{u})$$

$$\mathbf{F}(\mu, \tilde{\mathbf{U}}) = 0$$

with *steady* forcing

$$\mathbf{F}(\mu, 0) \neq 0.$$

In fact, the functions $\mathbf{u}(\omega(\varepsilon)t, \varepsilon)$, $\omega(\varepsilon)$ and $\mu(\varepsilon)$ which define the periodic bifurcating solution (Hopf's solution) are exactly the ones studied in Chapters

VII and VIII. We are interested in the loss of stability and secondary bifurcation of the solution $\mathbf{U}(\omega(\varepsilon)t, \varepsilon)$. In fact, we need not assume that $\mathbf{U}$ arises from a bifurcation. It is enough that $\mathbf{U}$ is a $T(\varepsilon)$-periodic solution of an autonomous equation, depending on a parameter.

The problem now under study, bifurcation of periodic solutions of autonomous problems, is very close to the problem of bifurcation of forced T-periodic problems which was studied in Chapters IX and X. We will show that the qualitative properties of secondary bifurcation of periodic solutions of autonomous problems and the properties of primary bifurcation of forced T-periodic problems are nearly the same. In both problems we find subharmonic bifurcation into nT-periodic solutions at rational points with* $n = 1, 2, 3, 4$ and, at other points, we get bifurcation into asymptotically quasi-periodic solutions, or when very special weak resonance conditions are satisfied, into subharmonic solutions with periods corresponding to integers $n \geq 5$, and the distributions of stability of the bifurcating solutions are the same in both problems.

But the two problems are not identical. In the autonomous problem, the outside world imposes data of "maximum symmetry," that is, steady data, so that solutions are indifferent to the choice of the time origin. In the forced T-periodic problem a definite pattern of temporal symmetry, T-periodicity, is imposed from the outside and the solutions are only indifferent to a shift in the origin of time by a period T. One consequence of this difference is that the subharmonic solutions which undergo secondary bifurcation from Hopf's solution (the autonomous case) have definite periods which (1) change with amplitude and (2) which are close to, but not exactly the same as the periods $nT(\varepsilon)$ ($n = 1, 2, 3, 4$) of the Hopf solution when $|\varepsilon| \neq 0$ is small. In the forced T-periodic problem the subharmonic solutions are exactly $\tau = nT$-periodic ($n = 1, 2, 3, 4$), where τ is independent of the amplitude ε.

A second consequence of this difference is technical and is associated with the fact that $\dot{\mathbf{u}}(s, \varepsilon)$ is always a solution with eigenvalue zero of the spectral problem (VIII.36), (VIII.38), for the stability of the Hopf solution. This property has the following significance. The bifurcating solutions $\{\mathbf{u}(s + \delta, \varepsilon), \mu(\varepsilon), \omega(\varepsilon)\}$ and $\{\mathbf{u}(s, \varepsilon), \mu(\varepsilon), \omega(\varepsilon)\}$ are equivalent to within a translation δ of the time origin. We call δ the *phase* of the bifurcating solution. The difference between two bifurcating solutions

$$\mathbf{\Theta}(s, \varepsilon, \delta) = \mathbf{u}(s + \delta, \varepsilon) - \mathbf{u}(s, \varepsilon)$$

satisfies

$$\begin{aligned} \omega(\varepsilon)\dot{\mathbf{\Theta}} &= \mathbf{f}(\mu(\varepsilon), \mathbf{u}(s + \delta, \varepsilon)) - \mathbf{f}(\mu(\varepsilon), \mathbf{u}(\delta, \varepsilon)) \\ &= \mathbf{f}_u(\mu(\varepsilon), \mathbf{u}(s, \varepsilon) | \mathbf{\Theta}) + O(\|\mathbf{\Theta}\|^2) \end{aligned}$$

and as $\delta \to 0$

$$\mathbf{\Theta}(s, \varepsilon, \delta) \sim \dot{\mathbf{u}}(s, \varepsilon)\delta.$$

* Here, n is the same as in Chapter IX.

So we can always go the other way and, starting with $\mathbf{u}(s, \varepsilon)$, construct a "bifurcating" solution $\mathbf{u}(s + \delta, \varepsilon)$ by pretending $\gamma(\varepsilon) = 0$ is an algebraically simple eigenvalue of $\mathbb{J}(\varepsilon)$ with eigenvector $\dot{\mathbf{u}}(s, \varepsilon)$. In treating true bifurcation problems it is necessary to avoid computing these phase shifts and the way we do it is to require it mathematically by insisting that true subharmonic bifurcating solutions should differ from phase shifts of $\mathbf{u}(s, \varepsilon)$. The mathematical condition for this, (XI.48), is most efficiently explained after establishing our method of constructing the bifurcation.

Notation

The notation for this chapter has much in common with the notation of Chapter IX. Some slight differences arise from the definitions

$$\hat{\omega}(\mu)t = s \quad (\text{see §XI.1})$$

and

$$\hat{\Omega}(\mu)t = s \quad (\text{see §XI.8}),$$

which require that we compute frequencies $\hat{\omega}$ and $\hat{\Omega}$. Some of the symbols which are also used in Chapter IX but have a slightly different meaning here are

$$J_0 \text{ and } J_0^* \qquad \text{in §XI.2}$$

$$\mathbb{J} \text{ and } \mathbb{J}^* \qquad \text{in §XI.4.}$$

The amplitude α of the bifurcating solution is defined by (XI.45).

$$(\cdot)_n = \frac{\partial^n(\cdot)}{\partial \mu^n} \quad \text{at } \mu = \mu_0.$$

$$(\cdot)^{(n)} = \frac{\partial^n(\cdot)}{\partial \alpha^n} \quad \text{at } \alpha = 0.$$

(ktlo) = known terms of lower order.

XI.1 Spectral Problems

In our analysis of secondary bifurcation of periodic solutions of autonomous problems we suppress, as far as possible, the idea that at a point of bifurcation one solution is primary and the other is secondary. For this reason we do not start with the evolution equation in local form; instead, we try to characterize the solution of the autonomous problem

$$\frac{d\mathbf{V}}{dt} = \mathbf{F}(\mu, \mathbf{V}) \tag{XI.2}$$

where $\mathbf{F}(\cdot, \cdot)$ is assumed to be smooth enough to allow what we do and $\mathbf{F}(\mu, 0)$ need not vanish. We shall study the bifurcation of periodic solutions of (XI.2). These solutions are of the form

$$\mathbf{V} = \hat{\mathbf{U}}(\hat{\omega}(\mu)t, \mu) = \hat{\mathbf{U}}(s, \mu) = \hat{\mathbf{U}}(s + 2\pi, \mu), \tag{XI.3}$$

where $\hat{\omega}(\mu)$ is the frequency. For example,

$$\hat{\mathbf{U}}(s, \mu) = \tilde{\mathbf{U}}(\mu) + \hat{\mathbf{u}}(s, \mu)$$

may be the Hopf solution of Chapter VIII: $\hat{\mathbf{u}}(s, \mu) = \mathbf{u}(s, \varepsilon)$, $\hat{\omega}(\mu) = \omega(\varepsilon)$ when $\mu = \mu(\varepsilon)$. This solution satisfies

$$\hat{\omega}(\mu)\frac{d\hat{\mathbf{U}}}{ds} = F(\mu, \hat{\mathbf{U}}). \tag{XI.4}$$

Note that the solution (XI.3) is parameterized with μ rather than ε.

Our aim is now to find the conditions under which subharmonic solutions bifurcate from (XI.3). To study bifurcation we need to analyze the spectral problem associated with the linearized theory of stability of (XI.3). This problem, reduced to local form, was studied in §VIII.4. But now we need to know more about the spectral problem and in particular we need to derive formulas for the strict loss of stability of the Hopf solution (XI.3) which will guarantee the existence of bifurcation (see §II.9).

To obtain the spectral problem we linearize (XI.2) around (XI.3):

$$\mathbf{V} = \hat{\mathbf{U}}(s, \mu) + \mathbf{v}(t), \qquad s = \hat{\omega}(\mu)t,$$

where

$$\frac{d\mathbf{v}}{dt} = \mathbf{F}_v(\mu, \hat{\mathbf{U}}(s, \mu)|\mathbf{v}).$$

Floquet theory then implies that we may ascertain the stability of $\hat{\mathbf{U}}(s, \mu)$ by study of the exponents $\gamma(\mu) = \xi(\mu) + i\eta(\mu)$ in the representation

$$\mathbf{v}(t) = e^{\gamma t}\mathbf{\Gamma}(s), \qquad \mathbf{\Gamma}(s) = \mathbf{\Gamma}(s + 2\pi).$$

The exponents are eigenvalues of the spectral problem

$$\gamma\mathbf{\Gamma} = -\hat{\omega}(\mu)\frac{d\mathbf{\Gamma}}{ds} + \mathbf{F}_v(\mu, \hat{\mathbf{U}}(s, \mu)|\mathbf{\Gamma}). \tag{XI.5}$$

We also have an adjoint eigenvalue problem

$$\bar{\gamma}\mathbf{\Gamma}^* = \hat{\omega}(\mu)\frac{d\mathbf{\Gamma}^*}{ds} + \mathbf{F}_v^*(\mu, \hat{\mathbf{U}}(s, \mu)|\mathbf{\Gamma}^*) \tag{XI.6}$$

associated with the scalar product

$$[\cdot, \cdot]_{2\pi} = \frac{1}{2\pi}\int_0^{2\pi} \langle \cdot, \cdot \rangle \, ds$$

where $\mathbf{F}_v^*(\mu, \hat{\mathbf{U}}(s, \mu)|\cdot)$ is the unique linear operator satisfying

$$\langle \mathbf{F}_v(\mu, \hat{\mathbf{U}}(s, \mu)|\mathbf{a}), \mathbf{b}\rangle = \langle \mathbf{a}, \mathbf{F}_v^*(\mu, \hat{\mathbf{U}}(s, \mu)|\mathbf{b})\rangle$$

for arbitrary vectors **a** and **b**. Recall that for vectors in $\mathbb{C}^n$, $\langle \mathbf{a}, \mathbf{b}\rangle = \mathbf{a}\cdot\bar{\mathbf{b}}$.

XI.2 Criticality and Rational Points

At criticality $\mu = \mu_0$,

$$\begin{aligned}
\xi(\mu_0) &= 0\\
\eta(\mu_0) &\stackrel{\text{def}}{=} \eta_0\\
\hat{\omega}(\mu_0) &\stackrel{\text{def}}{=} \omega_0\\
\hat{\mathbf{U}}(s, \mu_0) &\stackrel{\text{def}}{=} \mathbf{U}_0(s)\\
J_0 &\stackrel{\text{def}}{=} -\omega_0\frac{d}{ds} + \mathbf{F}_v(\mu_0, \mathbf{U}_0(s)|\cdot)\\
J_0^* &\stackrel{\text{def}}{=} \omega_0\frac{d}{ds} + \mathbf{F}_v^*(\mu_0, \mathbf{U}_0(s)|\cdot),
\end{aligned} \tag{XI.7}$$

where J_0 and J_0^* act on 2π-periodic functions of s. The spectral problems at criticality are

$$i\eta_0\mathbf{\Gamma}_0 = J_0\mathbf{\Gamma}_0 \tag{XI.8$_1$}$$

and

$$-i\eta_0\mathbf{\Gamma}_0^* = J_0^*\mathbf{\Gamma}_0^*. \tag{XI.8$_2$}$$

If the Floquet exponent $i\eta_0$ is an eigenvalue of J_0 at criticality, then $i(\eta_0 + l\omega_0)$, $l \in \mathbb{Z}$, is also an eigenvalue with eigenvector $\hat{\mathbf{\Gamma}}(s) = e^{-ils}\mathbf{\Gamma}_0(s) = \hat{\mathbf{\Gamma}}(s + 2\pi)$. The Floquet multiplier at criticality

$$\begin{aligned}
\lambda_0 &= \exp \gamma(\mu_0)T(\mu_0)\\
&= \exp\frac{i(\eta_0 + n\omega_0)2\pi}{\omega_0}\\
&= \exp\left(\frac{2\pi i\eta_0}{\omega_0}\right)
\end{aligned} \tag{XI.9}$$

maps repeated points on the imaginary axis of the complex γ-plane into unique points of the complex λ-plane. We may cover the unit circle of the λ-plane by restricting our considerations to the principal branch

$$0 \le \frac{\eta_0}{\omega_0} < 1 \tag{XI.10}$$

of the complex γ-plane.

We say that

$$\frac{\eta_0}{\omega_0} = \frac{m}{n} \tag{XI.11}$$

satisfying (XI.10) is in the set of rational points if m and n are integers and $m = 0$ when $n = 1$; otherwise $m \neq 0$. The Floquet multiplier at criticality is that nth root of unity when $\eta_0/\omega_0 = m/n$ is a rational point:

$$\lambda_0^n = (e^{2\pi i m/n})^n = 1. \tag{XI.12}$$

XI.3 Spectral Assumptions about J_0

The simplest and most typical situations which lead to subharmonic bifurcation into self-excited solutions of autonomous problems are similar to those leading to subharmonic bifurcations of forced T-periodic problems which were described in Chapter IX. In the autonomous case, however, it is necessary to accomodate for the fact that $\mathbf{\Gamma} = \dot{\mathbf{U}}(s, \mu)$ satisfies (XI.5) for all μ whenever $\gamma(\mu) = 0$. We showed, in the last theorem proved in §VIII.4, that $\gamma(\mu) = 0$ cannot always be an algebraically simple eigenvalue.

We may therefore formulate the simplest hypotheses about the eigenvalues $\gamma(\mu_0) = i\eta_0$ of J_0 as follows:

I. $i\eta_0 = 0$ is an isolated double eigenvalue of J_0; or
II. $i\eta_0 \neq 0$ is an isolated simple eigenvalue of J_0.

If we assume that $\eta_0/\omega_0 = m/n$, $m \neq 0$, we have (XI.8) in the form

$$\frac{i\omega_0 m}{n}\mathbf{\Gamma}_0 \equiv J_0\mathbf{\Gamma}_0 \tag*{(XI.13)$_1$}$$

and

$$\frac{-i\omega_0 m}{n}\mathbf{\Gamma}_0^* = J_0^*\mathbf{\Gamma}_0^*. \tag*{(XI.13)$_2$}$$

The solution $\mathbf{v}(t)$ of the linearized problem (XI.5) at criticality is then given by

$$\begin{aligned}\mathbf{v}(t) &= e^{\gamma(\mu_0)t}\mathbf{\Gamma}_0(s) = e^{i(m/n)\omega_0 t}\mathbf{\Gamma}_0(s)\\ &= e^{i(m/n)s}\mathbf{\Gamma}_0(s) \overset{\text{def}}{=} \mathbf{Z}(s) = \mathbf{Z}(s + 2\pi n)\end{aligned} \tag{XI.14}$$

XI.4 Spectral Assumptions about $\mathbb{J}$ in the Rational Case

The vector $\mathbf{Z}(s)$ satisfies the equation $\mathbb{J}\mathbf{Z} = 0$, where

$$\mathbb{J} \overset{\text{def}}{=} -\omega_0\frac{d}{ds} + \mathbf{F}_v(\mu_0, \mathbf{U}_0(s)|\cdot)$$

is a linear operator which acts on $2\pi n$-periodic functions of s, $n \in \mathbb{N}$, and $\mathbf{U}_0(s) \overset{\text{def}}{=} \mathbf{U}(s, \mu_0)$. The linear operator

$$\mathbb{J}^* = \omega_0 \frac{d}{ds} + \mathbf{F}_v^*(\mu_0, \mathbf{U}_0(s)|\cdot)$$

is adjoint to $\mathbb{J}$ relative to $[\cdot, \cdot]_{2\pi n}$; i.e., $[\mathbb{J}\mathbf{a}, \mathbf{b}]_{2\pi n} = [\mathbf{a}, \mathbb{J}^*\mathbf{b}]_{2\pi n}$ for arbitrary $2\pi n$-periodic vectors $\mathbf{a}(s)$ and $\mathbf{b}(s)$.

Assuming the validity of hypotheses (I) or (II) about J_0, there is at least the $2\pi n$-periodic, real-valued eigenfunction

$$\mathbf{Z}_0(s) = \dot{\mathbf{U}}_0(s) = \mathbf{Z}_0(s + 2\pi). \tag{XI.15$_1$}$$

If $\eta_0 \neq 0$, or if $\gamma(\mu_0) = 0$ is geometrically double, then

$$\mathbf{Z}_1(s) = e^{i(m/n)s}\mathbf{\Gamma}_0(s) = \mathbf{Z}_1(s + 2\pi n) \tag{XI.15$_2$}$$

belongs to the eigenvalue zero of $\mathbb{J}$; that is $\mathbb{J}\mathbf{Z}_0 = \mathbb{J}\mathbf{Z}_1 = 0$. If $\mathbf{Z}_1(s)$ is complex then

$$\mathbf{Z}_2(s) = \overline{\mathbf{Z}}_1(s) \tag{XI.15$_3$}$$

is also an eigenfunction of $\mathbb{J}$ belonging to the eigenvalue zero. The arguments given in §IX.7 and IX.8 apply here. $\mathbf{Z}_1(s)$ may be assumed to be real-valued when the multiplier λ_0 is; that is, when

$$\lambda_0 = 1, \qquad \frac{m}{n} = \frac{0}{1}$$

or

$$\lambda_0 = -1, \qquad \frac{m}{n} = \frac{1}{2}.$$

In all other cases, λ_0 is complex-valued.

In §IX.8 we wrote a lemma about $\mathbb{J}$ in the forced T-periodic problem. Now we wish to state the corresponding lemma about the autonomous problem. This lemma accounts for the fact that in the autonomous problem $\dot{\mathbf{U}}_0 = \mathbf{Z}_0$ is always in the null space of $\mathbb{J}$, $\mathbb{J}\mathbf{Z}_0 = 0$.

Assume that the hypotheses (I) or (II) about the eigenvalue of J_0 hold. Then, we may distinguish the following cases.

(a) $n = 1$ and zero is an index-one double (semi-simple) eigenvalue of $\mathbb{J} = J_0$. In this case we can find 2π-periodic vectors verifying the following equations:

$$\begin{aligned} &\mathbf{Z}_0(s) = \dot{\mathbf{U}}_0(s) = \mathbf{\Gamma}_{00}(s), \qquad \mathbf{Z}_0^* = \mathbf{\Gamma}_{00}^*(s) \\ &\mathbf{Z}_1(s) = \mathbf{\Gamma}_{01}(s), \qquad \mathbf{Z}_1^*(s) = \mathbf{\Gamma}_{01}^*(s), \end{aligned} \tag{XI.15$_4$}$$

so that $\mathbb{J}\mathbf{Z}_l = \mathbb{J}^*\mathbf{Z}_l^* = 0$,

$$[\mathbf{Z}_l, \mathbf{Z}_m^*]_{2\pi} = \delta_{lm}, \qquad l, m = 1, 2. \tag{XI.16$_1$}$$

The loss of stability of the solution $\hat{\mathbf{U}}(\cdot, \mu)$ under conditions specified in (a) is rather special. For example, in the case of Hopf bifurcation treated in §VIII.4 case (a) cannot occur unless $\omega_\mu(\mu_0) = 0$ and some additional conditions (specified under XI.27) are realized.

(b) $n = 1$ and zero is an index-two (not semi-simple) eigenvalue of $\mathbb{J} = J_0$ with one proper eigenvector $\mathbf{Z}_0 = \mathbf{\Gamma}_{00}$, one generalized eigenvector $\mathbf{Z}_1 = \mathbf{\Gamma}_{01}$, corresponding proper adjoint eigenvector $\mathbf{Z}_1^* = \mathbf{\Gamma}_{01}^*$, and generalized adjoint eigenvector $\mathbf{Z}_0^* = \mathbf{\Gamma}_{00}^*$ satisfying (XI.29) and (XI.30).

In §VIII.4 we gave some sufficient conditions, associated with $\omega_\mu(\mu_0) \neq 0$ for the realization of case (b).

(c) $n = 2$, zero is a semi-simple double eigenvalue of $\mathbb{J}$ with two real 4π-periodic eigenvectors

$$\mathbf{Z}_0(s) = \dot{\mathbf{U}}_0(s) = \mathbf{Z}_0(s + 2\pi) \tag{XI.16$_2$}$$

and

$$\mathbf{Z}_1(s) = e^{i(s/2)}\mathbf{\Gamma}_0(s) = \overline{\mathbf{Z}}_1(s) = \mathbf{Z}_1(s + 4\pi) \tag{XI.16$_3$}$$

satisfying (XI.16)$_1$.

(d) $n > 2$, zero is a semi-simple triple eigenvalue of $\mathbb{J}$ with one real 2π-periodic eigenvector $\mathbf{Z}_0(s) = \dot{\mathbf{U}}(s)$ and two complex-valued $2\pi n$-periodic eigenvectors

$$\mathbf{Z}_1(s) = e^{i(m/n)s}\mathbf{\Gamma}_0(s), \qquad \mathbf{\Gamma}_0(s) = \mathbf{\Gamma}_0(s + 2\pi) \tag{XI.16$_4$}$$

and $\overline{\mathbf{Z}}_1 = \mathbf{Z}_2$. In this case the eigenvectors satisfy biorthogonality conditions

$$[\mathbf{Z}_l, \mathbf{Z}_m^*]_{2\pi n} = \delta_{lm}, \qquad l, m = 0, 1, 2.$$

XI.5 Strict Loss of Stability at a Simple Eigenvalue of J_0

In all the problems of bifurcation which we have studied, a condition of strict loss of stability implies the existence of double-point bifurcation. The same type or strict crossing condition will suffice to guarantee the existence of subharmonic bifurcation of $2\pi/\hat{\omega}(\mu)$ periodic solutions.

In the present derivation of formulas expressing strict crossing we assume that $\gamma_0 = i\eta_0 = i(m/n)\omega_0$ is a simple eigenvalue of J_0. This assumption is typical when $n \neq 1$; when $n = 1$, $m = 0$ and $\gamma_0 = 0$ is a double eigenvalue of J_0. By a strict crossing we mean that the eigenvalue $\gamma(\mu) = \xi(\mu) + i\eta(\mu)$, whose real part changes sign at μ_0, satisfies

$$\xi_\mu(\mu_0) = \operatorname{Re} \gamma_\mu(\mu_0) \overset{\text{def}}{=} \operatorname{Re} \gamma_1 = \xi_1 > 0.$$

The equation governing γ_1 can be obtained by differentiating (XI.5) with respect to μ at μ_0:

$$\gamma_1\mathbf{\Gamma}_0 + \hat{\omega}_1\dot{\mathbf{\Gamma}}_0 + \gamma_0\mathbf{\Gamma}_1 = J_0\mathbf{\Gamma}_1 + \mathscr{J}\mathbf{\Gamma}_0, \qquad \mathbf{\Gamma}_1(s) = \mathbf{\Gamma}_1(s + 2\pi), \tag{XI.17$_1$}$$

where

$$\mathscr{I}(\cdot) \stackrel{\text{def}}{=} \mathbf{F}_{vv}(\mu_0, \mathbf{U}_0 | \hat{\mathbf{U}}_1 | \cdot) + \mathbf{F}_{v\mu}(\mu_0, \mathbf{U}_0 | \cdot). \tag{XI.17$_2$}$$

Equation (XI.17) is solvable for $\boldsymbol{\Gamma}_1(s)$ if the terms involving $\boldsymbol{\Gamma}_0$ are orthogonal to eigenvectors $\boldsymbol{\Gamma}_0^*$ solving

$$(J_0^* - \gamma_0)\boldsymbol{\Gamma}_0^* = 0, \qquad \bar{\boldsymbol{\Gamma}}_0^*(s) = \bar{\boldsymbol{\Gamma}}_0^*(s + 2\pi),$$

where

$$J_0^* = \omega_0 \frac{d}{ds} + \mathbf{F}_v^*(\mu_0, \mathbf{U}_0 | \cdot).$$

Since γ_0 is a simple eigenvalue of J_0 there is one eigenvector $\boldsymbol{\Gamma}_0$ belonging to γ_0 and one $\bar{\boldsymbol{\Gamma}}_0$ belonging to $\bar{\gamma}_0$. Similarly $\boldsymbol{\Gamma}_0^*$ belongs to $\bar{\gamma}_0$ and $\bar{\boldsymbol{\Gamma}}_0^*$ to γ_0 and

$$[\boldsymbol{\Gamma}_0, \boldsymbol{\Gamma}_0^*]_{2\pi} - 1 = [\boldsymbol{\Gamma}_0, \bar{\boldsymbol{\Gamma}}_0^*]_{2\pi} = 0. \tag{XI.18}$$

Equation (XI.17) is solvable if the inhomogeneous terms are orthogonal to $\boldsymbol{\Gamma}_0^*$:

$$\gamma_1 + \hat{\omega}_1[\dot{\boldsymbol{\Gamma}}_0, \boldsymbol{\Gamma}_0^*]_{2\pi} = [\mathscr{I}\boldsymbol{\Gamma}_0, \boldsymbol{\Gamma}_0^*]_{2\pi}.$$

Recalling that $\mathbf{Z}_1 = e^{i(m/n)s}\boldsymbol{\Gamma}_0$ and $\mathbf{Z}_1^* = e^{i(m/n)s}\boldsymbol{\Gamma}_0^*$, we compute

$$[\dot{\boldsymbol{\Gamma}}_0, \boldsymbol{\Gamma}_0^*]_{2\pi n} = [\dot{\boldsymbol{\Gamma}}_0 e^{i(m/n)s}, \mathbf{Z}_1^*]_{2\pi n} = \left[\left(\dot{\mathbf{Z}}_1 - \frac{im}{n}\mathbf{Z}_1\right), \mathbf{Z}_1^*\right]_{2\pi n}$$

$$= [\dot{\mathbf{Z}}_1, \mathbf{Z}_1^*]_{2\pi n} - \frac{im}{n}$$

and find that

$$\left(\gamma_1 - \frac{im}{n}\hat{\omega}_1\right) + \hat{\omega}_1[\dot{\mathbf{Z}}_1, \mathbf{Z}_1^*]_{2\pi n} = [\mathscr{I}\mathbf{Z}_1, \mathbf{Z}_1^*]_{2\pi n}. \tag{XI.19}$$

Equation (XI.19) applies when $n \neq 1$. When $n = 2$, $m/n = \frac{1}{2}$, $\mathbf{Z}_1$ and $\mathbf{Z}_1^*$ are real-valued, and

$$\eta_1 = \operatorname{Im} \gamma_1 = \frac{m}{n}\hat{\omega}_1 = \frac{\hat{\omega}_1}{2}$$

and

$$\xi_1 + \hat{\omega}_1[\dot{\mathbf{Z}}_1, \mathbf{Z}_1^*]_{4\pi} = [\mathscr{I}\mathbf{Z}_1, \mathbf{Z}_1^*]_{4\pi}. \tag{XI.20}$$

It can be shown, in fact, that

$$\eta_l = \frac{\hat{\omega}_l}{2}.$$

XI.6 Strict Loss of Stability at a Double Semi-Simple Eigenvalue of J_0

We now suppose that $\gamma_0 = 0$ is an algebraically double, semi-simple, double eigenvalue of J_0. Then there are two independent eigenfunctions $\mathbf{\Gamma}_{00} = \dot{\mathbf{U}}_0(s)$ and $\mathbf{\Gamma}_{01}$ satisfying $J_0\mathbf{\Gamma}_{01} = 0$ and two independent adjoint eigenfunctions $\mathbf{\Gamma}_{00}^*$ and $\mathbf{\Gamma}_{01}^*$ such that

$$[\mathbf{\Gamma}_{00}, \mathbf{\Gamma}_{00}^*]_{2\pi} = [\mathbf{\Gamma}_{01}, \mathbf{\Gamma}_{01}^*]_{2\pi} = 1 \tag{XI.21}$$

and

$$[\mathbf{\Gamma}_{00}, \mathbf{\Gamma}_{01}^*]_{2\pi} = [\mathbf{\Gamma}_{01}, \mathbf{\Gamma}_{00}^*]_{2\pi} = 0. \tag{XI.22}$$

Every eigenvector in the null space of J_0 may be formed as a linear combination of the independent ones

$$\mathbf{\Gamma}_0 = A\mathbf{\Gamma}_{00} + B\mathbf{\Gamma}_{01} \tag{XI.23}$$

To compute γ_1 in (XI.17), it is necessary to determine A and B. The values of A and B can be determined by the biorthogonality conditions required for the solvability of (XI.17).

Using (XI.23) we may write (XI.17) as

$$\gamma_1(A\mathbf{\Gamma}_{00} + B\mathbf{\Gamma}_{01}) + \hat{\omega}_1(A\dot{\mathbf{\Gamma}}_{00} + B\dot{\mathbf{\Gamma}}_{01}) = J_0\mathbf{\Gamma}_1 + \mathscr{J}(A\mathbf{\Gamma}_{00} + B\mathbf{\Gamma}_{01}). \tag{XI.24}$$

There is a special solution of (XI.24) which may be identified by differentiating (XI.5) with respect to μ at μ_0. This gives rise to (XI.24) with $\gamma_1 = B = 0$ and $A = 1$. Then

$$\hat{\omega}_1\dot{\mathbf{\Gamma}}_{00} = J_0\mathbf{\Gamma}_1 + \mathscr{J}\mathbf{\Gamma}_{00}$$

is solvable if

$$\hat{\omega}_1[\dot{\mathbf{\Gamma}}_{00}, \mathbf{\Gamma}_{0j}^*]_{2\pi} = [\mathscr{J}\mathbf{\Gamma}_{00}, \mathbf{\Gamma}_{0j}^*]_{2\pi}, \qquad j = 0, 1. \tag{XI.25}$$

and, using (XI.25) we find that (XI.24) is solvable if γ_1 is an eigenvalue of the matrix

$$\begin{bmatrix} 0 & [\mathscr{J}\mathbf{\Gamma}_{01} - \hat{\omega}_1\dot{\mathbf{\Gamma}}_{01}, \mathbf{\Gamma}_{00}^*]_{2\pi} \\ 0 & [\mathscr{J}\mathbf{\Gamma}_{01} - \hat{\omega}_1\dot{\mathbf{\Gamma}}_{01}, \mathbf{\Gamma}_{01}^*]_{2\pi} \end{bmatrix}. \tag{XI.26}$$

One eigenvalue is $\gamma_1^{(1)} = 0$; the other is then

$$\gamma_1^{(2)} = [\mathscr{J}\mathbf{\Gamma}_{01} - \hat{\omega}_1\dot{\mathbf{\Gamma}}_{01}, \mathbf{\Gamma}_{01}^*]_{2\pi}. \tag{XI.27}$$

In the present case our assumption that the loss of stability is strict implies that $\gamma_1^{(2)} > 0$. We remark that in the present case of a double semi-simple eigenvalue, the loss of stability of the bifurcated Hopf solution treated in §VIII.4 is associated with the special values

$$\hat{\omega}_1 = [\mathscr{J}\mathbf{\Gamma}_{00}, \mathbf{\Gamma}_{00}^*]_{2\pi} = [\mathscr{J}\mathbf{\Gamma}_{00}, \mathbf{\Gamma}_{01}^*]_{2\pi} = 0.$$

XI.7 Strict Loss of Stability at a Double Eigenvalue of Index Two*

At a double eigenvalue of index two we have two branches of eigenvalues $\gamma(\mu)$ with eigenfunction $\mathbf{\Gamma}(\mu)$ and $\tilde{\gamma}(\mu)$ with eigenfunction $\tilde{\mathbf{\Gamma}}(\mu)$ coming together at μ_0 in such a way that $\gamma(\mu_0) = \tilde{\gamma}(\mu_0) = 0$ and $\mathbf{\Gamma}_0 = \tilde{\mathbf{\Gamma}}_0 = \dot{\mathbf{U}}_0(s)$. In other words $\gamma = 0$ is a double eigenvalue of J_0 but there is only one eigenvector $\dot{\mathbf{U}}_0$ in the null space of J_0. We start at the point of degeneracy and seek to separate the eigenvalues by perturbing with μ. In fact we know from the start that one of the eigenvalues $\gamma(\mu) \equiv 0$ for μ near μ_0 and the other one, which is smooth in μ, controls stability.

At criticality $\gamma(\mu_0) = \gamma_0 = 0$,

$$\mathbf{\Gamma}_{00} = \dot{\mathbf{U}}_0,$$

$$\left.\begin{aligned} J_0\mathbf{\Gamma}_{00} &= 0 \\ J_0\mathbf{\Gamma}_{01} &= \mathbf{\Gamma}_{00} \end{aligned}\right\} \tag{XI.28}$$

$$\left.\begin{aligned} J_0^*\mathbf{\Gamma}_{01}^* &= 0 \\ J_0^*\mathbf{\Gamma}_{00}^* &= \mathbf{\Gamma}_{01}^* \end{aligned}\right\} \tag{XI.29}$$

$$[\mathbf{\Gamma}_{00}, \mathbf{\Gamma}_{00}^*]_{2\pi} = [\mathbf{\Gamma}_{01}, \mathbf{\Gamma}_{01}^*]_{2\pi} = 1 \tag{XI.30$_1$}$$

and

$$[\mathbf{\Gamma}_{01}, \mathbf{\Gamma}_{00}^*]_{2\pi} = [\mathbf{\Gamma}_{00}, \mathbf{\Gamma}_{01}^*]_{2\pi} = 0. \tag{XI.30$_2$}$$

We may assert that

$$\mathbf{\Gamma}(s, \mu) = \dot{\hat{\mathbf{U}}}(s, \mu), \qquad \gamma(\mu) = 0.$$

is the branch which is neutral with regard to stability. Stability of the Hopf solution is then determined by the sign of the second eigenvalue $\tilde{\gamma}(\mu)$ with eigenfunction $\tilde{\mathbf{\Gamma}}(\cdot, \mu)$. Of course, $\gamma(\mu_0) = \tilde{\gamma}(\mu_0) = 0$ and $\mathbf{\Gamma}(s, \mu_0) = \tilde{\mathbf{\Gamma}}(s, \mu_0) = \mathbf{\Gamma}_{00}$. The derivatives of $\tilde{\gamma}(\mu)$ and $\tilde{\mathbf{\Gamma}}(\cdot, \mu)$ at criticality satisfy

$$\tilde{\gamma}_1\mathbf{\Gamma}_{00} + \hat{\omega}_1\dot{\mathbf{\Gamma}}_{00} = J_0\tilde{\mathbf{\Gamma}}_1 + \mathcal{J}\mathbf{\Gamma}_{00}, \tag{XI.31$_1$}$$

$$\tilde{\gamma}_2\mathbf{\Gamma}_{00} + 2\tilde{\gamma}_1\tilde{\mathbf{\Gamma}}_1 + \hat{\omega}_2\dot{\mathbf{\Gamma}}_{00} + 2\hat{\omega}_1\dot{\tilde{\mathbf{\Gamma}}}_1 = J_0\tilde{\mathbf{\Gamma}}_2 + 2\mathcal{J}(\tilde{\mathbf{\Gamma}}_1) + \mathbf{m}(\mathbf{\Gamma}_{00}) \tag{XI.31$_2$}$$

where

$$\begin{aligned} \mathbf{m}(\mathbf{\Gamma}_{00}) \overset{\text{def}}{=} {} & \mathbf{F}_{vv}(\mu_0, \mathbf{U}_0|\hat{\mathbf{U}}_2|\mathbf{\Gamma}_{00}) + \mathbf{F}_{\mu\mu v}(\mu_0, \mathbf{U}_0|\mathbf{\Gamma}_{00}) \\ & + 2\mathbf{F}_{\mu vv}(\mu_0, \mathbf{U}_0|\hat{\mathbf{U}}_1|\mathbf{\Gamma}_{00}) + \mathbf{F}_{vvv}(\mu_0, \mathbf{U}_0|\hat{\mathbf{U}}_1|\hat{\mathbf{U}}_1|\mathbf{\Gamma}_{00}) \end{aligned}$$

* See §IV.4.4.2.

and all functions of s are 2π-periodic. The derivatives of the neutrally stable solution satisfy

$$\hat{\omega}_1 \dot{\mathbf{\Gamma}}_{00} = J_0 \mathbf{\Gamma}_1 + \mathscr{J} \mathbf{\Gamma}_{00}, \tag{XI.32$_1$}$$

$$\hat{\omega}_2 \dot{\mathbf{\Gamma}}_{00} + 2\hat{\omega}_1 \dot{\mathbf{\Gamma}}_1 = J_0 \mathbf{\Gamma}_2 + 2\mathscr{J} \mathbf{\Gamma}_1 + \mathbf{m}(\mathbf{\Gamma}_{00}). \tag{XI.32$_2$}$$

Application of the orthogonality condition (XI.30) to (XI.31) and (XI.32), using (XI.28) and (XI.29), leads at first order to

$$\tilde{\gamma}_1 + \hat{\omega}_1 [\dot{\mathbf{\Gamma}}_{00}, \mathbf{\Gamma}_{00}^*]_{2\pi} = [\tilde{\mathbf{\Gamma}}_1, \mathbf{\Gamma}_{01}^*]_{2\pi} + [\mathscr{J} \mathbf{\Gamma}_{00}, \mathbf{\Gamma}_{00}^*]_{2\pi}, \tag{XI.33}$$

$$\hat{\omega}_1 [\dot{\mathbf{\Gamma}}_{00}, \mathbf{\Gamma}_{01}^*]_{2\pi} = [\mathscr{J} \mathbf{\Gamma}_{00}, \mathbf{\Gamma}_{01}^*]_{2\pi}, \tag{XI.34}$$

$$\hat{\omega}_1 [\dot{\mathbf{\Gamma}}_{00}, \mathbf{\Gamma}_{00}^*]_{2\pi} = [\mathbf{\Gamma}_1, \mathbf{\Gamma}_{01}^*]_{2\pi} + [\mathscr{J} \mathbf{\Gamma}_{00}, \mathbf{\Gamma}_{00}^*]_{2\pi}. \tag{XI.35}$$

We may decompose the first derivatives of $\mathbf{\Gamma}(s, \mu)$, and $\tilde{\mathbf{\Gamma}}(s, \mu)$ uniquely as follows:

$$\mathbf{\Gamma}_1(s) = A_1 \mathbf{\Gamma}_{00}(s) + B_1 \mathbf{\Gamma}_{01}(s) + \boldsymbol{\chi}(s) \tag{XI.36$_1$}$$

and

$$\tilde{\mathbf{\Gamma}}_1(s) = \tilde{A}_1 \mathbf{\Gamma}_{00}(s) + \tilde{B}_1 \mathbf{\Gamma}_{01}(s) + \tilde{\boldsymbol{\chi}}(s), \tag{XI.36$_2$}$$

where $[\boldsymbol{\chi}, \mathbf{\Gamma}_{0l}^*]_{2\pi} = [\tilde{\boldsymbol{\chi}}, \mathbf{\Gamma}_{0l}^*]_{2\pi} = 0$ for $l = 0, 1$. The relation

$$\tilde{\gamma}_1 = \tilde{B}_1 - B_1 = [\tilde{\mathbf{\Gamma}}_1, \mathbf{\Gamma}_{01}^*]_{2\pi} - [\mathbf{\Gamma}_1, \mathbf{\Gamma}_{01}^*]_{2\pi} \tag{XI.37}$$

is implied by (XI.30, 33, 35, 36).

To complete the derivation of the strict crossing condition we must evaluate the constants on the right-hand side of (XI.37) using conditions for the solvability of $(\text{XI.31})_2$ and $(\text{XI.32})_2$. As a preliminary to this evaluation we derive the relation

$$J_0(\boldsymbol{\chi} - \tilde{\boldsymbol{\chi}}) = 0. \tag{XI.38$_1$}$$

To derive $(\text{XI.38})_1$, we subtract $(\text{XI.32})_1$ from $(\text{XI.31})_1$ and find that $J_0(\tilde{\mathbf{\Gamma}}_1 - \mathbf{\Gamma}_1) = \tilde{\gamma}_1 \mathbf{\Gamma}_{00}$. Then we use the decompositions $(\text{XI.36})_1$ and $(\text{XI.36})_2$ and simplify using (XI.28) and (XI.37). It follows from $(\text{XI.38})_1$ and the uniqueness of the decompositions that

$$\boldsymbol{\chi} = \tilde{\boldsymbol{\chi}}. \tag{XI.38$_2$}$$

We next form the scalar product of $(\text{XI.31})_2$ and $(\text{XI.32})_2$ with $\mathbf{\Gamma}_{01}^*$ and eliminate terms common to the two equations. After applying the bi-orthogonality conditions we get

$$2\tilde{\gamma}_1 \tilde{B}_1 + 2\hat{\omega}_1 [(\dot{\tilde{\mathbf{\Gamma}}}_1 - \dot{\mathbf{\Gamma}}_1), \mathbf{\Gamma}_{01}^*]_{2\pi} - 2[\mathscr{J}(\tilde{\mathbf{\Gamma}}_1 - \mathbf{\Gamma}_1), \mathbf{\Gamma}_{01}^*]_{2\pi} = 0. \tag{XI.39}$$

Upon application of (XI.36, 38, 32) this reduces to

$$\tilde{\gamma}_1 \tilde{B}_1 + (\tilde{B}_1 - B_1)\{\hat{\omega}_1 [\dot{\mathbf{\Gamma}}_{01}, \mathbf{\Gamma}_{01}^*]_{2\pi} - [\mathscr{J}(\mathbf{\Gamma}_{01}), \mathbf{\Gamma}_{01}^*]_{2\pi}\} = 0. \tag{XI.40}$$

It now follows from (XI.37) that

$$\tilde{B}_1 + \hat{\omega}_1[\dot{\mathbf{\Gamma}}_{01}, \mathbf{\Gamma}^*_{01}]_{2\pi} - [\mathscr{J}(\mathbf{\Gamma}_{01}), \mathbf{\Gamma}^*_{01}]_{2\pi} = 0 \tag{XI.41}$$

provided that $B_1 \neq \tilde{B}_1$. The possibility that $B_1 = \tilde{B}_1$ may be excluded since it implies that the solution $\gamma_1 = 0$ belongs to the neutral branch. Combining (XI.37) and (XI.41) we find that

$$\tilde{\gamma}_1 = -\hat{\omega}_1[\dot{\mathbf{\Gamma}}_{01}, \mathbf{\Gamma}^*_{01}]_{2\pi} + [\mathscr{J}(\mathbf{\Gamma}_{01}), \mathbf{\Gamma}^*_{01}]_{2\pi} - B_1,$$

where

$$B_1 = [\mathbf{\Gamma}_1, \mathbf{\Gamma}^*_{01}]_{2\pi}$$

may be written as

$$B_1 = [\dot{\hat{\mathbf{U}}}_1, \mathbf{\Gamma}^*_{01}]_{2\pi}$$

because $\mathbf{\Gamma}(\cdot, \mu) = \dot{\hat{\mathbf{U}}}(\cdot, \mu)$. By strict loss of stability we understand here that

$$0 < \tilde{\gamma}_1 = [\mathscr{J}\mathbf{\Gamma}_{01} - \hat{\omega}_1\dot{\mathbf{\Gamma}}_{01}, \mathbf{\Gamma}^*_{01}]_{2\pi} - [\dot{\hat{\mathbf{U}}}_1, \mathbf{\Gamma}^*_{01}]_{2\pi} \tag{XI.42}$$

where, using (XI.35),

$$[\dot{\hat{\mathbf{U}}}_1, \mathbf{\Gamma}^*_{01}]_{2\pi} = [\mathscr{J}\mathbf{\Gamma}_{00} - \hat{\omega}_1\mathbf{\Gamma}_{00}, \mathbf{\Gamma}^*_{00}]_{2\pi}.$$

Equation (XI.42) holds, not only in the present case of an index-two eigenvalue, but also for the semi-simple case treated in §XI.6. In the semi-simple case the second term of (XI.42) vanishes as a consequence of (XI.25).

XI.8 Formulation of the Problem of Subharmonic Bifurcation of Periodic Solutions of Autonomous Problems

In formulating the problem of subharmonic bifurcation of the Hopf solution (XI.3) it is convenient to map the periods of the Hopf solution and the subharmonic bifurcating solution into the same fixed domain. To explain this important point of convenience we note that the Hopf solution is given by (XI.3) as

$$\mathbf{V} = \hat{\mathbf{U}}(s, \mu) = \hat{\mathbf{U}}(s + 2\pi, \mu), \qquad s = \hat{\omega}(\mu)t$$

and we seek a subharmonic bifurcating solution of (XI.2)

$$\mathbf{V} = \hat{\boldsymbol{\psi}}(s, \mu) = \hat{\boldsymbol{\psi}}(s + 2\pi n, \mu), \qquad s = \hat{\Omega}(\mu)t$$

which is strictly $2\pi n$-periodic ($n \in \mathbb{N}$) in the reduced variable s, and is such that

$$\mathbf{U}_0(s) \stackrel{\text{def}}{=} \hat{\mathbf{U}}(s, \mu_0) = \hat{\boldsymbol{\psi}}(s, \mu_0)$$

with

$$\hat{\omega}(\mu_0) = \omega_0 = \hat{\Omega}(\mu_0) = \Omega_0.$$

We shall find that, in general, $\hat{\omega}(\mu) \neq \hat{\Omega}(\mu)$ so that $\hat{\boldsymbol{\psi}}(s, \mu) = \hat{\boldsymbol{\psi}}(\hat{\Omega}(\mu)t, \mu)$ is not generally $2\pi n/\hat{\omega}(\mu)$-periodic in t. The functions $\hat{\mathbf{U}}(s, \mu) = \hat{\mathbf{U}}(s + 2\pi, \mu)$ and $\hat{\omega}(\mu)$ satisfy (XI.3) and the functions $\hat{\boldsymbol{\psi}}(s, \mu) = \hat{\boldsymbol{\psi}}(s + 2\pi n, \mu)$ and $\hat{\Omega}(\mu)$ satisfy

$$\hat{\Omega}(\mu)\frac{d\hat{\boldsymbol{\psi}}}{ds} = \mathbf{F}(\mu, \hat{\boldsymbol{\psi}}(s, \mu)). \tag{XI.43}$$

The difference

$$\hat{\mathbf{Y}}(s, \mu) = \hat{\mathbf{U}}(s, \mu) - \hat{\boldsymbol{\psi}}(s, \mu) = \hat{\mathbf{Y}}(s + 2\pi n, \mu) \tag{XI.44}$$

is $2\pi n$-periodic in s, even though the functions defining this difference

$$\hat{\mathbf{U}}(\hat{\omega}(\mu)t, \mu) \quad \text{and} \quad \hat{\boldsymbol{\psi}}(\hat{\Omega}(\mu)t, \mu)$$

have different periods in t.

In the autonomous problem the ratio of the period $T(\hat{\Omega})$ of the bifurcated subharmonic solution to the period $T(\hat{\omega})$ of the given periodic solution is

$$\frac{T(\hat{\Omega})}{T(\hat{\omega})} = \frac{n\hat{\omega}(\mu)}{\hat{\Omega}(\mu)}, \qquad \hat{\omega}(\mu_0) = \hat{\Omega}(\mu_0).$$

Hence in general $T(\hat{\Omega})$ does not equal $nT(\hat{\omega})$,

XI.9 The Amplitude of the Bifurcating Solution

We may define an amplitude α for the bifurcating solution by any good linear functional of the difference $\hat{\mathbf{Y}}(s, \mu)$. Our choice is exactly the one used to study bifurcation of forced T-periodic solutions treated in §IX.10; that is,

$$a(\alpha) = [\hat{\mathbf{Y}}(s, \mu), \mathbf{Z}_1^*(s)]_{2\pi n} \tag{XI.45$_1$}$$

where $\mathbf{Z}_1^*(s) = \mathbf{Z}_1^*(s + 2\pi n)$ is given by (XI.15), (XI.16) where

$$\mathbb{J}\mathbf{Z}_1^* = 0$$

and

$$a(\alpha) = \alpha \qquad \text{when } \frac{m}{n} = \frac{0}{1}, \frac{1}{2} \tag{XI.45$_2$}$$

or

$$a(\alpha) = \alpha e^{i\phi(\alpha)} \quad \text{for } \frac{m}{n} \neq \frac{0}{1}, \frac{1}{2}. \tag{XI.45$_3$}$$

It is better to parameterize the bifurcating solution with α than with μ. The expansion of $\hat{\mathbf{Y}}(s, \mu)$ proceeds in powers of $\sqrt{|\mu - \mu_0|}$ whenever

$d\mu(0)/d\alpha = 0$ and $d^2\mu(0)/d\alpha^2 \neq 0$. In this case, as in the one in which $d\mu(0)/d\alpha \neq 0$, the series expansions of

$$\begin{aligned}
\mu &\stackrel{\text{def}}{=} \mu(\alpha) \\
\mathbf{U}(s, \alpha) &\stackrel{\text{def}}{=} \hat{\mathbf{U}}(s, \mu(\alpha)) \\
\omega(\alpha) &\stackrel{\text{def}}{=} \hat{\omega}(\mu(\alpha)) \\
\boldsymbol{\psi}(s, \alpha) &\stackrel{\text{def}}{=} \hat{\boldsymbol{\psi}}(s, \mu(\alpha)) \\
\Omega(\alpha) &\stackrel{\text{def}}{=} \hat{\Omega}(\mu(\alpha)) \\
\mathbf{Y}(s, \alpha) &\stackrel{\text{def}}{=} \hat{Y}(s, \mu(\alpha))
\end{aligned} \tag{XI.46}$$

are in integral powers of α. We adopt the following notation to distinguish between derivatives:

$$(\cdot)_n = \frac{\partial^n(\cdot)}{\partial \mu^n} \quad \text{at } \mu = \mu_0$$

$$(\cdot)^{(n)} = \frac{\partial^n(\cdot)}{\partial \alpha^n} \quad \text{at } \alpha = 0,$$

and seek a series solution for the functions $\boldsymbol{\psi}(s, \alpha) = \boldsymbol{\psi}(s + 2\pi n, \alpha)$, $\Omega(\alpha)$, and $\mathbf{Y}(s, \alpha) = \mathbf{Y}(s + 2\pi n, \alpha)$. On the other hand, the functions $\hat{\mathbf{U}}(s, \mu)$ and $\hat{\omega}(\mu)$ are given; they are assumed to be known from a previous computation of Hopf bifurcation. It follows, for example, that

$$\omega^{(1)} = \mu^{(1)}\hat{\omega}_1 \tag{XI.47}$$

etc., so that $(\cdot)^n$ may be calculated when $\mu^{(1)}, \ldots, \mu^{(n)}$ are known.

To insure that $\boldsymbol{\psi}(s, \alpha)$ is not merely a phase shift of $\mathbf{U}(s, \alpha)$, we require that $\mathbf{Y}(s, \alpha)$ should contain no part proportional to $\dot{\mathbf{U}}_0(s) = \mathbf{Z}_0(s)$; that is,

$$[\mathbf{Y}(s, \alpha), \mathbf{Z}_0^*]_{2\pi n} = 0. \tag{XI.48}$$

(A discussion of the principle behind (XI.48) is given in the introduction to this chapter.)

XI.10 Power-Series Solutions of the Bifurcation Problem

We turn next to the construction of the functions $\boldsymbol{\psi}(s, \alpha)$, $\mu(\alpha)$ and $\Omega(\alpha)$ in a series of powers of α:

$$\begin{bmatrix} \mathbf{U}(s, \alpha) - \mathbf{U}_0(s) \\ \boldsymbol{\psi}(s, \alpha) - \mathbf{U}_0(s) \\ \mathbf{Y}(s, \alpha) \\ \omega(\alpha) - \omega_0 \\ \Omega(\alpha) - \omega_0 \\ \mu(\alpha) - \mu_0 \end{bmatrix} = \sum_{n=1} \frac{\alpha^n}{n!} \begin{bmatrix} \mathbf{U}^{(n)}(s) \\ \boldsymbol{\psi}^{(n)}(s) \\ \mathbf{Y}^{(n)}(s) \\ \omega^{(n)} \\ \Omega^{(n)} \\ \mu^{(n)} \end{bmatrix}, \tag{XI.49}$$

where

$$\begin{aligned} \mathbf{U}(s, \alpha) &= \mathbf{U}(s + 2\pi, \alpha) \\ \boldsymbol{\psi}(s, \alpha) &= \boldsymbol{\psi}(s + 2\pi, \alpha) \\ \mathbf{Y}(s, \alpha) &= \mathbf{Y}(s + 2\pi, \alpha) \end{aligned} \tag{XI.50}$$

and

$$\begin{aligned} a^{(l)} &= [\mathbf{Y}^{(l)}, \mathbf{Z}_1^*]_{2\pi n}, \\ 0 &= [\mathbf{Y}^{(l)}, \mathbf{Z}_0^*]_{2\pi n} \qquad l > 0. \end{aligned} \tag{XI.51}$$

The coefficients in (XI.49) may be obtained by solving equations which arise from (XI.4) and (XI.43) by differentiation. The derivatives of the Hopf solution satisfy

$$\omega^{(1)}\dot{\mathbf{U}}_0 = \mathbb{J}\mathbf{U}^{(1)} + \mu^{(1)}\mathbf{F}_\mu(\mu_0, \mathbf{U}_0), \tag{XI.52$_1$}$$

$$\begin{aligned} \omega^{(2)}\dot{\mathbf{U}}_0 &+ 2\omega^{(1)}\dot{\mathbf{U}}^{(1)} \\ &= \mathbb{J}\mathbf{U}^{(2)} + 2\mu^{(1)}\mathbf{F}_{\mu v}(\mu_0, \mathbf{U}_0|\mathbf{U}^{(1)}) \\ &\quad + \mathbf{F}_{vv}(\mu_0, \mathbf{U}_0|\mathbf{U}^{(1)}|\mathbf{U}^{(1)}) \\ &\quad + (\mu^{(1)})^2\mathbf{F}_{\mu\mu}(\mu_0, \mathbf{U}_0) + \mu^{(2)}\mathbf{F}_\mu(\mu_0, \mathbf{U}_0), \end{aligned} \tag{XI.52$_2$}$$

$$\begin{aligned} \omega^{(3)}\dot{\mathbf{U}}_0 &+ 3\omega^{(2)}\dot{\mathbf{U}}^{(1)} + 3\omega^{(1)}\dot{\mathbf{U}}^{(2)} \\ &= \mathbb{J}\mathbf{U}^{(3)} + 3\mu^{(2)}\mathbf{F}_{v\mu}(\mu_0, \mathbf{U}_0|\mathbf{U}^{(1)}) + 3\mathbf{F}_{vv}(\mu_0, \mathbf{U}_0|\mathbf{U}^{(1)}|\mathbf{U}^{(2)}) \\ &\quad + \mathbf{F}_{vvv}(\mu_0, \mathbf{U}_0|\mathbf{U}^{(1)}|\mathbf{U}^{(1)}|\mathbf{U}^{(1)}) + \mu^{(3)}\mathbf{F}_\mu(\mu_0, \mathbf{U}_0) \\ &\quad + \mu^{(1)}\{3\mathbf{F}_{v\mu}(\mu_0, \mathbf{U}_0|\mathbf{U}^{(2)}) + \text{ktlo}\} \end{aligned} \tag{XI.52$_3$}$$

where ktlo = known terms of lower order,

$$\omega^{(l)}\dot{\mathbf{U}}_0 = \mathbb{J}\mathbf{U}^{(l)} + \mu^{(l)}\mathbf{F}_\mu(\mu_0, \mathbf{U}_0) + l\mu^{(l-1)}\mathbf{F}_{v\mu}(\mu_0, \mathbf{U}_0|\mathbf{U}^{(1)}) + (\text{ktlo}), \tag{XI.52$_4$}$$

where $\mathbf{U}^{(l)}(s) = \mathbf{U}^{(l)}(s + 2\pi)$ and $\mathbb{J}(\cdot) = -\omega_0\, d/ds + \mathbf{F}_v(\mu_0, \mathbf{U}_0|\cdot)$. The derivatives of the subharmonic bifurcating solution satisfy

$$\Omega^{(1)}\dot{\mathbf{U}}_0 = \mathbb{J}\boldsymbol{\psi}^{(1)} + \mu^{(1)}\mathbf{F}_\mu(\mu_0, \mathbf{U}_0) \tag{XI.53$_1$}$$

$$\begin{aligned} \Omega^{(2)}\dot{\mathbf{U}}_0 + 2\Omega^{(1)}\dot{\boldsymbol{\psi}}^{(1)} &= \mathbb{J}\boldsymbol{\psi}^{(2)} + 2\mu^{(1)}F_{\mu v}(\mu_0, \mathbf{U}_0|\boldsymbol{\psi}^{(1)}) \\ &\quad + F_{vv}(\mu_0, \mathbf{U}_0|\boldsymbol{\psi}^{(1)}|\boldsymbol{\psi}^{(1)}) + (\mu^{(1)})^2\mathbf{F}_{\mu\mu}(\mu_0, \mathbf{U}_0) \\ &\quad + \mu^{(2)}\mathbf{F}_\mu(\mu_0, \mathbf{U}_0), \end{aligned} \tag{XI.53$_2$}$$

$$\begin{aligned} \Omega^{(3)}\dot{\mathbf{U}}_0 &+ 3\Omega^{(2)}\dot{\boldsymbol{\psi}}^{(1)} + 3\Omega^{(1)}\dot{\boldsymbol{\psi}}^{(2)} \\ &= \mathbb{J}\boldsymbol{\psi}^{(3)} + 3\mu^{(2)}\mathbf{F}_{v\mu}(\mu_0, \mathbf{U}_0|\boldsymbol{\psi}^{(1)}) + 3\mathbf{F}_{vv}(\mu_0, \mathbf{U}_0|\boldsymbol{\psi}^{(1)}|\boldsymbol{\psi}^{(2)}) \\ &\quad + \mathbf{F}_{vvv}(\mu_0, \mathbf{U}_0|\boldsymbol{\psi}^{(1)}|\boldsymbol{\psi}^{(1)}|\boldsymbol{\psi}^{(1)}) + \mu^{(3)}\mathbf{F}_\mu(\mu_0, \mathbf{U}_0) \\ &\quad + \mu_1\{3F_{v\mu}(\mu_0, \mathbf{U}_0|\boldsymbol{\psi}^{(2)}) + \text{ktlo}\}, \end{aligned} \tag{XI.53$_3$}$$

$$\Omega^{(l)}\dot{\mathbf{U}}_0 = \mathbb{J}\boldsymbol{\psi}^{(l)} + l\mu^{(l-1)}\mathbf{F}_{v\mu}(\mu_0, \mathbf{U}_0|\boldsymbol{\psi}^{(1)}) + \mu^{(l)}\mathbf{F}_\mu(\mu_0, \mathbf{U}_0) + (\text{ktlo}) \tag{XI.53$_4$}$$

where $\boldsymbol{\psi}^{(l)}(s) = \boldsymbol{\psi}^{(l)}(s + 2\pi n)$.

Equations governing $\mathbf{Y}^{(n)} = \mathbf{U}^{(n)} - \boldsymbol{\psi}^{(n)}$ may be formed by subtraction. Recalling that $\dot{\mathbf{U}}_0 = \mathbf{Z}_0$ we find, using (XI.52)$_1$ and (XI.53)$_1$, that

$$(\omega^{(1)} - \Omega^{(1)})\mathbf{Z}_0 = \mathbb{J}\mathbf{Y}^{(1)}. \tag{XI.54}$$

Since $[\mathbb{J}\mathbf{Y}^{(1)}, \mathbf{Z}_0^*]_{2\pi n} = [\mathbf{Y}^{(1)}, \mathbb{J}^*\mathbf{Z}_0^*]_{2\pi n}$, we get

$$(\omega^{(1)} - \Omega^{(1)})[\mathbf{Z}_0, \mathbf{Z}_0^*]_{2\pi n} = \omega^{(1)} - \Omega^{(1)} = 0 \tag{XI.55}$$

whenever $\mathbb{J}^*\mathbf{Z}_0^* = 0$. Since this last equation holds except when zero is a double eigenvalue of J_0 of index two we have $\omega^{(1)} = \Omega^{(1)}$ in almost all the cases. In the exceptional case when $n = 1$ and zero is an index-two eigenvalue of J_0 we have, using (XI.51)$_1$ with $a^{(1)} = 1$, $\mathbf{Z}_0 = \mathbf{\Gamma}_{00}$, $\mathbf{Z}_0^* = \mathbf{\Gamma}_{00}^*$, $\mathbf{Z}_1^* = \mathbf{\Gamma}_{01}^*$, that

$$(\omega^{(1)} - \Omega^{(1)}) = [\mathbf{Y}^{(1)}, \mathbb{J}^*\mathbf{\Gamma}_{00}^*] = [\mathbf{Y}^{(1)}, \mathbf{\Gamma}_{01}^*] = 1 \tag{XI.56}$$

does not vanish.

It follows that in the general cases the bifurcating solutions are strictly subharmonic through terms of order α, and $\mathbf{Y}^{(1)}$ satisfies

$$\mathbb{J}\mathbf{Y}^{(1)} = 0. \tag{XI.57}$$

It is useful to replace $\boldsymbol{\psi}^{(1)}$ in (XI.53)$_2$ with $\mathbf{U}^{(1)} - \mathbf{Y}^{(1)}$:

$$\begin{aligned}\mathbf{F}_{vv}(\mu_0, \mathbf{U}_0|\boldsymbol{\psi}^{(1)}|\boldsymbol{\psi}^{(1)}) = {} & \mathbf{F}_{vv}(\mu_0, \mathbf{U}_0|\mathbf{U}^{(1)}|\mathbf{U}^{(1)}) \\ & - 2\mathbf{F}_{vv}(\mu_0, \mathbf{U}_0|\mathbf{U}^{(1)}|\mathbf{Y}^{(1)}) \\ & + \mathbf{F}_{vv}(\mu_0, \mathbf{U}_0|\mathbf{Y}^{(1)}|\mathbf{Y}^{(1)})\end{aligned} \tag{XI.58}$$

Then, after subtracting (XI.53)$_2$ from (XI.52)$_2$ using (XI.55), we get

$$\begin{aligned}&(\omega^{(2)} - \Omega^{(2)})\mathbf{Z}_0 + 2\omega^{(1)}\dot{\mathbf{Y}}^{(1)} \\ &\quad = \mathbb{J}\mathbf{Y}^{(2)} + 2\mu^{(1)}\mathbf{F}_{v\mu}(\mu_0, \mathbf{U}_0|\mathbf{Y}^{(1)}) + 2\mathbf{F}_{vv}(\mu_0, \mathbf{U}_0|\mathbf{U}^{(1)}|\mathbf{Y}^{(1)}) \\ &\quad\quad - \mathbf{F}_{vv}(\mu_0, \mathbf{U}_0|\mathbf{Y}^{(1)}|\mathbf{Y}^{(1)}).\end{aligned} \tag{XI.59}$$

Equations (XI.57) and (XI.59) do not hold in the exceptional case ($n = 1$) of a double eigenvalue zero of index two; the correct equations for this case are given in §XI.13.

XI.11 Subharmonic Bifurcation When $n = 2$

We are going to reduce this problem to the problem of bifurcation of $2T$-periodic solutions in the case of T-periodic forcing. The forced problem was studied in §IX.12. The problem with $n = 1$ which was also treated in the forced case in §XI.12 has some new features which will be treated in §XI.13 and XI.14. The subharmonic solutions for the autonomous problem are strictly subharmonic in the reduced variable s (see §XI.8). The computation of periodicity of this subharmonic solution

$$\boldsymbol{\psi}(s, \alpha) = \boldsymbol{\psi}(s + 2\pi n, \alpha), \qquad s = \Omega(\alpha)t$$

in real time t requires an ancillary computation to determine a new frequency $\Omega(\alpha)$.

When $n = 2$, $m = 1$ and $\mathbb{J}\mathbf{Z}_0 = \mathbb{J}\mathbf{Z}_1 = 0$ (see §XI.15). It follows that we may find $\mathbf{Y}(s, \alpha)$ in the decomposed form

$$\mathbf{Y}(s, \alpha) = \alpha\mathbf{Z}_1(s) + \boldsymbol{\chi}(s, \alpha), \tag{XI.60}$$

where (XI.48) is satisfied,

$$[\boldsymbol{\chi}(\cdot, \alpha), \mathbf{Z}_1^*]_{4\pi} = 0,$$

and

$$[\mathbf{Y}(\cdot, \alpha), \mathbf{Z}_1^*]_{4\pi} = \alpha.$$

It then follows from $(\text{XI.51})_1$ and $(\text{XI.45})_2$ that

$$[\mathbf{Y}^{(1)}, \mathbf{Z}_1^*]_{4\pi} = 1$$

and from (XI.57) that

$$\mathbf{Y}^{(1)} = \mathbf{Z}_1. \tag{XI.61}$$

To solve (XI.59) it is necessary that

$$[\mathbb{J}\mathbf{Y}^{(2)}, \mathbf{Z}_1^*]_{4\pi} = 0 \tag*{(XI.62)$_1$}$$

and

$$[\mathbb{J}\mathbf{Y}^{(2)}, \mathbf{Z}_0^*]_{4\pi} = 0. \tag*{(XI.62)$_2$}$$

We apply (XI.62) to (XI.59) and after setting

$$\mathbf{U}^{(1)} = \mu^{(1)}\hat{\mathbf{U}}_1 \quad \text{and} \quad \omega^{(1)} = \mu^{(1)}\hat{\omega}_1 \tag{XI.63}$$

we find, using (XI.16), that

$$2\mu^{(1)}[(\mathscr{J}\mathbf{Z}_1 - \hat{\omega}_1\dot{\mathbf{Z}}_1), \mathbf{Z}_1^*]_{4\pi} - [\mathbf{F}_{vv}(\mu_0, \mathbf{U}_0|\mathbf{Z}_1|\mathbf{Z}_1), \mathbf{Z}_1^*]_{4\pi} = 0, \tag*{(XI.64)$_1$}$$

where $\mathscr{J}$ is defined by $(\text{XI.17})_2$. Using (XI.20), we find that

$$2\mu^{(1)}\xi_1 - [\mathbf{F}_{vv}(\mu_0, \mathbf{U}_0|\mathbf{Z}_1|\mathbf{Z}_1), \mathbf{Z}_1^*]_{4\pi} = 0. \tag*{(XI.64)$_2$}$$

Since $\xi_1 > 0$, $(\text{XI.64})_2$ determines the value of $\mu^{(1)}$. In fact the second term of $(\text{XI.64})_2$ is in the form

$$[e^{is/2}\mathbf{F}_{vv}(\mu_0, \mathbf{U}_0|\boldsymbol{\Gamma}_0|\boldsymbol{\Gamma}_0), \boldsymbol{\Gamma}_0^*]_{4\pi} = 0$$

so that

$$\begin{aligned} \mu^{(1)} = 0, \qquad \mathbf{U}^{(1)}(s) = \mu^{(1)}\hat{\mathbf{U}}_1(s) = 0, \\ \Omega^{(1)} = \omega^{(1)} = \mu^{(1)}\hat{\omega}_1 = 0, \qquad \omega^{(2)} = \mu^{(2)}\hat{\omega}_1 \end{aligned} \tag{XI.65}$$

and

$$\boldsymbol{\psi}^{(1)} = -\mathbf{Y}^{(1)} = -\mathbf{Z}_1. \tag{XI.66}$$

It is convenient for a later application to write the equations for the present application with $n = 2$ in terms of $\mathbf{Y}^{(1)}$ rather than $\mathbf{Z}_1$. The same equations hold when $n > 2$, but then $\mathbf{Y}^{(1)}$ is a linear combination of $\mathbf{Z}_1$ and $\bar{\mathbf{Z}}_1$.

Using (XI.65) and (XI.66) we may simplify (XI.59):

$$(\mu^{(2)}\hat{\omega}_1 - \Omega^{(2)})\mathbf{Z}_0 = \mathbb{J}\mathbf{Y}^{(2)} - \mathbf{F}_{vv}(\mu_0, \mathbf{U}_0|\mathbf{Y}^{(1)}|\mathbf{Y}^{(1)}). \qquad (\text{XI.67})_1$$

Equation $(\text{XI.67})_1$ is solvable provided $(\text{XI.62})_2$ holds.

$$\Omega^{(2)} = \mu^{(2)}\hat{\omega}_1 + [\mathbf{F}_{vv}(\mu_0, \mathbf{U}_0|\mathbf{Y}^{(1)}|\mathbf{Y}^{(1)}), \mathbf{Z}_0^*]_{4\pi}. \qquad (\text{XI.67})_2$$

To compute $\mu^{(2)}$ and $\Omega^{(3)}$ we note that with the simplifications implied by (XI.65) and (XI.66), $(\text{XI.52})_3$ may be written as

$$\omega^{(3)}\mathbf{Z}_0 = \mathbb{J}\mathbf{U}^{(3)} + \mu^{(3)}\mathbf{F}_\mu(\mu_0, \mathbf{U}_0) \qquad (\text{XI.68})$$

and $(\text{XI.53})_3$ may be written as

$$\begin{aligned}\Omega^{(3)}\mathbf{Z}_0 - 3\Omega^{(2)}\mathbf{Y}_1 = {} & \mathbb{J}\boldsymbol{\psi}^{(3)} - 3\mu^{(2)}\mathbf{F}_{v\mu}(\mu_0, \dot{\mathbf{U}}_0|\mathbf{Y}^{(1)}) \\ & - 3\mathbf{F}_{vv}(\mu_0, \mathbf{U}_0|\mathbf{Y}^{(1)}|\mathbf{U}^{(2)} - \mathbf{Y}^{(2)}) \\ & - \mathbf{F}_{vvv}(\mu_0, U_0|\mathbf{Y}^{(1)}|\mathbf{Y}^{(1)}|\mathbf{Y}^{(1)}) \\ & + \mu^{(3)}\mathbf{F}_\mu(\mu_0, \mathbf{U}_0).\end{aligned} \qquad (\text{XI.69})$$

Note next that

$$\mathbf{U}^{(2)} = \mu^{(2)}\hat{\mathbf{U}}_1(s) \qquad (\text{XI.70})$$

and the difference between (XI.68) and (XI.69) becomes

$$\begin{aligned}&(\omega^{(3)} - \Omega^{(3)})\mathbf{Z}_0 + 3[\mathbf{F}_{vv}(\mu_0, \mathbf{U}_0|\mathbf{Y}^{(1)}|\mathbf{Y}^{(1)}), \mathbf{Z}_0^*]_{4\pi}\dot{\mathbf{Y}}^{(1)} \\ &\quad = \mathbb{J}\mathbf{Y}^{(3)} + 3\mu^{(2)}\{\mathscr{J}\mathbf{Y}^{(1)} - \hat{\omega}_1\dot{\mathbf{Y}}^{(1)}\} \\ &\quad\quad - 3\mathbf{F}_{vv}(\mu_0, \mathbf{U}_0|\mathbf{Y}^{(1)}|\mathbf{Y}^{(2)}) \\ &\quad\quad + F_{vvv}(\mu_0, \mathbf{U}_0|\mathbf{Y}^{(1)}|\mathbf{Y}^{(1)}|\mathbf{Y}^{(1)}).\end{aligned} \qquad (\text{XI.71})$$

Applying $(\text{XI.62})_1$ to (XI.71) we find, setting $\mathbf{Z}_1 = \mathbf{Y}^{(1)}$, that

$$\begin{aligned}\mu^{(2)}\xi_1 = {} & [F_{vv}(\mu_0, \mathbf{U}_0|\mathbf{Z}_1|\mathbf{Y}^{(2)}), \mathbf{Z}_1^*]_{4\pi} \\ & + [\mathbf{F}_{vv}(\mu_0, \mathbf{U}_0|\mathbf{Z}_1|\mathbf{Z}_1), \mathbf{Z}_0^*][\dot{\mathbf{Z}}_1, \mathbf{Z}_1^*]_{4\pi} \\ & - \tfrac{1}{3}[\mathbf{F}_{vvv}(\mu_0, \mathbf{U}_0|\mathbf{Z}_1|\mathbf{Z}_1|\mathbf{Z}_1), \mathbf{Z}_1^*]_{4\pi}.\end{aligned} \qquad (\text{XI.72})$$

In general, $\mu^{(2)} \neq 0$. Finally we note that $\mathbf{Y}^{(2)}(s)$ determined by the equation $(\text{XI.67})_1$ is 2π-periodic, as well as 4π-periodic; this fact follows from the fact that $\mathbf{Z}_0(s) = \mathbf{Z}_0(s + 2\pi)$ and $\mathbf{f}(s) \overset{\text{def}}{=} \mathbf{F}_{vv}(\mu_0, \mathbf{U}_0(s)|\mathbf{Z}_1(s)|\mathbf{Z}_1(s)) = e^{is}\mathbf{F}_{vv}(\mu_0, \mathbf{U}_0(s)|\boldsymbol{\Gamma}_0(s)|\boldsymbol{\Gamma}_0(s)) = \mathbf{f}(s + 2\pi)$. Hence when (XI.72) holds (XI.71) is solvable when $(\text{XI.62})_2$ holds. All the inhomogeneous terms in (XI.71) are of the form

$$e^{is/2}\boldsymbol{\zeta}(s) = e^{-is/2}\bar{\boldsymbol{\zeta}}(s), \qquad \boldsymbol{\zeta}(s) = \boldsymbol{\zeta}(s + 2\pi),$$

and the 4π-scalar products of these terms vanish. Thus

$$[\mathbb{J}\mathbf{Y}^{(3)}, \mathbf{Z}_0^*]_{4\pi} = \omega^{(3)} - \Omega^{(3)} = 0. \qquad (\text{XI.73})$$

The computation of higher-order terms is carried out in an identical fashion. At each step we need the strict crossing condition $\xi_1 \neq 0$ so we can solve for $\mu^{(k)}$.

XI.12 Subharmonic Bifurcation When $n > 2$

When $n \neq 1, 2$ we have (see (XI.15)) $\mathbb{J}\mathbf{Z}_0 = \mathbb{J}\mathbf{Z}_1 = \mathbb{J}\bar{\mathbf{Z}}_1 = 0$. It follows that we may find $\mathbf{Y}(s, \alpha)$ in the decomposed form

$$\mathbf{Y}(s, \alpha) = \alpha[e^{i\phi(\alpha)}\mathbf{Z}_1(s) + e^{-i\phi(\alpha)}\bar{\mathbf{Z}}_1(s)] + \boldsymbol{\chi}(s, \alpha), \tag{XI.74}$$

where (XI.48) is satisfied,

$$[\boldsymbol{\chi}(\cdot, \alpha), Z_1^*]_{2\pi n} = 0$$

and

$$[\mathbf{Y}(\cdot, \alpha), \mathbf{Z}_1^*]_{2\pi n} = \alpha e^{i\phi(\alpha)}.$$

It then follows from $(\text{XI.51})_1$ and $(\text{XI.45})_3$ that

$$[\mathbf{Y}^{(1)}, \mathbf{Z}_1^*]_{2\pi n} = e^{i\phi_0}$$

and from (XI.57) that

$$\begin{aligned}\mathbf{Y}^{(1)} &= e^{i\phi_0}\mathbf{Z}_1 + e^{-i\phi_0}\bar{\mathbf{Z}}_1\\ &= \exp i\left(\phi_0 + \frac{m}{n}s\right)\boldsymbol{\Gamma}_0 + \exp\left(-i\left(\phi_0 + \frac{m}{n}\right)s\right)\bar{\boldsymbol{\Gamma}}_0.\end{aligned} \tag{XI.75}$$

To solve (XI.59) it is necessary and sufficient that

$$[\mathbb{J}\mathbf{Y}^{(2)}, \mathbf{Z}_l^*]_{2\pi n} = [\mathbf{Y}^{(2)}, \mathbb{J}^*\mathbf{Z}_l^*]_{2\pi n} = 0, \qquad l = 0, 1, 2. \tag{XI.76}$$

We first apply (XI.76) to (XI.59) for $l = 1$. Since $\mathbf{Y}^{(2)}$ is real-valued we automatically have (XI.76) for $l = 2$. Then using (XI.63) we find

$$2\mu^{(1)}[\mathscr{J}\mathbf{Y}_1 - \hat{\omega}_1\dot{\mathbf{Y}}_1), \mathbf{Z}_1^*]_{2\pi n} - [\mathbf{F}_{vv}(\mu_0, \mathbf{U}_0|\mathbf{Y}_1|\mathbf{Y}_1), \mathbf{Z}_1^*]_{2\pi n} = 0. \tag{XI.77}$$

We may simplify (XI.77) by noting that if $\mathscr{L}$ is any linear, 2π-periodic operator and $\mathbf{Z}_1 = e^{i(m/n)s}\boldsymbol{\Gamma}_0(s)$, $\mathbf{Z}_1^* = e^{i(m/n)s}\boldsymbol{\Gamma}_0^*(s)$, then

$$[\mathscr{L}\bar{\mathbf{Z}}_1, \mathbf{Z}_1^*]_{2\pi n} = [e^{-2i(m/n)s}\mathscr{L}\bar{\boldsymbol{\Gamma}}_0, \boldsymbol{\Gamma}_0^*]_{2\pi n} = 0. \tag{XI.78}$$

We next replace $\mathbf{Y}^{(1)}$ in (XI.77) with the decomposition (XI.75) and utilize equation (XI.19) to reduce (XI.77) to

$$\begin{aligned}&2\mu^{(1)}\left(\gamma_1 - \frac{im}{n}\hat{\omega}_1\right)e^{i\phi_0} - e^{2i\phi_0}[\mathbf{F}_{vv}(\mu_0, \mathbf{U}_0|\mathbf{Z}_1|\mathbf{Z}_1), \mathbf{Z}_1^*]_{2\pi n}\\ &\quad - 2[\mathbf{F}_{vv}(\mu_0, \mathbf{U}_0|\mathbf{Z}_1|\bar{\mathbf{Z}}_1), \mathbf{Z}_1^*]_{2\pi n}\\ &\quad - e^{-2i\phi_0}[\mathbf{F}_{vv}(\mu_0, \mathbf{U}_0|\bar{\mathbf{Z}}_1|\bar{\mathbf{Z}}_1), \mathbf{Z}_1^*]_{2\pi n} = 0.\end{aligned} \tag{XI.79$_1$}$$

The second two terms in (XI.79)$_1$ vanish,

$$[e^{i(m/n)s}\mathbf{F}_{vv}(\mu_0, \mathbf{U}_0|\mathbf{\Gamma}_0|\mathbf{\Gamma}_0), \mathbf{\Gamma}_0^*]_{2\pi n} = 0$$

$$[e^{-i(m/n)s}\mathbf{F}_{vv}(\mu_0, \mathbf{U}_0|\mathbf{\Gamma}_0|\bar{\mathbf{\Gamma}}_0), \mathbf{\Gamma}_0^*]_{2\pi n} = 0,$$

and the last term vanishes unless $n = 3$

$$[e^{-3i(m/n)s}\mathbf{F}_{vv}(\mu_0, \mathbf{U}_0|\bar{\mathbf{\Gamma}}_0|\bar{\mathbf{\Gamma}}_0), \mathbf{\Gamma}_0^*]_{2\pi n} = \lambda_1\delta_{n3}.$$

Hence

$$2\mu^{(1)}\left(\gamma_1 - i\frac{m}{n}\hat{\omega}_1\right) - e^{-3i\phi_0}\lambda_1\delta_{n3} = 0, \tag{XI.79$_2$}$$

which is essentially the same as (IX.66). It gives $\mu^{(1)} = 0$ except in the case in which $n = 3$. The 6π-periodic solutions $\boldsymbol{\psi}(s, \alpha) = \boldsymbol{\psi}(s + 6\pi, \alpha)$ which bifurcate from $\hat{\mathbf{U}}(s, \mu(\alpha)) = \hat{\mathbf{U}}(s + 2\pi, \mu(\alpha))$, where $\alpha = 0$, have exactly the same properties as the $3T$-periodic solutions derived under §IX.14.

To compute the behavior of the bifurcating solutions in real time we must find the frequency $\Omega(\alpha)$. The second derivative Ω_2 of this frequency may be determined from (XI.59) using the condition that $[\mathbb{J}\mathbf{Y}^{(2)}, \mathbf{Z}_0^*]_{2\pi n} = 0$. The scalar products of the terms which are linear in $\mathbf{Y}_1$ are in the form $[e^{\pm i(m/n)s}a(s)]_{2\pi n} = 0$, where $a(s)$ is 2π-periodic and

$$\begin{aligned}[\mathbf{F}_{vv}(\mu_0, \mathbf{U}_0|\mathbf{Y}^{(1)}|\mathbf{Y}^{(1)}), \mathbf{Z}_0^*]_{2\pi n} \\ = 2[\mathbf{F}_{vv}(\mu_0, \mathbf{U}_0|\mathbf{Z}_1|\bar{\mathbf{Z}}_1), \mathbf{Z}_0^*]_{2\pi n} \\ = 2[\mathbf{F}_{vv}(\mu_0, \mathbf{U}_0|\mathbf{\Gamma}_0|\bar{\mathbf{\Gamma}}_0), \mathbf{Z}_0^*]_{2\pi n}.\end{aligned} \tag{XI.80}$$

It follows that for $n \in \mathbb{N}$, $n \neq 1, 2$ we have

$$\omega^{(2)} - \Omega^{(2)} = -[\mathbf{F}_{vv}(\mu_0, \mathbf{U}_0|\mathbf{\Gamma}_0|\bar{\mathbf{\Gamma}}_0), \mathbf{Z}_0^*]_{2\pi n}. \tag{XI.81}$$

We now assert that, apart from the calculation of the frequency $\Omega(\alpha)$ the 6π-periodic ($n = 3$) subharmonic solution of the reduced time s has all of the properties, including the stability properties (no stability for μ near μ_0) of the $3T$-periodic solutions computed under §IX.14.

We therefore turn our attention to the cases of subharmonic bifurcation in which $n \neq 1, 2, 3$. For all such cases we have $\mu^{(1)} = \mathbf{U}^{(1)} = \Omega^{(1)} = \omega^{(1)} = 0$. Equation (XI.71) also holds here, but with $\mathbf{Y}^{(1)} = e^{i\phi_0}\mathbf{Z}_1 + e^{-i\phi_0}\bar{\mathbf{Z}}_1$. This equation is solvable if (XI.76) holds for $\mathbf{Y}^{(3)}$ with $l = 0$ and $l = 1$. These biorthogonality conditions lead first with $l = 0$ to

$$\omega^{(3)} - \Omega^{(3)} = 0. \tag{XI.82}$$

The derivation of (XI.82) is similar to the derivation of (XI.73).

The second and third conditions $[\mathbb{J}\mathbf{Y}^{(3)}, \mathbf{Z}_1^*]_{2\pi n} = [\mathbb{J}\mathbf{Y}^{(3)}, \bar{\mathbf{Z}}_1^*]_{2\pi n} = 0$ for the solvability of (XI.71) lead to

$$\begin{aligned}\mu^{(2)}[\{\mathscr{J}\mathbf{Y}^{(1)} - \hat{\omega}_1\dot{\mathbf{Y}}^{(1)}\}, \mathbf{Z}_1^*]_{2\pi n} \\ = [\mathbf{F}_{vv}(\mu_0, \mathbf{U}_0|\mathbf{\Gamma}_0|\mathbf{\Gamma}_0), \mathbf{Z}_0^*]_{2\pi n}[\dot{\mathbf{Y}}^{(1)}, \mathbf{Z}_1^*]_{2\pi n} \\ + [\mathbf{F}_{vv}(\mu_0, \mathbf{U}_0|\mathbf{Y}^{(1)}|\mathbf{Y}^{(2)}), \mathbf{Z}_1^*]_{2\pi n} \\ - \tfrac{1}{3}[\mathbf{F}_{vvv}(\mu_0, \mathbf{U}_0|\mathbf{Y}^{(1)}|\mathbf{Y}^{(1)}|\mathbf{Y}^{(1)}), \mathbf{Z}_1^*]_{2\pi n}.\end{aligned} \tag{XI.83}$$

We next introduce (XI.75) into (XI.67) and find, using (XI.82),

$$
\begin{aligned}
&\mathbb{J}\mathbf{Y}^{(2)} + [\mathbf{F}_{vv}(\mu_0, \mathbf{U}_0 | \mathbf{\Gamma}_0 | \bar{\mathbf{\Gamma}}_0), \mathbf{Z}_0^*]_{2\pi} \mathbf{Z}_0 \\
&\quad - \exp 2i(\phi_0 + (m/n)s)\, \mathbf{F}_{vv}(\mu_0, \mathbf{U}_0 | \mathbf{\Gamma}_0 | \mathbf{\Gamma}_0) \\
&\quad - 2\mathbf{F}_{vv}(\mu_0, \mathbf{U}_0 | \mathbf{\Gamma}_0 | \bar{\mathbf{\Gamma}}_0) \\
&\quad - \exp(-2i(\phi_0 + (m/n)s))\, \mathbf{F}_{vv}(\mu_0, \mathbf{U}_0 | \bar{\mathbf{\Gamma}}_0 | \bar{\mathbf{\Gamma}}_0) = 0
\end{aligned}
$$

The decomposition (XI.74) now implies that

$$\mathbf{Y}^{(2)} = 2i\phi^{(1)}(e^{i\phi_0}\mathbf{Z}_1 - e^{-i\phi_0}\bar{\mathbf{Z}}_1) + \chi^{(2)}.$$

The solution of this equation which is orthogonal to $\mathbf{Z}_0^*$, $\mathbf{Z}_1^*$, and $\bar{\mathbf{Z}}_1^*$ is of the form

$$\chi^{(2)} = \zeta_0(s) + \exp 2i(\phi_0 + (m/n)s)\, \zeta_1(s) + \exp(-2i(\phi_0 + (m/n)s)\, \bar{\zeta}_1(s), \tag{XI.84}$$

where $\zeta_l(s) = \zeta_l(s + 2\pi)$, $l = 0, 1$ are periodic, to-be-determined, functions. Many terms in (XI.83) integrate to zero. Suppose

$$\mathbf{g} = e^{i(k/n)s}\hat{\mathbf{g}}(s), \quad \text{where } k = \pm m, \pm 3m.$$

Then

$$[\mathbf{g}, \mathbf{Z}_1^*]_{2\pi n} = \left[\exp\left(-i\left(\frac{m}{n} - \frac{k}{n}\right)s\right)\hat{\mathbf{g}}, \mathbf{\Gamma}_0^*\right]_{2\pi n} = 0 \tag{XI.85}$$

unless $k - m = rn$, where $r \in \mathbb{Z}$ and $0 \le m/n < 1$, $n \ge 4$. The only values $k = \pm m, \pm 3m$ leading to $r \in \mathbb{Z}$ are

$$k = m, n \quad \text{unrestricted}$$

and

$$k = -3m, \quad n = 4, \quad m = 1, 3.$$

We may therefore compute

$$[\dot{\mathbf{Y}}^{(1)}, \mathbf{Z}_1^*]_{2\pi} = e^{i\phi_0}\left\{i\frac{m}{n} + [\dot{\mathbf{\Gamma}}_0, \mathbf{\Gamma}_0^*]_{2\pi}\right\}$$

$$
\begin{aligned}
&[\mathbf{F}_{vv}(\mu_0, \mathbf{U}_0 | \mathbf{Y}^{(1)} | \mathbf{Y}^{(2)}), \mathbf{Z}_1^*]_{2\pi n} \\
&\quad = [\mathbf{F}_{vv}(\mu_0, \mathbf{U}_0 | \mathbf{Y}^{(1)} | \chi^{(2)}), \mathbf{Z}_1^*]_{2\pi n} \\
&\quad = e^{-3i\phi_0}[e^{-4i(m/n)s}\mathbf{F}_{vv}(\mu_0, \mathbf{U}_0 | \bar{\mathbf{\Gamma}}_0 | \bar{\zeta}_1), \mathbf{\Gamma}_0^*]_{2\pi n} \\
&\qquad + [\mathbf{F}_{vv}(\mu_0, \mathbf{U}_0 | \bar{\mathbf{\Gamma}}_0 | \zeta_1), \mathbf{\Gamma}_0^*]_{2\pi} \\
&\qquad + [\mathbf{F}_{vv}(\mu_0, \mathbf{U}_0 | \mathbf{\Gamma}_0 | \zeta_0 |), \mathbf{\Gamma}_0^*]_{2\pi},
\end{aligned}
$$

and

$$
\begin{aligned}
&[\mathbf{F}_{vvv}(\mu_0, \mathbf{U}_0 | \mathbf{Y}^{(1)} | \mathbf{Y}^{(1)} | \mathbf{Y}^{(1)}), \mathbf{Z}_1^*]_{2\pi n} \\
&\quad = e^{-3i\phi_0}[e^{-4i(m/n)s}\mathbf{F}_{vvv}(\mu_0, \mathbf{U}_0 | \bar{\mathbf{\Gamma}}_0 | \bar{\mathbf{\Gamma}}_0 | \bar{\mathbf{\Gamma}}_0), \mathbf{\Gamma}_0^*]_{2\pi n} \\
&\qquad + 3[\mathbf{F}_{vvv}(\mu_0, \mathbf{U}_0 | \bar{\mathbf{\Gamma}}_0 | \mathbf{\Gamma}_0 | \mathbf{\Gamma}_0), \mathbf{\Gamma}_0^*]_{2\pi n}
\end{aligned}
$$

Using these relations for the terms on the right-hand side of (XI.83) and simplifying the left-hand side with (XI.19), we find that

$$\mu^{(2)}\left(\gamma_1 - i\frac{m}{n}\hat{\omega}_1\right)e^{i\phi_0} = \lambda_2 e^{i\phi_0} = 0, \qquad n \geq 5 \tag{XI.86}$$

and

$$\mu^{(2)}\left(\gamma_1 - i\frac{m}{n}\hat{\omega}_1\right)e^{i\phi_0} = \lambda_2 e^{i\phi_0} + \lambda_3 e^{-3i\phi_0}, \qquad n = 4, \tag{XI.87}$$

where

$$\begin{aligned}
\lambda_2 &= [\mathbf{F}_{vv}(\mu_0, \mathbf{U}_0|\mathbf{\Gamma}_0|\mathbf{\Gamma}_0), \mathbf{Z}_0^*]_{2\pi}\left\{i\frac{m}{n} + [\dot{\mathbf{\Gamma}}_0, \mathbf{\Gamma}_0^*]_{2\pi}\right\} \\
&\quad + [\mathbf{F}_{vv}(\mu_0, \mathbf{U}_0|\bar{\mathbf{\Gamma}}_0|\zeta_1), \mathbf{\Gamma}_0^*]_{2\pi} + [\mathbf{F}_{vv}(\mu_0, \mathbf{U}_0|\mathbf{\Gamma}_0|\zeta_0), \mathbf{\Gamma}_0^*]_{2\pi} \\
&\quad - [\mathbf{F}_{vvv}(\mu_0, \mathbf{U}_0|\bar{\mathbf{\Gamma}}_0|\mathbf{\Gamma}_0|\mathbf{\Gamma}_0), \mathbf{\Gamma}_0^*]_{2\pi} \\
\lambda_3 &= [e^{-ims}\mathbf{F}_{vv}(\mu_0, \mathbf{U}_0|\bar{\mathbf{\Gamma}}_0|\bar{\zeta}_1), \mathbf{\Gamma}_0^*]_{2\pi} \\
&\quad - \tfrac{1}{3}[e^{-ims}\mathbf{F}_{vvv}(\mu_0, \mathbf{U}_0|\bar{\mathbf{\Gamma}}_0|\bar{\mathbf{\Gamma}}_0|\bar{\mathbf{\Gamma}}_0), \mathbf{\Gamma}_0^*]_{2\pi}, \qquad m = 1, 3.
\end{aligned}$$

Equation (XI.87) is in the form (IX.80), and (XI.86) is in the form (IX.101).

XI.13 Subharmonic Bifurcation When $n = 1$ in the Semi-Simple Case

We are now in the frame of case (a) of §XI.4. Zero is an index-one double (semi-simple) eigenvalue of $\mathbb{J} = J_0$ with two independent eigenvectors $\mathbf{\Gamma}_{00}$ and $\mathbf{\Gamma}_{01}$ and two independent adjoint eigenvectors $\mathbf{\Gamma}_{00}^*$ and $\mathbf{\Gamma}_{01}^*$ satisfying the biorthogonality conditions (XI.21). We seek a subharmonic solution $\boldsymbol{\psi}(s, \alpha) = \boldsymbol{\psi}(s + 2\pi, \alpha)$ and a frequency $\Omega(\alpha)$ in the series form given by (XI.49), where $\alpha = [\mathbf{Y}(s, \alpha), \mathbf{\Gamma}_{01}^*]_{2\pi}$ is the amplitude. Proceeding as in §XI.10, we find (XI.54) in the form

$$(\omega^{(1)} - \Omega^{(1)})\mathbf{\Gamma}_{00} = J_0\mathbf{Y}^{(1)}. \tag{XI.88}$$

Since $[J_0\mathbf{Y}^{(1)}, \mathbf{\Gamma}_{00}^*]_{2\pi} = 0$ and $[\mathbf{\Gamma}_{00}, \mathbf{\Gamma}_{00}^*]_{2\pi} = 1$, we find that

$$\Omega^{(1)} = \omega^{(1)} = \mu^{(1)}\hat{\omega}_1. \tag{XI.89}$$

Moreover, since $J_0\mathbf{Y}^{(1)} = 0$ and $\mathbf{\Gamma}_{00}$ and $\mathbf{\Gamma}_{01}$ are independent eigenvectors of J_0, we have $\mathbf{Y}^{(1)}$ as linear combination $C_1\mathbf{\Gamma}_{00} + C_2\mathbf{\Gamma}_{01}$. However, $(XI.51)_2$ (with $\mathbf{Z}_0^* = \mathbf{\Gamma}_{00}^*$) implies that $C_1 = 0$; $(XI.51)_1$ and $(XI.45)_2$ (with $\mathbf{Z}_1^* = \mathbf{\Gamma}_{01}^*$) imply that $C_2 = 1$. Hence

$$\mathbf{Y}^{(1)} = \mathbf{\Gamma}_{01}. \tag{XI.90}$$

Following again the line of equations in §XI.10, using (XI.63), (XI.89), and (XI.90), we find (XI.59) in the form

$$(\omega^{(2)} - \Omega^{(2)})\mathbf{\Gamma}_{00} = J_0\mathbf{Y}^{(2)} + 2\mu^{(1)}(\mathcal{J}\mathbf{\Gamma}_{01} - \hat{\omega}_1\dot{\mathbf{\Gamma}}_{01}) - \mathbf{F}_{vv}(\mu_0, \mathbf{U}_0|\mathbf{\Gamma}_{01}|\mathbf{\Gamma}_{01}). \tag{XI.91$_1$}$$

Necessary and sufficient conditions for the solvability of (XI.91) are that

$$[J_0\mathbf{Y}^{(2)}, \mathbf{\Gamma}^*_{00}]_{2\pi} = [J_0\mathbf{Y}^{(2)}, \mathbf{\Gamma}^*_{01}]_{2\pi} = 0. \tag{XI.91$_2$}$$

The second of these conditions gives

$$2\mu^{(1)}\gamma_1^{(2)} - [\mathbf{F}_{vv}(\mu_0, \mathbf{U}_0|\mathbf{\Gamma}_{01}|\mathbf{\Gamma}_{01}), \mathbf{\Gamma}^*_{01}]_{2\pi} = 0, \tag{XI.92}$$

where, according to (XI.27),

$$\gamma_1^{(2)} = [\mathcal{J}\mathbf{\Gamma}_{01} - \hat{\omega}_1\dot{\mathbf{\Gamma}}_{01}, \mathbf{\Gamma}^*_{01}]_{2\pi} > 0. \tag{XI.93}$$

So (XI.92) gives $\mu^{(1)}$. On the other hand, the other solvability condition leads to

$$\omega^{(2)} - \Omega^{(2)} = -[\mathbf{F}_{vv}(\mu_0, \mathbf{U}_0|\mathbf{\Gamma}_{01}|\mathbf{\Gamma}_{01}), \mathbf{\Gamma}^*_{00}]_{2\pi} + 2\mu^{(1)}[\mathcal{J}\mathbf{\Gamma}_{01} - \hat{\omega}_1\dot{\mathbf{\Gamma}}_{01}, \mathbf{\Gamma}^*_{00}]_{2\pi}.$$

To compute $\Omega^{(2)}$ we need the value of

$$\omega^{(2)} = \mu^{(2)}\hat{\omega}_1 + (\mu^{(1)})^2\hat{\omega}_2. \tag{XI.94}$$

We leave the specification of the algorithm for the computation of $\mu^{(2)}$ from the equation governing $\mathbf{Y}^{(3)}$ as an exercise for the reader.

XI.14 "Subharmonic" Bifurcation When $n = 1$ in the Case When Zero Is an Index-Two Double Eigenvalue of J_0

We turn next to case (b) of §XI.4. Zero is an index-two double eigenvalue of $\mathbb{J} = J_0$ with proper and generalized eigenvectors and adjoint eigenvectors satisfying (XI.28–30). We again seek a 2π-periodic bifurcating solution $\boldsymbol{\psi}(s, \alpha)$ and $\Omega(\alpha)$ in the series form given by (XI.49). The amplitude α is defined by

$$\alpha = [\mathbf{Y}(s, \alpha), \mathbf{\Gamma}^*_{01}]_{2\pi} \tag{XI.95}$$

Equation (XI.88) again governs at first order, but now

$$(\omega^{(1)} - \Omega^{(1)}) = [J_0\mathbf{Y}^{(1)}, \mathbf{\Gamma}^*_{00}]_{2\pi} = [\mathbf{Y}^{(1)}, J_0^*\mathbf{\Gamma}^*_{00}]_{2\pi} = [\mathbf{Y}^{(1)}, \mathbf{\Gamma}^*_{01}]_{2\pi} = 1. \tag{XI.96}$$

Since (XI.51)$_2$ requires that $[\mathbf{Y}^{(1)}, \mathbf{\Gamma}^*_{00}]_{2\pi} = 0$ we have

$$\Omega^{(1)} = \mu^{(1)}\hat{\omega}_1 - 1 \tag{XI.96}$$

and

$$\mathbf{Y}^{(1)} = \mathbf{\Gamma}_{01} = \mathbf{U}^{(1)} - \boldsymbol{\psi}^{(1)} = \mu^{(1)}\hat{\mathbf{U}}_1 - \boldsymbol{\psi}^{(1)}. \tag{XI.97}$$

The equation governing $\mathbf{Y}^{(2)}$ is now formed by subtracting $(\text{XI.53})_2$ from $(\text{XI.52})_2$ using (XI.96, 97) and (XI.63):

$$(\omega^{(2)} - \Omega^{(2)})\mathbf{\Gamma}_{00} + 2\dot{\boldsymbol{\psi}}_1 = J_0\mathbf{Y}^{(2)} + 2\mu^{(1)}\{\mathscr{J}\mathbf{\Gamma}_{01} - \hat{\omega}_1\dot{\mathbf{\Gamma}}_{01}\} - \mathbf{F}_{vv}(\mu_0, \mathbf{U}_0|\mathbf{\Gamma}_{01}|\mathbf{\Gamma}_{01}). \tag{XI.98}$$

This equation is solvable provided that $[J_0\mathbf{Y}^{(2)}, \mathbf{\Gamma}^*_{01}]_{2\pi} = 0$. Using (XI.97) and (XI.42) we find that

$$2\mu^{(1)}[\dot{\hat{\mathbf{U}}}_1, \mathbf{\Gamma}^*_{01}]_{2\pi} - 2[\dot{\mathbf{\Gamma}}_{01}, \mathbf{\Gamma}^*_{01}]_{2\pi} = 2\mu^{(1)}(\tilde{\gamma}_1 + [\dot{\hat{\mathbf{U}}}_1, \mathbf{\Gamma}^*_{01}]_{2\pi}) - [\mathbf{F}_{vv}(\mu_0, \mathbf{U}_0|\mathbf{\Gamma}_{01}|\mathbf{\Gamma}_{01}), \mathbf{\Gamma}^*_{01}]_{2\pi}.$$

Hence

$$2\mu^{(1)}\tilde{\gamma}_1 = [\mathbf{F}_{vv}(\mu_0, \mathbf{U}_0|\mathbf{\Gamma}_{01}|\mathbf{\Gamma}_{01}), \mathbf{\Gamma}^*_{01}]_{2\pi} - 2[\dot{\mathbf{\Gamma}}_{01}, \mathbf{\Gamma}^*_{01}]_{2\pi}. \tag{XI.99}$$

Since $\tilde{\gamma}_1 > 0$ by hypothesis, (XI.99) is solvable for $\mu^{(1)}$.

We compute the value of $\omega^{(2)} - \Omega^{(2)}$ by projecting (XI.98) with $\mathbf{\Gamma}^*_{00}$. Using (XI.95), we find that

$$[J_0\mathbf{Y}^{(2)}, \mathbf{\Gamma}^*_{00}]_{2\pi} = [\mathbf{Y}^{(2)}, J_0^*\mathbf{\Gamma}^*_{00}]_{2\pi} = [\mathbf{Y}^{(2)}, \mathbf{\Gamma}^*_{01}]_{2\pi} = 0,$$

$$\begin{aligned}\omega^{(2)} - \Omega^{(2)} &+ 2\mu^{(1)}[\dot{\hat{\mathbf{U}}}_1, \mathbf{\Gamma}^*_{00}]_{2\pi} - 2[\dot{\mathbf{\Gamma}}_{01}, \mathbf{\Gamma}^*_{00}]_{2\pi} \\ &= 2\mu^{(1)}[\{\mathscr{J}\mathbf{\Gamma}_{01} - \hat{\omega}_1\dot{\mathbf{\Gamma}}_{01}\}, \mathbf{\Gamma}^*_{00}]_{2\pi} \\ &\quad - [\mathbf{F}_{vv}(\mu_0, \mathbf{U}_0|\mathbf{\Gamma}_{01}|\mathbf{\Gamma}_{01}), \mathbf{\Gamma}^*_{00}]_{2\pi}.\end{aligned} \tag{XI.100}$$

$\Omega^{(2)}$ cannot be calculated from (XI.100) unless $\omega^{(2)} = \mu^{(2)}\hat{\omega}_1 + (\mu^{(1)})^2\hat{\omega}_2$ is known. So we need to determine $\mu^{(2)}$.

To find $\mu^{(2)}$, we first form the equation for $\mathbf{Y}^{(3)}$ by subtracting $(\text{XI.53})_3$ from $(\text{XI.52})_3$, using (XI.97):

$$\begin{aligned}(\omega^{(3)} &- \Omega^{(3)})\mathbf{\Gamma}_{00} + 3(\omega^{(2)} - \Omega^{(2)})\dot{\mathbf{U}}^{(1)} + 3\Omega^{(2)}\dot{\mathbf{\Gamma}}_{01} \\ &+ 3(\omega^{(1)} - \Omega^{(1)})\dot{\mathbf{U}}^{(2)} + 3\Omega^{(1)}\dot{\mathbf{Y}}^{(2)} \\ &= J_0\mathbf{Y}^{(3)} + 3\mu^{(2)}\mathbf{F}_{v\mu}(\mu_0, \mathbf{U}_0|\mathbf{\Gamma}_{01}) + 3\mathbf{F}_{vv}(\mu_0, \mathbf{U}_0|\mathbf{\Gamma}_{01}|\mathbf{U}^{(2)}) \\ &+ 3\mathbf{F}_{vv}(\mu_0, \mathbf{U}_0|\mathbf{U}^{(1)}|\mathbf{Y}^{(2)}) - 3\mathbf{F}_{vv}(\mu_0, \mathbf{U}_0|\mathbf{\Gamma}_{01}|\mathbf{Y}^{(2)}) \\ &+ 3\mathbf{F}_{vvv}(\mu_0, \mathbf{U}_0|\mathbf{U}^{(1)}|\mathbf{U}^{(1)}|\mathbf{\Gamma}_{01}) - 3\mathbf{F}_{vvv}(\mu_0, \mathbf{U}_0|\mathbf{U}^{(1)}|\mathbf{\Gamma}_{01}|\mathbf{\Gamma}_{01}) \\ &+ \mathbf{F}_{vvv}(\mu_0, \mathbf{U}_0|\mathbf{\Gamma}_{01}|\mathbf{\Gamma}_{01}|\mathbf{\Gamma}_{01}) + 3\mu^{(1)}\mathbf{F}_{v\mu}(\mu_0, \mathbf{U}_0|\mathbf{Y}^{(2)}) \\ &+ 3(\mu^{(1)})^2\mathbf{F}_{v\mu\mu}(\mu_0, \mathbf{U}_0|\mathbf{\Gamma}_{01}) + 6\mu^{(1)}F_{vv\mu}(\mu_0, \mathbf{U}_0|\mathbf{\Gamma}_{01}|\mathbf{U}^{(1)}) \\ &- 3\mu^{(1)}F_{vv\mu}(\mu_0, \mathbf{U}_0|\mathbf{\Gamma}_{01}|\mathbf{\Gamma}_{01}).\end{aligned} \tag{XI.101}$$

The unknowns in this equation are $\Omega^{(3)}$, $\mathbf{Y}^{(3)}$, and $\mu^{(2)}$. To identify the coefficient of $\mu^{(2)}$ we note that

$$\omega^{(1)} - \Omega^{(1)} = 1, \qquad \mathbf{U}^{(2)} = \mu^{(2)}\hat{U}_1 + \mu^{(1)^2}\hat{\mathbf{U}}_2$$

$$\Omega^{(2)} = \mu^{(2)}\hat{\omega}_1 + (\text{ktlo})$$

$$\begin{aligned}-3(\omega^{(1)} &- \Omega^{(1)})\dot{\mathbf{U}}^{(2)} - 3\Omega^{(2)}\dot{\mathbf{\Gamma}}_{01} + 3\mu^{(2)}\mathbf{F}_{v\mu}(\mu_0, \mathbf{U}_0|\mathbf{\Gamma}_{01}) \\ &+ 3\mathbf{F}_{vv}(\mu_0, \mathbf{U}_0|\mathbf{\Gamma}_{01}|\mathbf{U}^{(2)}) = 3\mu^{(2)}\{\mathscr{J}\mathbf{\Gamma}_{01} - \dot{\hat{\mathbf{U}}}_1 - \hat{\omega}_1\dot{\mathbf{\Gamma}}_{01}\} + (\text{ktlo}).\end{aligned} \tag{XI.102}$$

We may therefore write (XI.101) as

$$(\omega^{(3)} - \Omega^{(3)})\mathbf{\Gamma}_{00} = J_0\mathbf{Y}^{(3)} + 3\mu^{(2)}\{\mathscr{J}\mathbf{\Gamma}_{01} - \dot{\hat{\mathbf{U}}}_1 - \hat{\omega}_1\dot{\mathbf{\Gamma}}_{01}\} + (\text{ktlo}) \tag{XI.103}$$

Equation (XI.103) is solvable if $[J_0\mathbf{Y}^{(3)}, \mathbf{\Gamma}^*_{01}]_{2\pi} = 0$. Hence, using (XI.42) we find that

$$3\mu^{(2)}\tilde{\gamma}_1 + [(\text{ktlo}), \mathbf{\Gamma}^*_{01}]_{2\pi} = 0.$$

On the other hand, (XI.95) implies that

$$[J_0\mathbf{Y}^{(3)}, \mathbf{\Gamma}^*_{00}]_{2\pi} = [\mathbf{Y}^{(3)}, \mathbf{\Gamma}^*_{01}]_{2\pi} = 0,$$

so that

$$\omega^{(3)} - \Omega^{(3)} = 3\mu^{(2)}[\{\mathscr{J}\mathbf{\Gamma}_{01} - \dot{\hat{\mathbf{U}}}_1 - \hat{\omega}_1\dot{\mathbf{\Gamma}}_{01}\}, \mathbf{\Gamma}^*_{00}]_{2\pi} + [(\text{ktlo}), \mathbf{\Gamma}^*_{00}]_{2\pi}.$$

The computation of higher-order terms follows along similar lines.

XI.15 Stability of Subharmonic Solutions

The stability properties of the subharmonic solutions in the present autonomous case are the same as in the forced T-periodic case. When $n = 1$ and $n = 2$, subcritical bifurcating solutions are unstable and supercritical solutions are stable when $|\alpha|$ is small. The $6\pi(n = 3)$-periodic solution is locally unstable on both sides of criticality and the various possibilities for $n = 4$ which were discussed in IX.16 apply here as well. The proofs of these results are slightly more complicated than those given in Chapter IX. But all the results have been established in the more general context of maps by G. Iooss (*Bifurcation of Maps and Applications*, Amsterdam: North-Holland, 1979)). Here we shall indicate how these results may be obtained by a direct analysis of stability similar to the one given in Chapter IX.

Let $\boldsymbol{\phi}(t)$ be an arbitrarily small disturbance of $\boldsymbol{\psi}(s, \alpha)$. Then $\mathbf{V} = \boldsymbol{\psi}(s, \alpha) + \boldsymbol{\phi}(t)$, $s = \Omega(\alpha)t$, satisfies (XI.2) and the linearized equation for the disturbance $\boldsymbol{\phi}(t)$ is

$$\frac{d\boldsymbol{\phi}}{dt} = \mathbf{F}_v(\mu, \boldsymbol{\psi}|\boldsymbol{\phi}), \qquad \mu = \mu(\alpha).$$

Since $\boldsymbol{\psi}(s, \alpha) \in \mathbb{P}_{2\pi n}$ we may use Floquet theory to form a spectral problem for stability. Setting

$$\boldsymbol{\phi} = e^{\nu t}\boldsymbol{\zeta}(s), \qquad \boldsymbol{\zeta}(\cdot) \in \mathbb{P}_{2\pi n}$$

we find that

$$\nu\boldsymbol{\zeta} = -\Omega\frac{d\boldsymbol{\zeta}}{ds} + \mathbf{F}_v(\mu, \boldsymbol{\psi}|\boldsymbol{\zeta}) \overset{\text{def}}{=} \mathbb{J}\,\boldsymbol{\zeta} \tag{XI.104}$$

where

$$\mathbb{j}(\alpha) = -\Omega(\alpha)\frac{d}{ds} + \mathbf{F}_v(\mu(\alpha), \boldsymbol{\psi}(s, \alpha)|\cdot)$$

$$\text{domain } \mathbb{j}(\alpha) = \mathbb{P}_{2\pi n}$$

and since $(\Omega(\alpha), \boldsymbol{\psi}(s, \alpha), \mu(\alpha)) = (\omega_0, \mathbf{U}_0(s), \mu_0)$ when $\alpha = 0$,

$$\mathbb{j}(0) = \mathbb{J}.$$

We are interested in perturbing the eigenvalues such that

$$v(0) = 0.$$

We note that $\dot{\boldsymbol{\psi}}(s, \alpha)$ is a nullvector (an eigenvector with eigenvalue zero) of $\mathbb{j}(\alpha)$ for all α:

$$\mathbb{j}(\alpha)\dot{\boldsymbol{\psi}} = 0. \tag{XI.105}$$

We shall solve (XI.104) (find $\tilde{\boldsymbol{\zeta}}$ and v) by perturbation theory. The eigenfunction $\tilde{\boldsymbol{\zeta}}(s, \alpha)$ may be decomposed into a part $\dot{\boldsymbol{\psi}}(s, \alpha)$ lying on the null space of the operator $\mathbb{j}(\alpha)$ and is neutral with respect to stability and another part $\boldsymbol{\zeta}(s, \alpha)$ which determines the eigenvalues $v(\alpha)$ which change with α and determine stability:

$$\tilde{\boldsymbol{\zeta}}(s, \alpha) = C(\alpha)\left\{\boldsymbol{\zeta}(s, \alpha) + \frac{\tau(\alpha)}{v(\alpha)}\dot{\boldsymbol{\psi}}(s, \alpha)\right\}. \tag{XI.106}$$

The decomposition (XI.106) is not unique since (1) we have not yet required that $\boldsymbol{\zeta}(s, \alpha)$ should have no part proportional to $\dot{\boldsymbol{\psi}}(s, \alpha)$ and (2) $\tau(\alpha)$ and $C(\alpha)$ are arbitrary. To specify the decomposition more precisely we shall require that

$$\boldsymbol{\zeta}(s, 0) \stackrel{\text{def}}{=} \boldsymbol{\zeta}_0(s)$$

should contain no part proportional to

$$\dot{\boldsymbol{\psi}}(s, 0) = \mathbf{U}_0(s) = \mathbf{Z}_0.$$

We may realize this requirement by requiring that

$$[\boldsymbol{\zeta}, \mathbf{Z}_0^*]_{2\pi n} = 0. \tag{XI.107}$$

We may specify the decomposition (XI.106) completely if, in addition to (XI.107), we specify another condition fixing the amplitude of $\boldsymbol{\zeta}(s, \alpha)$. We may then find $v(\alpha)$ and $\tau(\alpha)$ by perturbation analysis. Since $\tilde{\boldsymbol{\zeta}}(s, \alpha)$ satisfies a linear equation the normalizing function $C(\alpha)$ is indeterminate and may be selected arbitrarily.

We find the equation for $\boldsymbol{\zeta}$ by substituting (XI.106) into (XI.104):

$$\tau\dot{\boldsymbol{\psi}} + v\boldsymbol{\zeta} = \mathbb{j}\boldsymbol{\zeta}, \qquad \boldsymbol{\zeta}(\cdot, \alpha) \in \mathbb{P}_{2\pi n}. \tag{XI.108}$$

When $\alpha = 0$, (XI.104) reduces to

$$\mathbb{J}\tilde{\zeta}_0 = 0.$$

Then (XI.108) reduces to

$$\tau_0 \dot{\mathbf{U}}_0 = \mathbb{J}\zeta_0. \qquad (\text{XI.109})_1$$

The first two derivatives of (XI.108) with respect to α at $\alpha = 0$ are

$$\tau^{(1)}\dot{\mathbf{U}}_0 + \tau_0 \dot{\boldsymbol{\psi}}^{(1)} + \nu^{(1)}\zeta_0 = \mathbb{J}\zeta^{(1)} + \mathring{\mathbb{J}}^{(1)}\zeta_0 \qquad (\text{XI.109})_2$$

and

$$\begin{aligned}\tau^{(2)}\dot{\mathbf{U}}_0 &+ 2\tau^{(1)}\dot{\boldsymbol{\psi}}^{(1)} + \tau_0 \dot{\boldsymbol{\psi}}^{(2)} + \nu^{(2)}\zeta_0 + 2\nu^{(1)}\zeta^{(1)} \\ &= \mathbb{J}\zeta^{(2)} + 2\mathring{\mathbb{J}}^{(1)}\zeta^{(1)} + \mathring{\mathbb{J}}^{(2)}\zeta_0,\end{aligned} \qquad (\text{XI.109})_3$$

where $\zeta^{(1)}(\cdot), \zeta^{(2)}(\cdot) \in \mathbb{P}_{2\pi n}$,

$$\mathring{\mathbb{J}}^{(1)}(\cdot) = -\Omega^{(1)}\frac{d(\cdot)}{ds} + \mu^{(1)}\mathbf{F}_{\mu v}(\mu_0, \mathbf{U}_0|\cdot) + \mathbf{F}_{vv}(\mu_0, \mathbf{U}_0|\boldsymbol{\psi}^{(1)}|\cdot)$$

and

$$\begin{aligned}\mathring{\mathbb{J}}^{(2)}(\cdot) = -\Omega^{(2)}\frac{d(\cdot)}{ds} &+ \mu^{(2)}\mathbf{F}_{\mu v}(\mu_0, \mathbf{U}_0|\cdot) + (\mu^{(1)})^2\mathbf{F}_{\mu\mu v}(\mu_0, \mathbf{U}_0|\cdot) \\ &+ 2\mu^{(1)}\mathbf{F}_{\mu vv}(\mu_0, \mathbf{U}_0|\boldsymbol{\psi}^{(1)}|\cdot) + \mathbf{F}_{vv}(\mu_0, \mathbf{U}_0|\boldsymbol{\psi}^{(2)}|\cdot) \\ &+ \mathbf{F}_{vvv}(\mu_0, \mathbf{U}_0|\boldsymbol{\psi}^{(1)}|\boldsymbol{\psi}^{(1)}|\cdot).\end{aligned}$$

We recall that $\dot{\mathbf{U}}_0(\cdot) \overset{\text{def}}{=} \mathbf{Z}_0$ and $[\mathbf{Z}_0, \mathbf{Z}_1^*]_{2\pi n} = 0$.

Let us consider the stability problem when $n = 2$. Then

$$\mu^{(1)} = 0, \qquad \Omega^{(1)} = \omega^{(1)} = \mu^{(1)}\hat{\omega}_1 = 0, \qquad \mathbf{U}^{(1)}(s) = \mu^{(1)}\hat{\mathbf{U}}_1(s) = 0$$

$$\boldsymbol{\psi}_1(s) = -\mathbf{Y}^{(1)}(s) = -\mathbf{Z}_1(s), \quad \Omega^{(2)} = \mu^{(2)}\hat{\omega}_1 + [\mathbf{F}_{vv}(\mu_0, \mathbf{U}_0|\mathbf{Z}_1|\mathbf{Z}_1), \mathbf{Z}_0^*]_{4\pi}$$

$$\mathbf{U}^{(2)}(s) = \mu^{(2)}\hat{\mathbf{U}}_1(s), \qquad \tau_0 = 0.$$

The condition (XI.107) plus a normalizing condition $[\zeta_0, Z_1^*] = 1$ gives

$$\zeta_0(s) = \mathbf{Z}_1(s) \quad \text{and} \quad \mathring{\mathbb{J}}^{(1)}\zeta_0 = -F_{vv}(\mu_0, \mathbf{U}_0|\mathbf{Z}_1|\mathbf{Z}_1).$$

Equation $(\text{XI.109})_2$ is solvable if $[\mathbb{J}\zeta^{(1)}, \mathbf{Z}_1^*]_{4\pi} = [\mathbb{J}\zeta^{(1)}, \mathbf{Z}_0^*]_{4\pi} = 0$. The first of these gives (see equation below $(\text{XI.64})_2$)

$$\nu^{(1)} = -[\mathbf{F}_{vv}(\mu_0, \mathbf{U}_0|\mathbf{Z}_1|\mathbf{Z}_1), \mathbf{Z}_1^*]_{4\pi} = 0$$

and the second gives

$$\tau^{(1)} = -[\mathbf{F}_{vv}(\mu_0, \mathbf{U}_0|\mathbf{Z}_1|\mathbf{Z}_1), \mathbf{Z}_0^*]_{4\pi} = \omega^{(2)} - \Omega^{(2)}.$$

Then $(\text{XI.109})_2$ may be reduced to

$$\mathbb{J}\zeta^{(1)} = \mathbf{F}_{vv}(\mu_0, \mathbf{U}_0|\mathbf{Z}_1|\mathbf{Z}_1) - [\mathbf{F}_{vv}(\mu_0, \mathbf{U}_0|\mathbf{Z}_1|\mathbf{Z}_1), \mathbf{Z}_0^*]_{4\pi}\mathbf{Z}_0. \qquad (\text{XI.110})$$

Comparing (XI.110) with $(XI.67)_2$ we conclude that

$$\zeta^{(1)} = \mathbf{Y}^{(2)}. \tag{XI.111}$$

We may now rewrite $(XI.109)_3$ using the relations specified in the paragraph above:

$$\begin{aligned}\tau^{(2)}\mathbf{Z}_0 &+ 3[\mathbf{F}_{vv}(\mu_0, \mathbf{U}_0|\mathbf{Z}_1|\mathbf{Z}_1), \mathbf{Z}_0^*]_{4\pi}\dot{\mathbf{Z}}_1 + \nu^{(2)}\mathbf{Z}_1 \\ &= \mathbb{J}\zeta^{(2)} - 3\mathbf{F}_{vv}(\mu_0, \mathbf{U}_0|\mathbf{Z}_1|\mathbf{Y}^{(2)}) + \mu^{(2)}\{\mathscr{J}\mathbf{Z}_1 - \hat{\omega}_1\dot{\mathbf{Z}}_1\} \\ &+ \mathbf{F}_{vvv}(\mu_0, \mathbf{U}_0|\mathbf{Z}_1|\mathbf{Z}_1|\mathbf{Z}_1).\end{aligned} \tag{XI.112}$$

Projecting (XI.112) with $\mathbf{Z}_1^*$ and comparing the result with (XI.72) we find that

$$\nu^{(2)} = -2\mu^{(2)}\xi_1$$

and

$$\nu(\alpha) = -\mu^{(2)}\xi_1\alpha^2 + O(\alpha^4). \tag{XI.113}$$

Equation (XI.113) shows that subcritical solutions are unstable and supercritical solutions are stable when α is small. Projecting (XI.112) with $\mathbf{Z}_0^*$ we find, after a short computation like that leading to (XI.73), that $\tau^{(2)} = 0$.

Turning next to the subharmonic with $n = 3$ we note that

$$\Omega^{(1)} = \omega^{(1)} = \mu^{(1)}\hat{\omega}_1, \qquad \mathbf{U}^{(1)}(s) = \mu^{(1)}\hat{\mathbf{U}}_1(s)$$

$$\boldsymbol{\psi}^{(1)}(s) = \mu^{(1)}\hat{U}_1(s) - \mathbf{Y}^{(1)}, \qquad \mathbf{Y}^{(1)} = e^{i\phi_0}\mathbf{Z}_1 + e^{-i\phi_0}\overline{\mathbf{Z}}_1$$

$$\mathbf{Z}_1 = e^{i(m/3)s}\boldsymbol{\Gamma}_0(s), \qquad \boldsymbol{\Gamma}_0(s) \in \mathbb{P}_{2\pi}, \qquad m = 1 \quad \text{or} \quad 2, \quad \text{and} \quad \tau_0 = 0.$$

The vector ζ_0 is a combination of independent nullvectors of $\mathbb{J}$ and the condition (XI.107) implies that there are complex constants C_0 and C_1 such that

$$\zeta_0 = C_0\mathbf{Z}_1 + C_1\overline{\mathbf{Z}}_1. \tag{XI.114}$$

We may simplify $(XI.109)_2$, using the relations specified above:

$$\begin{aligned}\tau^{(1)}\mathbf{Z}_0 &+ \nu^{(1)}C_0\mathbf{Z}_1 + \nu^{(1)}C_1\overline{\mathbf{Z}}_1 \\ &= \mathbb{J}\zeta^{(1)} + C_0\mu^{(1)}\{\mathscr{J}\mathbf{Z}_1 - \hat{\omega}_1\dot{\mathbf{Z}}_1\} + C_1\{\mathscr{J}\overline{\mathbf{Z}}_1 - \hat{\omega}_1\dot{\overline{\mathbf{Z}}}_1\} \\ &\quad - C_0\mathbf{F}_{vv}(\mu_0, \mathbf{U}_0|e^{i\phi_0}\mathbf{Z}_1 + e^{-i\phi_0}\overline{\mathbf{Z}}_1|\mathbf{Z}_1) \\ &\quad - C_1\mathbf{F}_{vv}(\mu_0, \mathbf{U}_0|e^{i\phi_0}\mathbf{Z}_1 + e^{-i\phi_0}\overline{\mathbf{Z}}_1|\overline{\mathbf{Z}}_1).\end{aligned}$$

Projecting this first with $\mathbf{Z}_1^*$, and then with $\overline{\mathbf{Z}}_1^*$ we find, using (XI.19) and (XI.79), that

$$\nu^{(1)}C_0 = C_0\mu^{(1)}\left(\gamma_1 - \frac{im}{3}\hat{\omega}_1\right) - C_1\lambda_1 e^{-i\phi_0}$$

$$\nu^{(1)}C_1 = C_1\mu^{(1)}\left(\bar{\gamma}_1 + \frac{im}{3}\hat{\omega}_1\right) - C_0\bar{\gamma}_1 e^{i\phi_0}.$$

So $\nu_1^{(1)}$ and $\nu_2^{(1)}$ are the eigenvalues of the matrix

$$\tilde{\mathbf{m}} = \begin{bmatrix} \mu^{(1)}\left(\gamma_1 - \dfrac{im}{3}\hat{\omega}_1\right) & -\lambda_1 e^{-i\phi_0} \\ -\bar{\lambda}_1 e^{i\phi_0} & \mu^{(1)}\left(\bar{\gamma}_1 + \dfrac{im}{3}\hat{\omega}_1\right) \end{bmatrix},$$

where $\mu^{(1)}$ is given by $(\text{XI.79})_2$,

$$2\mu^{(1)}\left(\gamma_1 - \frac{im}{3}\hat{\omega}_1\right) = \lambda_1 e^{-3i\phi_0},$$

with the same λ_1 as given there. The analysis of stability is exactly the same as in the forced case studied in §IX.14 (see (IX.74)):

$$\nu_1^{(1)} + \nu_2^{(1)} = \operatorname{tr} \tilde{\mathbf{m}} = 2\mu^{(1)}\xi_1 > 0$$

because

$$\mu^{(1)} = \frac{\frac{1}{2}|\lambda_1|}{|\gamma_1 - (im/3)\hat{\omega}_1|}, \qquad \xi_1 > 0$$

and

$$\begin{aligned} \nu_1^{(1)}\nu_2^{(1)} &= (\mu^{(1)})^2\left|\gamma_1 - \frac{im}{3}\hat{\omega}_1\right|^2 - |\lambda_1|^2 \\ &= -3(\mu^{(1)})^2\left|\gamma_1 - \frac{im}{3}\hat{\omega}_1\right|^2 < 0 \end{aligned}$$

and one of the eigenvalues of $\tilde{\mathbf{m}}$ is positive and the other is negative. It follows that one of the two eigenvalues

$$\begin{bmatrix} \nu_1(\varepsilon) \\ \nu_2(\varepsilon) \end{bmatrix} = \alpha \begin{bmatrix} \nu_1^{(1)} \\ \nu_2^{(1)} \end{bmatrix} + O(\alpha^2)$$

is positive on each side of criticality. So the 6π-periodic (in s) bifurcating solution is unstable for both positive and negative values of α when $|\alpha|$ is small.

We leave the demonstration of the other stability results asserted in §XI.14 as a demanding exercise which will test the understanding of devoted students.

XI.16 Summary of Results about Subharmonic Bifurcation in the Autonomous Case

Suppose one of spectral assumptions (I) and (II) of §XI.3 holds with $\eta_0/\omega_0 = m/n$ along with the strict crossing conditions of §XI.5–7.

i. When $n = 1$ a single, one-parameter (ε) family of $2\pi/\Omega(\varepsilon)$-periodic solutions of (XI.2) bifurcates on both sides of criticality. When $n = 2$ a single, one-parameter (ε) family of $4\pi/\Omega(\varepsilon)$-periodic solutions of (XI.2) bifurcates on one side of criticality. Supercritical ($\mu(\varepsilon) > 0$) bifurcating solutions are stable; subcritical ($\mu(\varepsilon) < 0$) bifurcating solutions are unstable.

ii. When $n = 3$ a single, one-parameter family of $6\pi/\Omega(\varepsilon)$-periodic solutions of (XI.2) bifurcates and is unstable on both sides of criticality.

iii. When $n = 4$ and $|\lambda_3| > |\gamma_1 - \frac{1}{4}im\hat{\omega}_1||\text{Im}(\lambda_2/(\gamma_1 - \frac{1}{4}im\hat{\omega}_1))|$, λ_2 and λ_3 being defined under (XI.87), $m = 1$ or 3, $\gamma_1 - \frac{1}{4}im\hat{\omega}_1$ satisfying (XI.19), then two one-parameter (ε) families of $8\pi/\Omega(\varepsilon)$-periodic solutions of (XI.2) bifurcate. If $|\lambda_2| < |\lambda_3|$, one of the two bifurcating solutions bifurcates on the subcritical side ($\mu < 0$) and the other on the supercritical side ($\mu > 0$), and both solutions are unstable. If $|\lambda_2| > |\lambda_3|$ the two solutions bifurcate on the same side of criticality and at least one of the two is unstable. The stability of the other solution depends on the details of the problem.

iv. When $n \geq 5$ and $\text{Im}\,(\lambda_2/(\gamma_1 - (im/n)\hat{\omega}_1)) \neq 0$, λ_2 being defined under (XI.87) or when $n = 4$ and the inequality of (iii) is not realized, there is in general no small-amplitude $2n\pi/\Omega(\varepsilon)$-periodic solution of (XI.2) near criticality. In all cases $\Omega(\varepsilon)$ is such that $\Omega(0) = \omega_0$, so the bifurcating solutions have periods *close to* a multiple of $2\pi/\hat{\omega}(\mu)$.

XI.17 Bifurcation of a Torus in Autonomous Nonresonant Cases

Consider the autonomous equation (XI.2) and suppose further that $\mathbf{V} = \hat{\mathbf{U}}(\hat{\omega}(\mu)t, \mu)$ is a $2\pi/\hat{\omega}(\mu)$-periodic solution of (XI.2). In earlier sections of this chapter we found the subharmonic solutions of (XI.2) which bifurcate from $\hat{\mathbf{U}}$. Now we are interested in solutions which bifurcate when the condition of strong resonance $\eta_0/\omega_0 = m/n$, $n = 1, 2, 3, 4$ is *not* satisfied, while other spectral assumptions about $J(\mu)$ are the same as in §IX.3. In this case, as in the forced periodic case studied in Chapter X, we obtain a torus of asymptotically quasi-periodic solutions.

The spectral problem for the stability of $\hat{\mathbf{U}}(\tau, \mu)$ is given by (XI.5) and the adjoint spectral problem by (XI.6). At $\mu = \mu_0$ we have on the imaginary axis the simple eigenvalues $\pm i\eta_0 + ki\omega_0$ and $0 + k'i\omega_0$ with any k, k' in $\mathbb{Z}$. Then for μ near μ_0 we have a pair of simple eigenvalues $\gamma(\mu)$, $\bar{\gamma}(\mu)$ of $J(\mu)$. Moreover, zero is still a simple eigenvalue of $J(\mu)$ associated with the eigenfunction $\hat{\mathbf{U}}(\tau, \mu) \stackrel{\text{def}}{=} \mathbf{\Gamma}_0(\tau, \mu)$. Let $\mathbf{\Gamma}(\tau, \mu)$ be the eigenfunction of (XI.5) associated with $\gamma(\mu)$. The adjoint eigenfunctions satisfying (XI.6) are $\mathbf{\Gamma}^*(\tau, \mu)$, $\bar{\mathbf{\Gamma}}^*(\tau, \mu)$ and $\mathbf{\Gamma}_0^*(\tau, \mu)$. By computations like those given in Exercise X.1, we find the following biorthogonality relations, for all τ:

$$\langle \mathbf{\Gamma}(\tau, \mu), \mathbf{\Gamma}^*(\tau, \mu) \rangle = \langle \mathbf{\Gamma}_0(\tau, \mu), \mathbf{\Gamma}_0^*(\tau, \mu) \rangle = 1 \tag{XI.115}$$

$$\begin{aligned} \langle \mathbf{\Gamma}(\tau, \mu), \mathbf{\Gamma}_0^*(\tau, \mu) \rangle &= \langle \bar{\mathbf{\Gamma}}(\tau, \mu), \mathbf{\Gamma}_0^*(\tau, \mu) \rangle \\ &= \langle \mathbf{\Gamma}(\tau, \mu), \bar{\mathbf{\Gamma}}^*(\tau, \mu) \rangle = 0. \end{aligned} \tag{XI.116}$$

We next split $\mathbf{V}$ as follows:

$$\mathbf{V} = \hat{\mathbf{U}}(\tau, \mu) + \mathbf{Z}, \tag{XI.117}$$

where

$$\langle \mathbf{Z}, \mathbf{\Gamma}_0^*(\tau, \mu)\rangle = 0, \tag{XI.118}$$

and where $\mathbf{Z}$ and τ are to-be-determined functions of (t, μ). In the case of strong resonance, studied in previous sections, $\tau = s$, $\mathbf{V} = \hat{\boldsymbol{\psi}}$, and $\mathbf{Z} = -\mathbf{Y}$. We wish to reduce this problem to the form (X.8), (X.9) which may be treated by the method of averaging used in §X.2.

Equation (XI.2) takes the form*

$$\begin{aligned}\dot{\tau}\left[\frac{d\hat{\mathbf{U}}}{d\tau} + \frac{d\mathbf{Z}}{d\tau}\right] &= \mathbf{F}(\mu, \hat{\mathbf{U}}(\tau, \mu) + \mathbf{Z})\\ &= \mathbf{F}[\mu, \hat{\mathbf{U}}(\tau, \mu)] + \mathbf{F}_V(\mu, \hat{\mathbf{U}}(\tau, \mu)|\mathbf{Z}) + \mathbf{N}(\mu, \tau, \mathbf{Z}),\end{aligned} \tag{XI.119}$$

where $\|\mathbf{N}(\mu, \tau, \mathbf{Z})\| = O(\|\mathbf{Z}\|^2)$, and since $\mathbf{F}[\mu, \hat{\mathbf{U}}(\tau, \mu)] = \hat{\omega}(\mu)\, d\hat{\mathbf{U}}(\tau, \mu)/d\tau$, we get

$$(\dot{\tau} - \hat{\omega})\frac{d\hat{\mathbf{U}}}{d\tau} + \dot{\tau}\frac{d\mathbf{Z}}{d\tau} = \mathbf{F}_v(\mu, \hat{\mathbf{U}}|\mathbf{Z}) + \mathbf{N}(\mu, \tau, \mathbf{Z}), \tag{XI.120}$$

hence

$$(\dot{\tau} - \hat{\omega})\left(\frac{d\hat{\mathbf{U}}}{d\tau} + \frac{d\mathbf{Z}}{d\tau}\right) = J(\mu)\mathbf{Z} + \mathbf{N}(\mu, \tau, \mathbf{Z}). \tag{XI.121}$$

Now differentiating (XI.118) with respect to τ, we find that

$$\left\langle \frac{d\mathbf{Z}}{d\tau}, \mathbf{\Gamma}_0^*(\tau, \mu)\right\rangle = -\langle \mathbf{Z}, \dot{\mathbf{\Gamma}}_0^*(\tau, \mu)\rangle. \tag{XI.122}$$

We project (XI.121) with $\mathbf{\Gamma}_0^*(\tau, \mu)$, using (XI.122), and find that

$$(\dot{\tau} - \hat{\omega})[1 - \langle \mathbf{Z}, \dot{\mathbf{\Gamma}}_0^*(\tau, \mu)\rangle] = \langle \mathbf{N}(\mu, \tau, \mathbf{Z}), \mathbf{\Gamma}_0^*(\tau, \mu)\rangle. \tag{XI.123}$$

Hence

$$\dot{\tau} = \hat{\omega}(\mu) + \frac{\langle \mathbf{N}(\mu, \tau, \mathbf{Z}), \mathbf{\Gamma}_0^*(\tau, \mu)\rangle}{1 - \langle \mathbf{Z}, \dot{\mathbf{\Gamma}}_0^*(\tau, \mu)\rangle}. \tag{XI.124}$$

Now, solving (XI.121) for $d\mathbf{Z}/d\tau$, using (XI.124), we obtain

$$\begin{aligned}\frac{d\mathbf{Z}}{d\tau} = \left(\hat{\omega}(\mu) + \frac{\langle \mathbf{N}(\mu, \tau, \mathbf{Z}), \mathbf{\Gamma}_0^*(\tau, \mu)\rangle}{1 - \langle \mathbf{Z}, \dot{\mathbf{\Gamma}}_0^*(\tau, \mu)\rangle}\right)^{-1}&\left(\mathbf{F}_v(\mu, \hat{\mathbf{U}}(\tau, \mu)|\mathbf{Z})\right.\\ &\left. + \mathbf{N}(\mu, \tau, \mathbf{Z}) - \frac{\langle \mathbf{N}(\mu, \tau, \mathbf{Z}), \mathbf{\Gamma}_0^*(\tau, \mu)\rangle}{1 - \langle \mathbf{Z}, \dot{\mathbf{\Gamma}}_0^*(\tau, \mu)\rangle}\frac{d\hat{\mathbf{U}}(\tau, \mu)}{d\tau}\right),\end{aligned} \tag{XI.125}$$

* In what follows, $\dot{\tau} \overset{\text{def}}{=} d\tau/dt$, while $\dot{\hat{\mathbf{U}}} = d\hat{\mathbf{U}}/d\tau$, $\dot{\mathbf{Z}} = d\mathbf{Z}/d\tau$.

which is of the form of (X.1) supplemented by the auxiliary equation (XI.118). The linearized operator here is

$$-\frac{d}{d\tau} + (\hat{\omega}(\mu))^{-1}\mathbf{F}_v(\mu, \hat{\mathbf{U}}(\tau, \mu)|\cdot) = \frac{1}{\hat{\omega}(\mu)} J(\mu). \qquad \text{(XI.126)}$$

The effect of the auxiliary condition (XI.118) is to suppress the zero eigenvalue (and the related eigenvalues on the imaginary axis) of $J(\mu)$.

Now, as in Chapter X, we decompose $\mathbf{Z}$:

$$\mathbf{Z} = z\mathbf{\Gamma}(\tau, \mu) + \bar{z}\bar{\mathbf{\Gamma}}(\tau, \mu) + \mathbf{W}, \qquad \text{(XI.127)}$$

where

$$\langle \mathbf{W}, \mathbf{\Gamma}^*(\tau, \mu)\rangle = \langle \mathbf{W}, \mathbf{\Gamma}_0^*(\tau, \mu)\rangle = 0, \qquad \text{(XI.128)}$$

$$z = \langle \mathbf{Z}, \mathbf{\Gamma}^*(\tau, \mu)\rangle. \qquad \text{(XI.129)}$$

Equation (XI.125) then leads to the system

$$\frac{dz}{d\tau} = \frac{\gamma(\mu)}{\hat{\omega}(\mu)} z + b(\tau, \mu, z, \bar{z}, \mathbf{W}) \qquad \text{(XI.130)}$$

$$\frac{d\mathbf{W}}{d\tau} = \frac{1}{\hat{\omega}(\mu)} \mathbf{F}_v(\mu, \hat{\mathbf{U}}(\tau, \mu)|\mathbf{W}) + \mathbf{B}(\tau, \mu, z, \bar{z}, \mathbf{W}) \qquad \text{(XI.131)}$$

where

$$|b(\tau, \mu, z, \bar{z}, \mathbf{W})| + \|\mathbf{B}(\tau, \mu, z, \bar{z}, \mathbf{W})\| = O[(|z| + \|\mathbf{W}\|)^2]. \qquad \text{(XI.132)}$$

By assumption the numbers $\gamma(\mu)/\hat{\omega}(\mu)|_{\mu=0} = i\eta_0/\omega_0$ do not satisfy to conditions of strong resonance and the Floquet exponants for the $\mathbf{W}$ part are all on the left-hand side of the complex plane. So, applying the results of Chapter X, we have the solutions of (XI.130, 131) on the bifurcated invariant torus in the following form:

$$z(\tau) = \rho(\tau) \exp i\left[\frac{\eta_0}{\omega_0}\tau + \theta(\tau)\right] + \sum_{p+q=2}^{N} \gamma'_{pq}(\tau, \mu)[\rho(\tau)]^{p+q} \exp(p-q)i\left[\frac{\eta_0}{\omega_0}\tau + \theta(\tau)\right] + O(\varepsilon^{N+1}) \qquad \text{(XI.133)}$$

$$\mathbf{W}(\tau) = \sum_{p+q=2}^{N} \mathbf{\Gamma}'_{pq}(\tau, \mu)[\rho(\tau)]^{p+q} \exp(p-q)i\left[\frac{\eta_0}{\omega_0}\tau + \theta(\tau)\right] + O(\varepsilon^{N+1}), \qquad \text{(XI.134)}$$

where, as in Chapter X, the truncation number N is unrestricted and $\rho(\tau)$, $\theta(\tau)$, μ are parameterized by ε:

$$\begin{aligned} \mu &= \mu_2\varepsilon^2 + \mu_4\varepsilon^4 + \cdots + O(\varepsilon^{N+1}) \\ \rho(\tau) &= \varepsilon + \varepsilon^{n-3}\rho_{n-3}(\theta) + \varepsilon^{n-2}\rho_{n-2}(\theta) + \cdots + O(\varepsilon^{N+1}) \qquad \text{(XI.135)} \\ \theta(\tau) + \varepsilon^{n-4}h_{n-4}[\theta(\tau)] &+ \varepsilon^{n-3}h_{n-3}[\theta(\tau)] + \cdots = \varepsilon^2\Omega(\varepsilon^2)\tau + \chi(\tau, \varepsilon) \end{aligned}$$

The n in (XI.135) is defined by $\exp(2\pi i n\eta_0/\omega_0) = 1$, and ρ_l and h_l are $2\pi/n$-periodic functions of zero mean value which may be determined as in §X.9, 10 in the form of polynomials of $e^{\pm in\theta}$. χ contains secular terms in τ and

$$\left|\frac{\partial \chi}{\partial \tau}\right| = O(\varepsilon^N)$$

uniformly in τ.

Equations (XI.117), (XI.127), (XI.133–135) *determine the trajectories*, parameterized by τ, with initial data for θ given by $\chi(0, \varepsilon)$. But we need $\tau(t)$ to obtain the law of movement on these trajectories.

XI.18 Asymptotically Quasi-Periodic Solutions on the Bifurcated Torus

We may now replace $\mathbf{Z}$ in (XI.124) by $\mathbf{Z}(\tau)$ of (XI.127) to determine $\tau(t)$ which gives the behavior of the solution on the trajectories lying on the torus. After following the path laid out in §X.11, we may write $(\text{XI.135})_3$ in the form

$$\frac{\eta_0}{\omega_0}\tau + \theta(\tau) + \varepsilon^{n-4}H_{n-4}\left(\tau, \frac{\eta_0}{\omega_0}\tau + \theta(\tau)\right) + \cdots$$
$$= \left[\frac{\eta_0}{\omega_0} + \varepsilon^2\Omega(\varepsilon^2)\right]\tau + \chi(\tau, \varepsilon), \qquad \text{(XI.136)}$$

where the H_l's satisfy $H_l(\cdot + 2\pi, \cdot) = H_l(\cdot, \cdot + 2\pi) = H_l(\cdot, \cdot)$. Hence, inverting (XI.136), we have

$$\frac{\eta_0}{\omega_0}\tau + \theta(\tau) = \left[\frac{\eta_0}{\omega_0} + \varepsilon^2\Omega(\varepsilon^2)\right]\tau$$
$$+ \varepsilon^{n-4}G_{n-4}\left(\tau, \left[\frac{\eta_0}{\omega_0} + \varepsilon^2\Omega(\varepsilon^2)\right]\tau + \chi_{10}\right) + \cdots + \chi_1(\tau, \varepsilon),$$
$$\text{(XI.137)}$$

where the G_l are 2π-periodic with respect to both arguments and $\chi_{10} = \chi_1(0, \varepsilon)$, $|\partial\chi_1(\tau, \varepsilon)/\partial\tau| = O(\varepsilon^N)$. We need only treat the case in which G_l has a finite Fourier decomposition in the second argument. Higher harmonics occur only in higher-order terms in ε and we truncate at ε^{N-1}.

In the same way, following methods of analysis used in §X.9, we find that

$$\rho(\tau) = \varepsilon + \varepsilon^{n-3}R_{n-3}\left(\tau, \frac{\eta_0}{\omega_0}\tau + \theta(\tau)\right) + \cdots + O(\varepsilon^{N+1}), \quad \text{(XI.138)}$$

where the R_l are 2π-periodic with respect to both arguments. We next replace $\rho(\tau)$ and $(\eta_0/\omega_0)\tau + \theta(\tau)$ by their expressions (XI.138), (XI.137) in (XI.133), (XI.134). Equation (XI.124) may then be expressed as

$$\frac{d\tau}{dt} = \hat{\omega}[\mu(\varepsilon)] + \sum_{k\geq 1}^{N} \varepsilon^k T_k\left(\tau, \left[\frac{\eta_0}{\omega_0} + \varepsilon^2\Omega(\varepsilon^2)\right]\tau + \chi_{10}\right) + O(\varepsilon^{N+1}) + \chi_2(\tau, \varepsilon), \tag{XI.139}$$

where the T_k's are 2π-periodic with respect to both arguments and $|\partial\chi_2(\tau, \varepsilon)/\partial\tau| = O(\varepsilon^{N+1})$. In general, there are secular terms in χ_2, but the T_k's have finite Fourier decompositions* (same proof as for the G_l's), all higher harmonics going into the terms $O(\varepsilon^{N+1})$.

It is also not hard to see that

$$\left[\hat{\omega}[\mu(\varepsilon)] + \sum_{k\geq 1}^{N} \varepsilon^k T_k(\tau, \tilde{\Omega}(\varepsilon^2)\tau)\right]^{-1} = \omega_0^{-1} + \sum_{k\geq 1}^{N} \varepsilon^k S_k(\tau, \tilde{\Omega}(\varepsilon^2)\tau) + O(\varepsilon^{N+1}), \tag{XI.140}$$

where S_k are 2π-periodic with respect to both variables and

$$\tilde{\Omega}(\varepsilon^2) \stackrel{\text{def}}{=} \frac{\eta_0}{\omega_0} + \varepsilon^2\Omega(\varepsilon^2).$$

When $\tau = 0(1)$, the secular terms in χ_2 are bounded, and $|\chi_2| = O(\varepsilon^{N+1})$, (XI.139) may then be integrated:

$$\omega_0^{-1}\tau + \sum_{k\geq 1}^{N} \varepsilon^k \int_0^\tau S_k(s, \tilde{\Omega}(\varepsilon^2)s)\, ds + \chi_3(\tau, \varepsilon) = t - t_0, \tag{XI.141}$$

where $|X_3| = O(\varepsilon^{N+1}(1 + \tau + \tau^2))$. To compute the integrals we may write

$$S_k(s, \tilde{\Omega}(\varepsilon^2)s) = \sum_{l_1, l_2} S_{kl_1l_2} \exp i(l_1 + \tilde{\Omega}l_2)s,$$

and if $l_1 + \tilde{\Omega}l_2 \neq 0$ for all l_1, l_2 in the summation, then

$$\int_0^\tau S_k(s, \tilde{\Omega}s)\, ds = \sum_{l_1, l_2} \frac{S_{kl_1l_2}}{i(l_1 + \tilde{\Omega}l_2)} (\exp i(l_1 + \tilde{\Omega}l_2)\tau - 1). \tag{XI.142}$$

If, for some l_1 and l_2, $l_1 + \tilde{\Omega}l_2 = 0$, then (XI.142) is not valid and the integration in (XI.141) has to be done in another way. Since the summation over l_2 in (XI.142) is finite, we may compute the integrals as in (XI.142) provided only that $l_1 + \tilde{\Omega}l_2 \neq 0$. For infinite summation we could expect to encounter small denominators and divergent series even when $\tilde{\Omega}(\varepsilon^2)$ is irrational.

Equations (XI.141) and (XI.142) show that truncated solutions which suppress higher-order terms (which may be secular) are in the form

$$\mathbf{V} \approx \mathbf{V}(\omega t, \omega\tilde{\Omega}t), \tag{XI.143}$$

* In the second argument.

where $\mathbf{V}$ is 2π-periodic with respect to both arguments. To determine ω, we change variables to transform the right-hand-side of (XI.139) into a function which is constant through terms of order ε^{N+1}. Of course $\omega = \omega_0 + O(\varepsilon)$, and

$$\tilde{\Omega} = \frac{\eta_0}{\omega_0} + \varepsilon^2\Omega(\varepsilon^2).$$

The rotation number of the Poincaré map (see §X.15) is given here by

$$\hat{\rho}(f) = \frac{\eta_0}{\omega_0} + \varepsilon^2\Omega(\varepsilon^2) + O(\varepsilon^N).$$

For approximate solutions (XI.143) the number $\hat{\rho}$ is given by the ratio of frequencies. The comments of §X.15 do not have the same force here because an irrational $\hat{\rho}$ is not enough to guarantee that the solutions on the torus are quasi-periodic. We do get quasi-periodic solutions when $\hat{\rho}$ is not too well approximated by rationals. Fortunately most numbers ("most" defined in the sense of Lebesgue measure) have the required property and for these the discussion of Chapter X is valid.

XI.19 Strictly Quasi-Periodic Solutions on the Bifurcated Torus*

The problem of bifurcation simplifies in many physical problems governed by (VIII.1) which are invariant under certain groups of transformations. These simplifications can operate in problems of partial differential equations in which the invariance to a group of transformations involving space and time leads to a Hopf bifurcation into wave-like solutions. To study the bifurcation of the wave-like solution it is then both convenient and possible to reduce the problem strictly, and not just asymptotically, to an autonomous one. Steady solutions of the autonomous problem are then periodic solutions of the original problem. But if the autonomous problem governing the bifurcation of the wave-like solution undergoes a Hopf bifurcation again, the bifurcation solutions will be strictly doubly periodic with two frequencies and a rotation number, all analytic in ε. Suppose, for example, we have a partial differential equation for a bifurcating scalar field $v(r, \theta, z, t)$ which is invariant to rotations around z in cylindrical coordinates (r, θ, z). And suppose further that $v = 0$ undergoes Hopf bifurcation into a rotating wave $v = \tilde{v}(r, \theta - \omega t, z)$. The problem governing the bifurcation of $\tilde{v}(r, \phi, z)$

* This section is motivated by recent results on bifurcation of rotating waves (see Rand, David. Dynamics and symmetry: predictions for modulated waves in rotating fluids, Arch. Rational Mech. Anal. (to appear in 1981), Michael Renardy, Bifurcation of Rotating Waves, Arch. Rational Mech. Anal. (to appear in 1981).

$\phi = \theta - \omega t$, is *autonomous*, and if it undergoes another Hopf bifurcation the new solution will be of the form $u(r, \phi, z, t)$, T-periodic in t. The bifurcated solution $u(r, \phi, z, t) = \tilde{u}(r, \theta - \omega t, z, t)$ will live on a two-dimensional torus of doubly periodic flows with two frequencies $(2\pi/T, \omega)$ and rotation number, all analytic in ε.

Invariance of systems of differential equations in $\mathbb{R}^n$ to groups of rotations in phase space also leads to big simplifications.

EXAMPLE. Consider the evolution of the three-dimensional vector $\mathbf{x} = (x, y, z)$, $\mathbf{x} \in \mathbb{R}^3$, where

$$\frac{d\mathbf{x}}{dt} = \mathbf{f}(\mu, \mathbf{x}) \tag{XI.144}$$

and $\mathbf{f}(\mu, \mathbf{R}_\theta \mathbf{x}) = \mathbf{R}_\theta \mathbf{f}(\mu, x)$ is invariant to rotations of angle θ about the y axis,

$$[\mathbf{R}_\theta] = \begin{bmatrix} \cos\theta & -\sin\theta & 0 \\ \sin\theta & \cos\theta & 0 \\ 0 & 0 & 1 \end{bmatrix}.$$

To simplify the study of bifurcation we introduce the *generator* $\mathbf{S}_\omega$ of the *group* $\mathbf{R}_{\omega t}(t \in \mathbb{R})$ which is defined through the linear evolution problem

$$\frac{d\mathbf{Z}}{dt} = \omega\mathbf{k} \wedge \mathbf{Z} \overset{\text{def}}{=} \mathbf{S}_\omega \mathbf{Z}, \qquad \mathbf{Z} \in \mathbb{R}^3, \tag{XI.145}$$

where ω is real constant and $\mathbf{k}$ is unit vector along the z axis. The general solution of (XI.145) is

$$\mathbf{Z}(t) = e^{\mathbf{s}}\omega^t \mathbf{Z}(0) = \underline{R}_{\omega t} Z(0) \in \mathbb{P}_{2\pi/\omega}.$$

Returning now to (XI.144), we set

$$\mathbf{x}(t) = e^{\mathbf{s}_\omega t}\mathbf{y}(t) = \mathbf{R}_{\omega t}\mathbf{y}(t),$$

where

$$\frac{d\mathbf{y}}{dt} = \mathbf{f}(\mu, \mathbf{y}) - \mathbf{S}_\omega \mathbf{y}. \tag{XI.146}$$

Steady solutions of (XI.146) correspond to periodic solutions $\mathbf{x}(t) = \mathbb{R}_{\omega t}\mathbf{y} \in \mathbb{P}_{2\pi/\omega}$ of (XI.144). Periodic solutions of (XI.146) correspond to doubly periodic solutions of (XI.144) which are, in fact, quasi-periodic for most ω.

EXERCISE

XI.1 Consider example VII.1 on page 134 and show that (VII.47) is invariant under rotations $\mathbf{R}_\theta$ of the (x, y) plane. Compute Hopf bifurcation along the lines laid out in the example.

XI.2 Consider evolution problems of the form $\dot{\mathbf{u}} = \mathbf{f}(t, \mu, \delta, \underline{u})$, where $\underline{f}(t, \mu, \delta, 0) = 0$, depending on a parameter δ perturbing the problem of this chapter. Suppose further that we have a Floquet multiplier $\lambda = e^{\sigma(\mu, \delta)T}$, where $\sigma(\mu, \delta) = i\omega_0 + \mu\sigma_\mu + \delta\sigma_\delta + 0[|\mu|^2 + |\delta|^2]$ and $\text{Re}\ \sigma_\mu \neq 0$.

(1) Compute σ_μ and σ_δ in terms of scalar products.
(2) Assume $\lambda(0, 0)$ is simple and real. Show that $\lambda(0, 0) = 1$ or -1. Then show that $\lambda(\mu, \delta)$ is still real when $|\mu| + |\delta|$ is small. Compute the critical values $\mu^{(c)}(\delta) = \delta\mu_\delta + 0(\delta^2)$ for which $\lambda[\mu^{(c)}(\delta), \delta] = 1$ or -1. Show that when μ crosses $\mu^{(c)}(\delta)$, δ being fixed, $\lambda(\mu, \delta)$ crosses the unit circle through 1 or -1, respectively.
(3) Assume that $\lambda(0, 0)$ is simple and $\omega_0 = 2\pi m/nT$ with $0 < m/n < 1$ (as in this chapter) and $n \geq 3$. Compute the critical values $\mu^{(c)}(\delta) = \delta\mu_\delta + 0(\delta^2)$ for which $|\lambda[\mu^{(c)}(\delta), \delta]| = 1$. What is the value of $\lambda[\mu^{(c)}(\delta), \delta]$ on the unit circle? Show that the condition $[\lambda(0, 0)]^n = 1$ *does not persist* for $\delta \neq 0$. *Answer*:

$$\arg \lambda[\mu^{(c)}(\delta), \delta] = \omega_0 + \delta \frac{(\text{Im}\ \sigma_\delta\ \text{Re}\ \sigma_\mu - \text{Im}\ \sigma_\mu\ \text{Re}\ \sigma_\delta)}{\text{Re}\ \sigma_\mu} + O(\delta^2).$$

Index